Deformations of Mathematical Structures

TABLE OF CONTENTS

Part II. COMPLEX ANALYTIC GEOMETRY

edited by P. Dolbeault (Paris), J. Ławrynowicz (Łódź), and E. Vesentini (Pisa)

* Invited Speaker

FOREWORD

These Proceedings contain selected papers by the speakers invited to the Seminar on Deformations, organized in 1985/87 by Julian Ławrynowicz (Łódź), whose most fruitful parts took place in 1986 in Lublin during the 3rd Finnish-Polish Summer School in Complex Analysis [in cooperation with O. Martio (Jyväskylä)] held simultaneously with the 9th Conference on Analytic Function in Poland [in cooperation with S. Dimiev (Sofia), P. Dolbeault (Paris), K. Spallek (Bochum), and E. Vesentini (Pisa)]. The Lublin session of the Seminar, organized jointly with S. Dimiev and K. Spallek, was preceded by a session organized by them at Druzhba (near Varna) in 1985 and followed by a similar session at Druzhba in 1987.

The collection contains 31 papers connected with deformations of mathematical structures in the context of complex analysis with physical applications: (quasi)conformal deformation uniformization, potential theory, several complex variables, geometric algebra, algebraic geometry, foliations, Hurwitz pairs, and Hermitian geometry. They are research papers in final form: no version of them will be submitted for publication elsewhere. In contrast to the previous volume (*Seminar on Deformations, Proceedings, Łódź-Warsaw 1982/84*, ed. by J. Ławrynowicz, Lecture Notes in Math. 1165, Springer, Berlin-Heidelberg--New York-Tokyo 1985, X + 331 pp.) open problems are not published as separate research notes, but are included in the papers.

The papers were recommended for publication by the Advisory Committee formed by the members of the Organizing Committees of the 3rd Finnish-Polish Summer School in Complex Analysis [B. Bojarski (Warszawa), Z. Charzyński (Łódź), J. Krzyż (Lublin), J. Ławrynowicz (Łódź), O. Lehto (Helsinki), P. Lounesto (Helsinki), O. Martio (Jyväskylä), J. Siciak (Kraków), T. Sorvali (Joensuu), K. Suominen (Helsinki), O. Tammi (Helsinki), and E. Złotkiewicz (Lublin) and of the 9th Conference on Analytic Functions [C. Andreian Cazacu (București), Z. Charzyński, P. Dolbeault, F.W. Gehring (Ann Arbor, MI), A.A. Gonchar (Moscow), J. Górski (Katowice), L. Iliev (Sofia), J. Krzyż, J. Ławrynowicz, O. Lehto, J. Siciak, K. Spallek, K. Strebel (Zürich), W. Tutschke (Halle/Saale), E. Vesentini, and E. Złotkiewicz] and - in addition - it contained J. Bingener (Regensburg), S. Dimiev (Sofia), H. Grauert (Göttingen), J. Leiterer (Berlin, GDR), S. Łojasiewicz (Kraków), and S. Walczak (Łódź).

The preparation of the Proceedings was possible thanks to the help of the Łódź Society of Sciences and Arts, especially to its President, Professor Witold Śmiech. The organizers express also their gratitude to the Kluwer Academic Publishers for its kind consent to publish the Proceedings as an "out of series" work. Finally, the organizers wish to thank warmly Mrs. A. Marciniak, an English philologist, for improving the language style and Mrs. E. Gałuszka and her for typing the paper.

Łódź, December 1987 Julian Ławrynowicz

Part I

PROCEEDINGS OF THE THIRD FINNISH-POLISH SUMMER SCHOOL IN COMPLEX ANALYSIS

edited by

Julian Ławrynowicz

Institute of Mathematics
Polish Academy of Sciences
Łódź, Poland

and

Olli Martio

Institute of Mathematics
University of Jyväskylä
Jyväskylä, Finland

SOME DIFFERENTIAL OPERATORS IN REAL AND COMPLEX GEOMETRY

Jerzy Kalina and Antoni Pierzchalski
Institute of Mathematics, Polish Academy of Sciences, PL - 90-136 Łódź, Poland
Institute of Mathematics, University of Łódź, PL - 90-237 Łódź, Poland

ABSTRACT. The famous Laplace-Beltrami operator Δ acting on differential forms on a Riemannian manifold M determines in some sense the geometry of M. For example the Hodge decomposition theorem implies that in the compact case

$$X(M) = \text{Trace } e^{-t\Delta_{even}} - \text{Trace } e^{-t\Delta_{odd}},$$

where $X(M)$ is the Euler characteristic of M and $\Delta_{even} = \Delta|_{\bigoplus_{p\ even} \Lambda^p(M)}$, $\Delta_{odd} = \Delta|_{\bigoplus_{p\ odd} \Lambda^p}$. Of course the theory of the operator in the complex case is much richer. We are going to give a short review of the theory of the Laplace-Beltrami operator on compact complex manifolds. In particular, the Hodge decomposition and its applications will be given. The case of a compact Kähler manifold will also be mentioned. Some other elliptic operators essentially connected with the geometry of M will be introduced. One of them is the so-called Ahlfors-Laplacian S^*S acting on 1-forms. S is the Ahlfors' operator which arises naturally in the theory of quasiconformal deformations of M. S^*S is strongly influenced by the geometry of M. It behaves nicely both in the real and in the complex cases. Before passing to the operators some necessary information from the theory of real and complex geometry will be given.

INTRODUCTION

Geometry of a manifold depends on a metric structure. The metric structure is ussually prescribed by a so-called Riemannian scalar product, i.e. by a symmetric and positive definite two form. In what follows we will assume that M is a Riemannian manifold of dimension m with a Riemannian scalar product g.

The aim of our survey paper is to show that the metric structure induces in a very natural way some linear differential operators. The operators depend essentially on the geometry of M, i.e. on g. On the other hand, the operators could determine that geometry to some extent.

J. Ławrynowicz (ed.), Deformations of Mathematical Structures, 3–28.

The most famous is the classical Laplace-Beltrami operator Δ (in short: Laplacian) acting on forms. A major part of the paper is devoted to this operator: first in the real, next in the complex cases. In the complex case the theory is especially interesting: there are three different Laplacians. They coincide with one another when the metric structure is compatible with the complex one, i.e. when a manifold is Kählerian.

The Ahlfors-Laplace operator which arises naturally in conformal or quasi-conformal geometry will also be mentioned.

The subject of our paper is real and complex geometry. The question is how to join them. Our attitude is the following: Equip step by step a (real) manifold with richer and richer structure starting from a Riemannian structure and resulting in a Kählerian one.

A Kählerian manifold may, therefore, be regarded as a Riemannian manifold of dimension m equipped with a tensor field J of the type $(1,1)$ satisfying the condition $J^2 = -\mathrm{id}$ (it implies orientability and even dimension $m = 2n$) and some integrability condition (it implies that M is locally biholomorphic with $\mathbb{C}^n$). Moreover, we require that g is hermitian in the sense that $g(JX, JY) = g(X, Y)$ and that the fundamental form ω defined by $\omega(X, Y) = g(JX, Y)$ is closed, i.e. $d\omega = 0$.

The main theorem of the paper is the Hodge decomposition theorem in the real and complex cases and its applications. The importance of the theorem for real and complex geometry follows from a variety of its consequences. For example, some topological invariants such as Euler characteristic, Betti and Hodge numbers can be checked with the help of the theorem (see: Sections 6 and 10).

All manifolds and mappings in question are assumed to be smooth, i.e. of the class C^∞. For simplicity we confine ourselves to compact and oriented manifolds.

2. LAPLACIAN ON A RIEMANNIAN MANIFOLD

Let M be a Riemannian manifold of dimension m with a Riemannian scalar product g. There exists a unique connection ∇ on M, the so-called Riemannian connection for g, which is torsion free and for which g is parallel ($\nabla g = 0$).

The curvature tensor R of a connection ∇ is defined by

$$R(X, Y)Z = \nabla_X \nabla_Y Z - \nabla_Y \nabla_X Z - \nabla_{[X, Y]} Z$$

for arbitrary vector fields X, Y, Z. The Ricci tensor is the trace of it:

$$\mathrm{Ric}(X, Y) = \mathrm{trace}(Z \to R(Z, X)Y) \tag{2.1}$$

TM and T^*M denote the tangent and cotangent bundles, respectively. The scalar product g may be extended onto the whole tensor algebra of M, in particular, onto the exterior algebra

$$\Lambda = \Lambda T^*M = \bigoplus_{p=1}^{m} \Lambda^p T^*M .$$

For example, if $x \in M$, and $v, w \in T_x^*M = \Lambda^1 T_x^*M$, we put

$$g(v, w) = g(v^{\#}, w^{\#}),$$

where $v^{\#}, w^{\#}$ are tangent vectors dual to v and w, respectively. If, now, $v_1 \wedge \ldots \wedge v_p, w_1 \wedge \ldots \wedge w_p \in \Lambda^p T_x^*M$, then

$$g(v_1 \wedge \ldots \wedge v_p, w_1 \wedge \ldots \wedge w_p) = \sum_{\pi} \varepsilon(\pi)\, g(v_1, w_{\pi_1}) \ldots g(v_p, w_{\pi_p}),$$

where the sum is taken over all permutations $\pi = (\pi_1, \ldots, \pi_p)$ of $\{1, \ldots, p\}$. Assuming that the spaces Λ^p and Λ^q are orthogonal for $p \neq q$ we extend g onto the whole Λ. The use of the same letter g for the extended product should not be confusing.

Now, for any two forms $\lambda, \mu \in C^\infty(\Lambda)$ we define the global scalar product $< , >$ by

$$<\lambda, \mu> = \int_M g(\lambda, \mu) \mathrm{Vol}_M , \tag{1}$$

where Vol_M is the volume of the metric g (M has been assumed to be compact and oriented).

Now, take one of the simplest and most natural first order differential operators $d: C^\infty(\Lambda) \to C^\infty(\Lambda)$. We are interested in finding its formal adjoint d^*. To this aim let us introduce the Hodge-star homomorphism $*: \Lambda^p \to \Lambda^{n-p}$. It is uniquely determined by the condition:

$$g(\lambda, \mu)\mathrm{Vol}_M = \lambda \wedge * \mu \qquad \lambda, \mu \in C^\infty(\Lambda^p). \tag{2}$$

The homomorphism $*$ may be extended onto the whole algebra Λ.

The properties given follow directly from definition (2):

$$**|_{\Lambda^p} = (-1)^{p(n-p)} \tag{a}$$

and

$$g(*\lambda, *\mu) = g(\lambda, \mu). \tag{b}$$

By (1) and (2) we get

$$<\lambda, \mu> = \int_M \lambda \wedge * \mu, \qquad \lambda, \mu \in C^\infty(\Lambda^p). \tag{3}$$

When applying the Stokes theorem to $<d\lambda, \nu>$ we get, by (3), (a) and (b), so that

$$<d\lambda, \nu> = <\lambda, \bar{\omega} * d * \omega \nu>,$$

where ω and $\bar{\omega}$ restricted to Λ^p denote multiplication by $(-1)^p$ and $(-1)^{p(n-p)}$, respectively.

Consider the operator

$$d^* = \bar{\omega} * d * \omega. \tag{4}$$

Then d^* is formally adjoint to d in the sense that

$$< d\lambda, \nu > = < \lambda, d^*\nu > . \tag{5}$$

It is a first-order linear differential operator on M.

By definition (5), properties (a) and (b) of $*$ and by the well--known properties of d one can easily check the following properties of d^*:

(i) $d^*: C^\infty(\Lambda^p) \to C^\infty(\Lambda^{p-1})$
in particular, $d^*f = 0$ on functions f,

(ii) $d^*d^* = 0$,

(iii) $d^*\lambda = (-1)^{n(p+1)+1} * d * \lambda, \qquad \lambda \in C^\infty(\Lambda^p)$
in particular, if n is even,

(iii) $d^* = - * d *$.

Now, using d and d^* we can build the Laplace-Beltrami operator (Laplacian) as follows:

$$\Delta = d^*d + dd^*. \tag{6}$$

THEOREM 2.1. *Laplacian* Δ *is a second-order linear differential operator and it has the following properties:*

1° $\Delta: C^\infty(\Lambda^p) \to C^\infty(\Lambda^p)$,
while $\Delta = -\sum_i \partial^2/(\partial x^i)^2$ *on functions in* $M = \mathbb{R}^n$,

2° $*\Delta = \Delta *$,

3° $\Delta = (d + d^*)^2$,

4° $< \Delta\lambda, \mu > = < \lambda, \Delta\mu >$,

5° *If* $\Delta\mu = 0$, *then* $< \Delta\lambda, \mu > = 0$,

6° $\Delta\lambda = 0$ *if and only if* $d\lambda = 0 \wedge d^*\lambda = 0$.

Proof. 1° is evident. 2° is numerical. 3° is a consequence of the fact that $d^2 = 0$ and $d^{*2} = 0$. 4°: By (5) and (6), we get $< \Delta\lambda, \mu > = <(d^*d + dd^*)\lambda, \mu > = < d^*d\lambda, \mu > + < dd^*\lambda, \mu > = <\lambda, d^*d\mu > + < \lambda, dd^*\mu > = < \lambda, \Delta\mu >$. 5° is a consequence of 4°. 6°: The implication "$\Leftarrow$" is evident, so if $\Delta\lambda = 0$ then $0 = < \Delta\lambda, \mu > = <(d^*d + dd^*)\lambda, \lambda > = <d\lambda, d\lambda > + <d^*\lambda, d^*\lambda >$. It implies that $d\lambda = 0$ and $d^*\lambda = 0$, and completes the proof.

3. AHLFORS-LAPLACIAN ON A RIEMANNIAN MANIFOLD

There are many linear differential operators D which appear naturally in differential geometry. Generally, they are of the form

(3.1) $\quad D = \text{polynomial} \quad (\nabla, R\otimes, g\otimes, \text{contractions}).$

There arise questions of a dependence of a given operator D on the geometry of M. One of them is the following: How does D transform itself under a deformation of the Riemannian metric g ? Of course, there is a great variety of deformations: different for various parts of geometry. The conformal geometry, for instance, investigates deformations of the form $\bar{g} = \Omega^2 g$, where Ω is a positive function, and is interested in determining differential operators of the form (3.1) which are "conformally invariant (covariant)". On the other hand, they "produce" invariants which, to some extent, can determine the geometry of the manifold. For details we refer to [4, 5].

Let us take the example of the Ahlfors-Laplacian. In the papers [1,2] Ahlfors introduced and investigated the operator S acting on vector fields Z in $\mathbb{R}^m$. SZ is the matrix field of the form

$$SZ_{jk} = Z^j_{|k} + Z^k_{|j} + \frac{2}{m}\,\mathrm{tr}(Z^m_{|n})\,\delta_{jk}.$$

It is related to the so-called quasiconformal deformations in $\mathbb{R}^n$, namely, its norm is a measure of quasiconformality of Z [1]. The concept of S is generalized onto a Riemannian manifold [13]. If we consider S acting on 1-forms rather than on vector fields we get

$$S\alpha = 2\,\nabla^s \alpha + \frac{2}{m}\, d^*\alpha\, g,$$

where ∇^s is the symmetrized covariant differentiation: $(\nabla^s_\alpha)(X,Y) = \frac{1}{2}(\nabla_X \alpha(Y) + \nabla_Y \alpha(X))$. $S\alpha$ is a symmetric trace free tensor field of the type (0,2). If $\mathfrak{M}$ denotes the space of all such fields then $S\colon C^\infty(\Lambda^1) \to \mathfrak{M}$ is a first order linear operator. Its kernel consists of all conformal Killing 1-forms.

For $\phi \in \mathfrak{M}$ define

$$S^*\phi = 2\,\delta\,\phi,$$

where $\delta\phi$ is the 1-form given locally by $\delta\phi_j = -\nabla^i \phi_{ij}$. One can prove (see [13]) that S^* is formally adjoint to S in the following sense:

$$<S\alpha, \phi> = <\alpha, S^*\phi>, \qquad \alpha \in C^\infty(\Lambda^1), \quad \phi \in \mathfrak{M}.$$

Consequently, the Ahlfors-Laplacian

$$S^*S\colon C^\infty(\Lambda^1) \to C^\infty(\Lambda^1)$$

is a linear second order differential operator of the form (3.1). S^*S is related to the Laplace-Beltrami operator Δ by the formula

$$S^*S\,\alpha = -4\,\mathrm{Ric}\cdot\alpha + 2\,\Delta\alpha + \frac{2n-4}{n}\,dd^*\,\alpha,$$

where Ric is the Ricci tensor defined by (2.1). The formula expresses the dependence of S^*S on the geometry of M (Ricci term). On the other hand, S^*S "produces" local invariants listed in [5].

4. ELLIPTIC OPERATORS

The operators d, d^*, Δ, S, S^*, S^*S are examples of linear differential operators on M. Let us recall a general definition.

Let E and F be two vector bundles over M. A linear differential operator P from E into F is a linear mapping

$$P: C^\infty(E) \to C^\infty(F)$$

such that for every $\alpha \in C^\infty(E)$

$$\text{support}\, P\alpha \subset \text{support}\, \alpha .$$

The definition (cf. [9]) implies in a coordinate neighbourhood

$$P\alpha(x) = \sum_{|\gamma| \leq k} a_\gamma(x)\,(D^\gamma \alpha)(x).$$

The integer k is said to be the order of P, i.e. the order of P is the highest order of derivatives appearing in P.

It is interesting that many of the properties of P depend only on these derivatives. In order to refer to them we introduce the notion of the (principal) symbol.

Let P be a linear differential operator of order k. Let $x \in M$ and $\alpha \in C^\infty(E)$ with $v = \alpha(x)$. Let $w \in T^*_x M$. Take $f \in C^\infty(M)$ such that $f(p) = 0$, $df(p) = w$. The symbol $\sigma_P(w, x)\colon E_x \to F_x$ is the following homeomorphism:

$$\sigma_P(w, x)v = \frac{i^k}{k!}\,P(f^k\alpha)(x).$$

We say that P is elliptic if its symbol is injective at each point x.

Example 1. $\sigma_d(w)v = i\,w \wedge v$, $\sigma_{d^*}(w)v = -i\,w \mathbin{\llcorner} v$ (interior product), $\sigma_\Delta = w \wedge (w \mathbin{\llcorner} v) + w \mathbin{\llcorner} (w \wedge v)$.

Consequently, Δ is elliptic.

Example 2. $\sigma_S(w)\alpha = i(w \otimes \alpha + \alpha \otimes w - \frac{2}{m}\, g(w, \alpha) g)$, $\sigma_{S^*}(w)\phi = -2\,i\,w \mathbin{\llcorner} \phi$, $\sigma_{S^*S} = 2((w \mathbin{\llcorner} w)\alpha + w \mathbin{\llcorner} (\alpha \otimes w) - \frac{2}{m}(w \mathbin{\llcorner} \alpha)\, w \mathbin{\llcorner} g)$.

Consequently, S^*S is elliptic.

Operators Δ and S^*S point out at a general rule of building elliptic operators.

Assume that $P: C^\infty(E) \to C^\infty(F)$ and $Q: C^\infty(F) \to C^\infty(G)$ are first order differential operators and E, F, G are vector bundles equipped with a Riemannian (Hermitian) scalar product.

THEOREM 4.1 ([12]). *If the following symbol sequence*

$$E_x \xrightarrow{\sigma_P(w,x)} F_x \xrightarrow{\sigma_Q(w,x)} G_x$$

is exact at F_x *for every* $w \in T^*_x M$, $w \neq 0$, *then the second order operator*

$$L = PP^* + Q^*Q : C^\infty(F) \to C^\infty(F)$$

is elliptic at x.

In the special case $P = 0$ we get

COROLLARY. *If* $\sigma_Q(w, x)$ *is injective for every* $w \in T^*_x M$, $w \neq 0$, *then* Q^*Q *is elliptic.*

Observe that in the case where $E = \Lambda^{p-1}$, $F = \Lambda^p$, $G = \Lambda^{p+1}$ and P, Q are exterior differentiation, we derive, by Theorem 4.1, ellipticity of Δ, while Corollary implies ellipticity of S^*S.

5. HODGE DECOMPOSITION THEOREM AND ITS CONSEQUENCES.

Ellipticity of a linear differential operator has many interesting consequences. A decomposition type theorem is one of the most important.

Let us consider a simple example of a linear map $A: \mathbb{R}^n \to \mathbb{R}^m$. Then the equation $Ax = y$ has a solution if and only if y is orthogonal to $\ker A^*$. Indeed, $z \perp \operatorname{Im} A$ if and only if $\langle z, Ax \rangle = 0$ for all x. This is equivalent to $\langle A^*z, x \rangle = 0$ for all x which just means that $A^*z = 0$. We may sum it up in the form

$$\mathbb{R}^m = \operatorname{Im} A \oplus \ker A^*.$$

An analogous theorem (often called the Fredholm alternative) may be obtained for any linear elliptic differential operator on a compact manifold. We are going to present it in the special case of the Laplace -Beltrami operator Δ.

Before doing this let us state the following definitions:

A form ω is called harmonic if $\Delta\omega = 0$.

Denote by H^p the kernel of Δ restricted to $C^\infty(\Lambda^p)$, i.e. the vector space of all harmonic p-forms on M:

$$H^p = \{\omega \in C^\infty(\Lambda^p) : \Delta\omega = 0\}.$$

THEOREM 5.1 (The Hodge decomposition). *For each integer* p, $0 \leq p \leq n$

$$C^\infty(\Lambda^p) = \Delta(C^\infty(\Lambda^p)) \oplus H^p. \tag{1}$$

The decomposition is orthogonal.

Outline of the proof. First prove that H^p is of finite dimension, if not - H^p would consist an infinite orthonormal sequence $\{e_j\}$.

By Theorem 6.6 in [14], the sequence should consist a Cauchy subsequence, what is impossible, since $\|e_i - e_j\| = \sqrt{2}$, for $i \neq j$.

Since Δ is a selfadjoint operator $\Delta\alpha$ is orthogonal to every harmonic form. Consequently $\Delta(C^\infty(\Lambda^p)) \subset H^{p\perp}$. In order to prove the decomposition formula (1) it is enough to prove that $H^{p\perp} \subset \Delta(C^\infty(\Lambda^p))$. To this aim, take $\alpha \in H^{p\perp}$ and construct a linear functional ℓ on $\Delta(C^\infty(\Lambda^p))$ by setting $\ell(\Delta\beta) = \langle\alpha, \beta\rangle$ for all $\beta \in C^\infty(\Lambda^p)$. It is well defined, for if $\Delta\beta_1 = \Delta\beta_2$ then, by linearity of Δ, $\beta_1 - \beta_2 \in H^p$ and therefore $\langle\alpha, \beta_1 - \beta_2\rangle = 0$. It could be proved (using e.g. estimates of the Theorem 6.6 of [14]) that this linear functional ℓ is bounded on $\Delta(C^\infty(\Lambda^p))$. By the Hahn-Banach theorem it extends to a bounded linear functional on $C^\infty(\Lambda^p)$.

In this way we obtain a bounded linear functional $\ell: C^\infty(\Lambda^p) \to \mathbb{R}$ satisfying the condition $\ell(\Delta^*\beta) = \langle\alpha, \beta\rangle$ for all $\beta \in C^\infty(\Lambda^p)$ (observe that $\Delta^* = \Delta$). Each such functional ℓ is called a weak solution to the equation $\Delta\omega = \alpha$. Using now the Regularity Theorem (cf. [14], Theorem 6.5) saying that for every weak solution ℓ there exists $\omega \in C^\infty(\Lambda^p)$ such that $\ell(\beta) = \langle\omega\ \beta\rangle$ we deduce that $\Delta\omega = \alpha$, which completes the proof.

COROLLARY 5.2.

$$C^\infty(\Lambda^p) = d^*d(C^\infty(\Lambda^p)) \oplus dd^*(C^\infty(\Lambda^p)) \oplus H^p; \tag{2}$$

the decomposition is orthogonal.

P r o o f. By Theorem 5.1, equality (2) follows from the definition of Δ: $\Delta = d^*d + dd^*$ and the equality $\langle dd^*\alpha, d^*d\beta\rangle = \langle d^*\alpha, d^*(dd)\beta\rangle = 0$.

COROLLARY 5.3.

$$C^\infty(\Lambda^p) = d^*(C^\infty(\Lambda^{p+1})) \oplus d(C^\infty(\Lambda^{p-1})) \oplus H^p;$$

the decomposition is orthogonal.

P r o o f. By Corollary 5.2 it is enough to prove that $d^*(C^\infty(\Lambda^{p+1})) \cap d(C^\infty(\Lambda^{p-1})) = \emptyset$. Indeed, if $\alpha = d^*\alpha_1 = d^*\alpha_2$ where $\alpha_1 \in C^\infty(\Lambda^{p+1})$ and $\alpha_2 \in C^\infty(\Lambda^{p-1})$ then $\langle\alpha, \alpha\rangle = \langle d^*\alpha_1, d\alpha_2\rangle = \langle\alpha_1, dd\alpha_2\rangle = 0$, what implies $\alpha = 0$.

6. CONSEQUENCES OF THE HODGE DECOMPOSITION

Theorem 5.1 has many important consequences. Let us record few of them.

First of all we can answer the question of solvability of the equation $\Delta\omega = \alpha$:

PROPOSITION 6.1. *The equation* $\Delta\omega = \alpha$ *has a solution* $\omega \in C^\infty(\Lambda^p)$ *if and only if* α *is a* p*-form orthogonal to* H^p. *If* ω_1 *and* ω_2 *are two such solutions then* $\omega_1 - \omega_2$ *is harmonic.*

P r o o f. The first part follows directly from Theorem 5.1. The other one is a consequence of the linearity of Δ.

Now arises the question how to get a solution ω for a given α. It needs the construction of the Green operator:

PROPOSITION 6.2. *There exist a linear operator* $G\colon C^\infty(\Lambda^p) \to H^{p\perp}$ *defined by*

$$G(\lambda) = \omega \quad \textit{if and only if} \quad \Delta\omega = \lambda - H(\lambda) \tag{1}$$

where $H(\lambda)$ *denotes the harmonic component of* λ *in the Hodge decomposition.* G *is the Green operator for* Δ.

Proof. Take $\lambda \in C^\infty(\Lambda^p)$. By Theorem 5.1, $\lambda - H(\lambda) \in \Delta(C^\infty(\Lambda^p))$. By Proposition 6.1 there exist a unique $\omega \in H^{p\perp}$ such that $\Delta\omega = \lambda - H(\lambda)$. Put $G(\lambda) = \omega$ for completing the proof.

PROPOSITION 6.3. *If a linear operator* $A\colon C^\infty(\Lambda^p) \to C^\infty(\Lambda^q)$ *commutes with* Δ *i.e.* $\Delta A = \Delta A$ *then it commutes with* G. *In particular* G *commutes with* d, d^*, $*$ *and* Δ.

Proof. By (1), the operator G may be defined as follows

$$G = (\Delta | H^{p\perp})^{-1} \circ \pi_p, \tag{2}$$

where π_p denotes the projection of $C^\infty(\Lambda^p)$ into $H^{p\perp}$. By the assumption $A\Delta = \Delta A$, we get $A(H^p) \subset H^q$, which, by Theorem 5.1, implies A that $A(H^{p\perp}) \subset H^{q\perp}$. Consequently,

$$A \circ \pi_p = \pi_q \circ A \tag{3}$$

and, if we restrict ourselves to $H^{p\perp}$,

$$A \circ (\Delta | H^{p\perp}) = (\Delta | H^{q\perp}) \circ A,$$

or equivalently,

$$\Delta | (H^{q\perp})^{-1} \circ A = A \circ (\Delta | H^{p\perp})^{-1}. \tag{4}$$

Now $A \circ G = G \circ A$ follows from (2), (3) and (4).

Define p-th (de Rham) cohomology group $\mathcal{H}^p$ of M as follows. $\mathcal{H}^p = \{\mathrm{Ker}(d\colon C^\infty(\Lambda^p) \to C^\infty(\Lambda^{p+1}))\}/\{\mathrm{Im}(d\colon C^\infty(\Lambda^p) \to C^\infty(\Lambda^{p+1}))\}$. $\mathcal{H}^p$ is then a real vector space: the quotient space of the real vector space of closed p-forms modulo the subspace of exact p-forms. Elements of $\mathcal{H}^p$ (cosets) are called cohomology classes.

PROPOSITION 6.4. *The spaces* $\mathcal{H}^p$ *and* H^p *are isomorphic to each other. In particular, in each cohomology class there is precisely one harmonic form.*

Proof. Since, in the space of the forms, both addition and multiplication by numbers commute with d it is enough to prove the second part of the assertion. Let $\omega \in C^\infty(\Lambda^p)$. By Corollary 5.2 and the definition of the Green operator G, we get

$$\omega = d^*d\,G(\omega) + dd^*\,G(\omega) + H(\omega).$$

Since G commutes with d (cf. Proposition 6.3), we have

$$\omega = d^* G(d\omega) + dd^* G(\omega) + H(\omega).$$

If we assume that ω is p-closed form $(d\omega = 0)$, then

$$\omega - H(\omega) = dd^* G(\omega),$$

i.e. ω and $H(\omega)$ differ by an exact form, therefore, they are in the same cohomology class. On the other hand, if two harmonic forms differ by an exact form α, i.e. if $\omega_1 - \omega_2 = d\beta$, then they have to be equal, since, by Corollary 5.3, $\omega_1 - \omega_2$ is orthogonal to $d\beta$.

Remark. Cohomology groups $\mathcal{H}^p$ of a manifold M do not depend on the Riemannian structure, while the spaces H^p do. Since any differentiable manifold may be equipped with a Riemannian scalar product, we can see that the de Rham cohomology groups of a compact orientable manifold are all finite dimensional. In particular, we derive the following

COROLLARY 6.5. *The Euler characteristic*

$$\chi(M) =: \sum_{p=0}^{m} (-1)^p \dim \mathcal{H}^p$$

is finite.

Let us express the Euler characteristic by the spectrum of Δ. Consider two spaces

$$C^\infty(\Lambda)^+ = \bigoplus_{p\text{-even}} C^\infty(\Lambda^p) \quad \text{and} \quad C^\infty(\Lambda)^- = \bigoplus_{p\text{-odd}} C^\infty(\Lambda^p).$$

Then $C^\infty(\Lambda) = C^\infty(\Lambda)^+ \oplus C^\infty(\Lambda)^-$. Let us introduce the first order linear operator

$$P = d + d^*_{|C^\infty(\Lambda)^+} : \quad C^\infty(\Lambda)^+ \to C^\infty(\Lambda)^-$$

and its formal adjoint

$$P^* = d^* + d_{|C^\infty(\Lambda)^-} : C^\infty(\Lambda)^- \to C^\infty(\Lambda)^+.$$

Consequently,

$$P^*P = \Delta_{|C^\infty(\Lambda)^+} \quad \text{and} \quad PP^* = \Delta_{|C^\infty(\Lambda)^-}$$

and, therefore

$$\dim \mathrm{Ker}(P^*P) - \dim \mathrm{Ker}(PP^*) = \sum_{p=1}^{m} (-1)^p \dim H^p.$$

By Proposition 6.4 and the definition of the Euler characteristic $\chi(M)$ (cf. Corollary 6.5), we have

$$\chi(M) = \dim \mathrm{Ker}(\Delta_{\text{even}}) - \dim \mathrm{Ker}(\Delta_{\text{odd}}),$$

where $\Delta_{even} = \Delta|_{C^\infty(\Lambda)^+}$, $\Delta_{odd} = \Delta|_{C^\infty(\Lambda)^-}$.

Let $0, \ldots, 0 < \lambda_1 < \lambda_2 < \ldots$ be ordered eigenvalues of Δ_{even} listed according to their multiplicities and $0, \ldots, 0 < \mu_1 < \mu_2 \leq \ldots$ ordered eigenvalues of Δ_{odd}. Observe that

$$\lambda_1 = \mu_1, \quad \lambda_2 = \mu_2, \quad \ldots$$

Indeed, if $x \neq 0$ is an eigenvalue of $\Delta_{even} = P^*P$ then $P^*P\,\omega = \lambda\,\omega$ for some $\omega \neq 0$. Applying P to the both sides of the last equality we get $PP^*(P\omega) = \lambda(P\omega)$. Since $P\omega \neq 0$, λ is an eigenvalue of $\Delta_{odd} = PP^*$. Moreover, the multiplicity of λ for Δ_{odd} is at least as great as for Δ_{even}. The symmetry of reasoning completes the proof of the observation.

If, for an arbitrary real number t, we let

$$\mathrm{Trace}(e^{-t\,\Delta_{even}}) = \sum_j e^{-t\,\lambda_j} \quad \text{and} \quad \mathrm{Trace}(e^{-t\,\Delta_{odd}}) = \sum_j e^{-t\,\mu_j},$$

where both sums include also zero eigenvalue according to their multiplicites, then we have proved

PROPOSITION 6.6. (cf. [3])

$$\chi(M) = \mathrm{Trace}(e^{-t\,\Delta_{even}}) - \mathrm{Trace}(e^{-t\,\Delta_{odd}}).$$

It is well known that $\chi(M)$ is topologically invariant. Proposition 6.6 shows that it can be described by the spectrum of the differential operator Δ.

The spectrum of Δ can tell us much more: it determines to some extent the geometry of M. The relationship between the geometry of a manifold and the spectrum of the Laplace-Beltrami operator was investigated and explored extensively (spectral geometry). The subject is very wide so it could be discussed in a separate paper. We can only mention that the spectrum of Δ determines the dimension and the volume of the compact manifold. It also says whether M is of constant scalar curvature, constant sectional curvature, is Einstein, etc. Also, in the complex case it says, e.g., whether M is Kähler . For details we refer to [6] and the literature listed there.

7. COMPLEX STRUCTURES ON VECTOR SPACES

In this section we shall give some linear-algebraic results which will be applied to tangent spaces of manifolds in subsequent sections.

A *complex structure* on a real vector space V is a linear endomorphism $J: V \to V$ such that $J^2 = -\mathrm{id}$. A real vector space V with a complex structure J can be turned into a complex vector space as follows:

$$(a + i\,b)X := aX + bJX, \quad \text{for } X \in V \text{ and } a, b \in \mathbb{R}.$$

To make a complex structure exist on a real vector space V it is necessary (and, in fact, sufficient) that $\dim_{\mathbb{R}} V = 2n$. If the space V has real dimension $2n$ then V, considered as a complex vector space, has complex dimension n.

Example. Let us take into account the complex vector space $\mathbb{C}^n$ of n-tuplex of complex number $Z = (z^1, \ldots, z^n)$. If we set

$$z^k = x^k + i\,y^k, \qquad x^k, y^k \in \mathbb{R}, \qquad k = 1, \ldots, n,$$

then $\mathbb{C}^n$ can be identified with real vector space $\mathbb{R}^{2n}$; in fact this identification is $\mathbb{R}$-isomorphism between $\mathbb{R}^{2n}$ and $\mathbb{C}^n$. On $\mathbb{C}^n$ we have the linear endomorphism J_o given by

$$J_o Z = i\,Z, \qquad Z \in \mathbb{C}^n.$$

The complex structure of $\mathbb{R}^{2n}$, obtained from J_o by the above $\mathbb{R}$-isomorphism, maps $(x^1, \ldots, x^n, y^1, \ldots, y^2)$ into $(-y_1, \ldots, -y_n, x^1, \ldots, x^n)$ and is called the canonical complex structure of $\mathbb{R}^{2n}$. Its matrix, in terms of natural basis for $\mathbb{R}^{2n}$, has the form

$$J_o = \begin{pmatrix} 0 & I_n \\ -I_n & 0 \end{pmatrix},$$

where I_n is the identity matrix of degree n.

Let V be a real vector space and V^* its dual. A complex structure J on V induces a complex structure on V^* denoted also by J and given by

$$J\,Y^*(X) = Y^*(J\,X) \quad \text{for} \quad X \in V \quad \text{and} \quad Y^* \in V^*.$$

Let V be n-dimensional real vector space. By V^c we denote its *complexification*, i.e. $V^c = V \otimes_{\mathbb{R}} \mathbb{C}$. Then V becomes a real subspace of V^c in a natural manner. Any vector Z from V^c will be denoted by $Z = X + i\,Y$, where X and Y are taken from V. In the space V^c we have the endomorphism

$$Z = X + i\,Y \to X - i\,Y, \qquad X, Y \in V,$$

called the *complex conjugation*. The complex conjugation can be extended in a natural way on $T^r_s(V^c)$ (the space of tensor of the type (r,s) over V^c).

Assume that V is a real $2n$-dimensional vector space with a complex structure J. Then J can be extended to a complex structure on V^c by the formulae

$$J(X + i\,Y) := J\,X + i\,J\,Y \quad \text{for} \quad X, Y \in V.$$

It is easy to verify that J is $\mathbb{C}$-linear on V^c, where multiplication by i on V^c is given by the expression $i\,Z = i(X + i\,Y) = -Y + i\,X$.
Let set

$$V^{1,0} = \{Z \in V^c : JZ = iZ\} \quad \text{and} \quad V^{0,1} = \{Z \in V^c : JZ = -iZ\}.$$

We have the following

PROPOSITION 7.1.

(1) $V^{1,0} = \{X - iJX : X \in V\}$ *and* $V^{0,1} = \{X + iJX : X \in V\}$;

(2) $V^c = V^{1,0} \oplus V^{0,1}$ *(as complex subspaces)*;

(3) *The complex conjugation defines a real isomorphism between* $V^{1,0}$ *and* $V^{0,1}$.

Proof. It is evident.

The decomposition of V^{*c} in the direct sum induces the decomposition of exterior algebra ΛV^{*c}. We denote by $\Lambda^{p,q}V^{*c}$ the subspace of ΛV^{*c} spanned by $\alpha \wedge \beta$, where $\alpha \in \Lambda^p V_{1,0}$ and $\beta \in \Lambda^q V_{0,1}$. It is evident that the exterior algebra may be decomposed:

$$\Lambda^{*c} = \sum_{r=0}^{n} \Lambda^r V^{*c} \quad \text{with} \quad \Lambda^r V^{*c} = \sum_{p+q=r} \Lambda^{p,q} V^{*c}.$$

The complex conjugation in V^{*c} gives a real isomorphism between $\Lambda^{p,q} V^{*c}$ and $\Lambda^{q,p} V^{*c}$.

8. HERMITIAN INNER PRODUCTS

Let V be a real vector space with a complex structure J. A *Hermitian inner product* on V is an inner product h such that

$$h(JX, JY) = h(X, Y) \quad \text{for} \quad X, Y \in V.$$

Let h be a Hermitian inner product in a real vector space V with a complex structure J. Then h can be extended uniquely to the complex symmetric bilinear form denoted also by h on V^c by the following expression:

$$h(X + iY, Z + iW) = h(x, Z) + i\,h(Y, Z) + i\,h(X, W) - h(Y, W).$$

This extension has the following properties:

(i) $h(\bar{Z}, \bar{W}) = h(Z, W)$ for $Z, W \in V^c$,

(ii) $h(Z, \bar{Z}) > 0$ for non-zero $Z \in V^c$,

(iii) $h(Z, \bar{W}) = 0$ for $Z \in V^{1,0}$ and $W \in V^{0,1}$.

It is easy to check that the form

$$\omega(X, Y) = h(X, JY) \quad \text{for} \quad X, Y \in V$$

is skew-symmetric and so it is an element of $\Lambda^2 V^*$. The form ω can be uniquely extended to the skew-symmetric bilinear form on V^c, denoted also by ω, in the same manner as it was done for h. By an easy calculation we verify that

$$\omega(X \pm JX, Y \pm JY) = 0 \quad \text{for } X, Y \in V.$$

Therefore $\omega \in \Lambda^{1,1} V^{*c}$ and

$$\omega(X - iJX, Y + iJY) = 2(\omega(X, Y) - i\,h(X, Y)).$$

If $\{e_1, \ldots, e_n\}$ is a $\mathbb{C}$-basis for $V^{1,0}$ and $\{e_1^*, \ldots, e_n^*\}$ the basis for $V_{1,0}$, and if we set

$$h_{j\bar{k}} = h(e_j, \bar{e}_k) \quad \text{for } j, k = 1, 2, \ldots, n,$$

then $h_{j\bar{k}} = \bar{h}_{k\bar{j}}$ and $\omega = -2i \sum_{j,k=1}^{n} h_{j\bar{k}}\, e_j^* \wedge \bar{e}_k^*$.

The form ω is usually called the *Kähler form* (or *fundamental form*). It also is easily seen that the Kähler form is J-invariant.

9. COMPLEX AND ALMOST COMPLEX MANIFOLDS

It is known that the tangent space of a complex manifold M in each point has the canonical structure of a complex vector space. We sometimes investigate only the structure T_aM for each $a \in M$ instead of the complex structure underlying M. This leads to the class of manifolds richer than the class of complex manifolds.

Let M be C^∞-manifold of $\dim_{\mathbb{R}} M = 2n$. Assume that for every $a \in M$ on T_aM we have a structure of $\mathbb{C}$-vector space. We say that this structure depends smoothly on a point if for every $a \in M$ there exist a neighbourhood $U \ni a$ and a set of complex-valued 1-forms $\{\omega_i\}_{1 \le i \le n}$ on U such that the mapping

$$T_xM \to \mathbb{C}^n, \quad x \in M,$$

given by

$$T_xM \ni X \to (\omega_1(x)(X), \ldots, \omega_n(x)(X)) \in \mathbb{C}^n,$$

is $\mathbb{C}$-isomorphism (with respect to the given complex structure in T_xM). We have the following definition:

We say that a real manifold M is given an *almost complex structure* if every T_xM, $x \in M$, is given a complex structure which depends smoothly on $x \in M$. Forms $\omega_1, \ldots, \omega_n$, which locally give us the complex structure, are called *structural forms*.

In general structural forms are not closed. Let us denote by J_a, $a \in M$, the R-linear endomorphism of T_aM given by $X \to \sqrt{-1}\,X$ (multiplication by $\sqrt{-1}$ is taken with respect to a given $\mathbb{C}$-vector structure in T_aM). For a smooth vector field X on M we define the smooth vector field JX by

$$J X(a) = J_a X(a), \quad \text{for} \quad a \in M.$$

We can see that J is a tensor of the type (1,1) which fulfils $J^2 = -\mathrm{id}$. This tensor is sometimes called an almost complex structure and (M, J) an *almost complex manifold.*

PROPOSITION 9.1. *Every almost complex manifold is of even dimension and orientable.*

Proof. Choose a covering $\{U_j\}_{j \in I}$ of M such that in each U_j there exists a system of structural forms $\{\omega_1^j, \ldots, \omega_n^j\}$, where $2n = \dim_{\mathbb{R}} M$. Put

$$\omega^{(j)} = (i/2)^n \omega_1^j \wedge \bar{\omega}_1^{(j)} \wedge \ldots \wedge \omega_n^{(j)} \wedge \bar{\omega}_n^{(j)}.$$

This form is non-zero on U_j. We note that if $\{\omega_1, \ldots, \omega_n\}$ and $\{\omega_1', \ldots, \omega_n'\}$ are two systems of such forms on U and U', respectively, then there exist complex-valued functions $a_{\mu\nu}$ on $U \cap U'$ for $\mu, \nu = 1, 2, \ldots, n$ such that

$$\omega_\nu = \Sigma\, a_{\mu\nu}\, \omega_\mu' \quad \text{on} \quad U \cap U'.$$

Let $D = \det(a_{\mu\nu})$. Then

$$\omega_1 \wedge \bar{\omega}_1 \wedge \ldots \wedge \omega_n \wedge \bar{\omega}_n = |D|^2\, \omega_1' \wedge \bar{\omega}_1' \wedge \ldots \wedge \omega_n' \wedge \bar{\omega}_n'$$

and $|D|^2 > 0$. It means that

$$\omega^{(j)} = f_{jj'}\, \omega^{(j')} \quad \text{on} \quad U_j \cap U_{j'},$$

where $f_{jj'} > 0$. When using a partition of unity we can define in the standard way a non-zero $2n$-form on M, which proves that M is orientable. The fact that M has the even dimension follows from the existence of almost complex structure.

The orientation of an almost complex manifold M given in the proof above is called the *natural orientation.*

Let us consider the space $\mathbb{C}^n$ of n-tuples of complex numbers $(z^1, \ldots, z^n)$ with $z^i = x^i + i\, y^i$, $i = 1, 2, \ldots, n$. With respect to the coordinate system $(x^1, \ldots, x^n, y^1, \ldots, y^n)$ we define an almost complex structure on $\mathbb{C}^n$ by

$$J_o(\partial/\partial x^j) = -\partial/\partial y^j, \quad J(\partial/\partial y^j) = \partial/\partial x^j \quad \text{for} \quad j = 1, 2, \ldots, n.$$

The question arises: Is there any connection of this complex structure with holomorphic mappings? The answer is in the following

PROPOSITION 9.2. *A mapping of an open subset of* $\mathbb{C}^n$ *into* $\mathbb{C}^m$ *preserves the almost complex structures of* $\mathbb{C}^n$ *and* $\mathbb{C}^m$, *i.e.,* $f_* \circ J_o = J_o \circ f_*$ *if and only if* f *is holomorphic.*

Proof. Simple computation.

If M is a complex manifold then we may locally transport the complex structure from $\mathbb{C}^n$ by means of a chart and observe that by virtue of Proposition 9.2 this transport does not depend on the chart and in consequence, we obtain the global almost complex structure on M which comes from the complex structure of an underlying manifold. Such an almost complex structure is called *complex structure*. If $f: M \to M'$, where M and M' are complex manifolds, then the mapping f is holomorphic if and only if f preserves the complex structures of M and M', i.e. $f_* \circ J = f_* \circ J'$. If M is an almost complex manifold then

$$T_x^c(M) = T_x^{1,0}(M) \oplus T_x^{0,1}(M).$$

Elements of $T_x^c(M)$ are called complex vectors.
Similarly, we have the dual decomposition

$$T_x^{*c}(M) = T_x^{*1,0}(M) \oplus T_x^{*0,1}(M)$$

and, consequently,

$$\Lambda T_x^{*c}(M) = \sum_{r=0}^{n} \Big(\sum_{p+q=r} \Lambda^{p,q} T_x^{*c}(M)\Big).$$

If $\{\omega_1, \ldots, \omega_n\}$ is a set of structural forms in a neighbourhood of $x \in M$, then $\omega_1(x), \ldots, \omega_n(x)$ form $\mathbb{C}$-basis for $T_x^{*1,0}(M)$. Similarly, the basis for $\Lambda^{p,q} T_x^{*c}(M)$ is given by

$$\omega_{j_1}(x) \wedge \ldots \wedge \omega_{j_p}(x) \wedge \bar{\omega}_{k_1}(x) \wedge \ldots \wedge \bar{\omega}_{k_q}(x),$$

where $1 \le j_1 \le \ldots \le j_p \le n$ and $1 \le k_1 < \ldots < k_q \le n$. On passing to sections, we obtain

$$A^r = \sum_{p+q=r} A^{p,q},$$

where $A^{p,q} = \{\phi \in A^r : \phi(x) \in \Lambda^{p,q} T_x^{*c}(M) \quad \text{for} \quad x \in M\}$.
The differentials $d\omega_j$ are 2-forms and so

$$d\omega_j = \eta_j + \eta_j' + \eta_j'',$$

where η_j, η_j' and η_j'' the types (2,0), (1,1) and (0,2), respectively. We get

COROLLARY 9.3. *If* ω *is of the type* (p, q), *then* $d\omega$ *is the sum of four forms of the types* $(p-1, q+2)$, $(p, q+1)$, $(p+1, q)$ *and* $(p+2, q-1)$, *respectively. Besides, components of the types* $(p-1, q+2)$ *and* $(p+2, q-1)$ *equal zero if and only if all forms* η'' *in the above decomposition are equal to zero.*

An almost complex structure on a manifold M is called *integrable* if for any set $\{\omega_1, \ldots, \omega_n\}$ of structural forms their differentials $d\omega_j$ do not contain terms of the type (0,2).

Let Ω be 2-form on an almost complex manifold M. Define a new 2-form by

$$S_\Omega(X, Y) = \Omega(X, Y) + i\,\Omega(JX, Y) + i\,\Omega(X, JY) - \Omega(JX, JY)$$

for $X, Y \in TM$.

Next, we can easily extend S_Ω to the complex valued 2-form S_Ω on T^cM in the same way as h was extended. If Ω is of the type (2,0) or (1,1) then $S_\Omega(X, Y) = 0$ for every X, Y, but if Ω is of the type (0,2) then $S_\Omega(X, Y) = 2\Omega(X, Y)$. When applying these observations to $\Omega = d\omega_\nu$, where ω_ν is a structural form and using the following expressions:

$$d\omega_\nu(X, Y) = X\,\omega_\nu(Y) - Y\,\omega_\nu(X) - \omega_\nu([X, Y])$$

and

$$\omega_\nu(JX) = i\,\omega_\nu(X),$$

we get

COROLLARY 9.4. *An almost complex structure on* M *is integrable if and only if for every vector field* X *and* Y *on* M *the following equality is fulfilled:*

$$[X, Y] + J[JX, Y] + J[X, JY] - [JX, JY] = 0.$$

Remark. We can easily show that the quantity

$$N(X, Y) = [X, Y] + J[JX, Y] + J[X, JY] - [JX, JY],$$

for $X, Y \in TM$, is the tensor field of the type (1,2) which is called the *torsion* of J. Our Corollary says that an almost complex structure is integrable if it has no torsion.

Let $(x^1, \dots, x^{2n})$ be a local coordinate system in M. By setting $X = \partial/\partial x^j$ and $Y = \partial/\partial x^k$ in the equation defining N we can see that the components of N are of the form

$$N^i_{jk} = 2\sum_{h=1}^{2n} (J^h_j\,\partial_h J^i_k - J^h_k\,\partial_h J^i_j - J^i_h\,\partial_j J^h_k + J^i_h\,\partial_k J^h_j),$$

where $\partial_h = \partial/\partial x^h$. We observe that the complex structure has no torsion because all terms $\partial_h J^i_j$ equal 0 in a complex local coordinate system. The question arises whether a sufficiently smooth differentiable almost complex manifold satisfying $N = 0$ is, in fact, a complex analytic manifold. For the answer we have

THEOREM (Newlender and Nirenberg [10, 11]). *An almost complex structure is a complex structure if it is integrable, i.e. if it has no torsion.*

Proof. For the proof in the case of the real analytic category see [8].

Let M be an almost complex manifold. A Riemannian metric h on M is said be *Hermitian* if it satisfies

$$h(JX, JY) = h(X, Y) \quad \text{for} \quad X, Y \in TM.$$

An almost complex manifold (a complex manifold, resp.) with a Hermitian metric h is called an *almost Hermitian manifold* (a *Hermitian manifold*, resp.).

We can easily see that any paracompact almost complex manifold admits a Hermitian metric. It is enough to take any Riemannian metric g and set

$$h(X, Y) = g(X, Y) + g(JX, JY) \quad \text{for any vectors } X \text{ and } Y.$$

By the previous section we see that a Hermitian metric h can be extended to $\mathbb{C}$-linear tensor on $T^c M$, also denoted by h, such that

(i) $h(\bar{Z}, \bar{W}) = \overline{h(Z, W)}$ for $X, Y \in TM$;

(ii) $h(Z, \bar{Z}) > 0$ for any non-zero complex vector Z;

(iii) $h(Z, \bar{W}) = 0$ for any vector field Z of the type $(1,0)$ and for any vector W of the type $(0,1)$.

The *fundamental 2-form* (or *Kähler form*) ω associated with J is defined as follows:

$$\omega(X, Y) = h(JX, Y) \quad \text{for} \quad X, Y \in TM.$$

It can be extended to $T^c M$ by bilinearity (see the previous section). The extended form ω is then of the type $(1,1)$. Consider two vectors $V = (X - iJX) \in T^{1,0}M$ and $W = (Y + iJY) \in T^{0,1}M$. We have

$$h(X - iJX, Y + iJY) = 2h(X, Y) - 2i\,\omega(X, Y).$$

It implies that the Riemannian structure g is equal to $1/2 \operatorname{Re} h$ and the fundamental form ω equals $i/2 \operatorname{Im} h$. When identifying $T^{0,1}M$ with $\bar{T}^{1,0}M$, we can also treat h as a Hermitian tensor on the holomorphic tangent bundle $T^{1,0}M \times T^{1,0}M \ni (V, W) \mapsto h(V, \bar{W})$. Obviously,

(a) $h(V, \bar{W}) = \overline{h(W, V)}$;

(b) $h(aV, \overline{bW}) = a\bar{b}\, h(V, \bar{W})$;

(c) $h(V, \bar{V}) \geq 0$.

If a real $(1,1)$-form ω is given, we can reconstruct a tensor h from it, but there is no reason for it to be positive definite. If this is the case, h is the Hermitian structure and ω is called positive. In local coordinates the real metric will be defined by $h(\partial/\partial x^j, \partial/\partial x^k) = h_{jk}$ and $ds^2 = \Sigma\, h_{jk}\, dx^j \otimes dx^k$, which means that $h(X^j\, \partial/\partial x^j, X^k\, \partial/\partial x^k) = \Sigma\, h_{jk}\, X^j X^k$. Similarly, in the complex case, we call j and $\bar{j}$ the indices of the coordinates z^j and $\bar{z}^j$ $(j = 1, 2, \ldots, n)$ and J, K the systems of coordinates $(j, \bar{j})$, and we set

$$h_{JK} = h(\partial/\partial z^J, \partial/\partial z^K).$$

The above conditions imply that $h_{jk} = h_{\bar{j}\bar{k}} = 0$, $ds^2 = \Sigma h_{j\bar{k}}(dz^j \otimes d\bar{z}^k +$

$+ d\bar{z}^k \otimes dz^j)$, and $\omega = -2i \sum h_{j\bar{k}} dz^j \wedge d\bar{z}^k$.

Remark. We often describe ds^2 only by its action on $T^{1,0}M$ as $ds^2 = \sum h_{j\bar{k}} dz^j \otimes d\bar{z}^k$. To describe it on TM, the complex conjugate term should be odd. That is why some authors use $ds^2 = 2\sum h_{j\bar{k}} dz^j \otimes d\bar{z}^k$ as a notation. Another way of expressing locally the Hermitian structure is defining a coframe for h, i.e. an n-tuple $(\phi^1, \ldots, \phi^n)$ of forms of the type $(1,0)$, orthonormal with respect to h. This is always possible (locally) by the diagonalization process. Thus, we have

$$ds^2 = \sum \phi^j \otimes \bar{\phi}^j \quad (+\sum \bar{\phi}^j \otimes \phi^j)$$
$$\omega = -2i \sum \phi^j \wedge \bar{\phi}^j.$$

Finally, we will note that a Hermitian structure induces a volume element on M.

The *canonical volume element* on (M, h) is defined in local coordinates by

$$V_h = \det(h_{j\bar{k}})\ dx^1 \wedge \ldots \wedge dx^{2n}, \quad \dim_{\mathbb{R}} M = 2n.$$

The change of variable formulae implies that V_h is globally defined. Obviously,

$$\omega^n = (2/i)^n n! \det(h_{j\bar{k}}) dz^1 \wedge d\bar{z}^1 \wedge \ldots \wedge dz^n \wedge d\bar{z}^n =$$
$$(-4)^n n! \det(h_{j\bar{k}}) dx^1 \wedge dy^1 \wedge \ldots \wedge dx^n \wedge dy^n = (-4)^n n!\ V_h.$$

The concept of the compatibility of the complex and Riemannian structures leads to the concept of Kähler metric.

A Hermitian metric on a complex manifold is called *Kähler metric* if the fundamental 2-form is closed. A complex manifold which admits a Kähler metric is called *Kähler manifold*. It is easy to verify that any complex submanifold of a Kähler manifold is a Kähler manifold. The Kähler form on it is simply the pull-back of the orginal one. From the above we can see that any algebraic manifold is a Kähler manifold.

A Kähler metric has an interesting property, namely: osculation with the Euclidean metric of the second order. We say that the metric h osculates the Euclidean metric at the second order if for every point $z_0 \in M$ there exists in a neighbourhood of z_0 a system of coordinates $z = (z^j)$ such that

$$ds^2 = \sum (\delta_{ij} + g_{ij}) dz^i \otimes d\bar{z}^j,$$

where $g_{ij}(z_0) = 0$ and $d\, g_{ij}(z_0) = 0$. For the proof of this fact see [7].

10. HARMONIC THEORY ON COMPACT COMPLEX MANIFOLDS

Let M be a compact connected Hermitian manifold of complex dimension $n=2m$. We wish to find canonical representatives for the Dolbeault cohomology groups $H^{p,q}_{\bar\partial}(M)$. To do this, we copy the development in the real case replacing d by $\bar\partial$ and adding complex conjugate signes in various places.

First, the Hermitian metric on M induces a Hermitian inner product $(\cdot\ ,\ \cdot)_x$ on each space $\Lambda^{p,q}T^{*c}_x(M)$ and hence it induces a "global" Hermitian inner product on $A^{p,q}(M)$ by

$$<\psi,\eta>=\int_M(\psi(x),\eta(x))_x dV(x)\ , \qquad \psi,\eta\in A^{p,q}$$

making $A^{p,q}$ into a complex pre-Hilbert space. The star operator $*$: $A^{p,q}\to A^{m-p\ m-q}$ is defined by the requirement that $\psi(x)\wedge *\eta(x)=(\psi(x),\eta(x))_x dV(x)$. By the above it is easy to verify that $**\eta=(-1)^{p+q}\eta$ for every $\eta\in A^{p,q}$. We shall begin with the following

PROPOSITION 10.1. *The formal adjoint* $\bar\partial^*: A^{p,q}\to A^{p,q-1}$ *of* $\bar\partial: A^{p,q-1}\to A^{p,q}$ *is given by*

$$\bar\partial^*=-*\bar\partial *\ .$$

Proof. By applying the Stokes theorem we get

$$<\bar\partial\phi,\psi>=\int_M\bar\partial\phi\wedge *\psi=\int_M d\phi\wedge *\psi=(-1)^{p+q}\int_M\phi\wedge d(*\psi)=$$
$$(-1)^{p+q}\int_M\phi\wedge\bar\partial *\psi=(-1)^{p+q}(-1)^{2n-p-q+1}\int_M\phi\wedge *(*\bar\partial *)\psi=$$
$$-<\phi,*\bar\partial *\psi>.$$

The $\bar\partial$-*Laplacian* $\Delta_{\bar\partial}: A^{p,q}\to A^{p,q}$ is defined by $\Delta_{\bar\partial}=\bar\partial^*\bar\partial+\bar\partial\bar\partial^*$. A (p,q)-form ψ is said to be $\bar\partial$ harmonic if $\Delta_{\bar\partial}\psi=0$.

Let us denote by $H^{p,q}=Z^{p,q}/B^{p,q}$ the (p,q)-cohomology group in the sense of Dolbeault. Now let a Dolbeault cohomology class $[\psi]\in H^{p,q}$ be represented by (p,q)-form ψ. Then the other $\bar\partial$-closed (p,q)-form in this cohomology is the form $\psi+\bar\partial\eta$, where $\eta\in A^{p,q-1}$. Thus, the set of $\bar\partial$-closed (p,q)-form in a given cohomology class is an affine subspace $S\subset A^{p,q}$. A natural question is whether there exists which has a minimum norm in S. Since $A^{p,q}$ is not complete there may not be such an element. To study this we find a criterion for ψ to have a minimum norm. First note that $\|\psi+\bar\partial\eta\|^2=\|\psi\|^2+\|\bar\partial\eta\|^2+2\,\mathrm{Re}<\psi,\bar\partial\eta>$.

PROPOSITION 10.2. *A $\bar\partial$-closed form ψ is of minimum norm within its Dolbeault cohomology class if and only if* $\bar\partial^*\psi=0$.

Proof. If $\bar\partial^*\psi=0$ then for every $\eta\in A^{p,q-1}$ we have

$$\|\psi+\bar\partial\eta\|^2=\|\psi\|^2+\|\bar\partial\eta\|^2+2\,\mathrm{Re}<\psi,\bar\partial\eta>=$$
$$\|\psi\|^2+\|\bar\partial\eta\|^2+2\,\mathrm{Re}<\bar\partial^*\psi,\eta>=\|\psi\|^2+\|\eta\|^2\geq\|\psi\|^2.$$

Therefore ψ has a minimal norm in its cohomology class. Conversely, if ψ has a minimal norm then, using 1-parameter variation of its norm, we get

$$0 = d/dt\,\|\psi + t\,\bar{\partial}\,\eta\|^2_{|t=0} = 2\,\mathrm{Re}\langle\psi, \bar{\partial}\,\eta\rangle$$

and

$$0 = d/dt\,\|\psi + i\,t\bar{\partial}\,\eta\|^2_{|t=0} = 2\,\mathrm{Im}\langle\psi, \bar{\partial}\,\eta\rangle,$$

for every $\eta \in A^{p,q-1}$. Hence $\langle\psi, \bar{\partial}\,\eta\rangle = \langle\bar{\partial}^*\psi, \eta\rangle = 0$ for every $\eta \in A^{p,q-1}$, therefore $\bar{\partial}^*\psi = 0$.
As in the real case we have, similarly,

PROPOSITION 10.3. $\Delta_{\bar{\partial}}\psi = 0$ *if and only if* $\bar{\partial}^*\psi = 0$ *and* $\bar{\partial}\psi = 0$.

Proof. $\langle\Delta_{\bar{\partial}}\psi, \psi\rangle = \|\bar{\partial}^*\psi\|^2 + \|\bar{\partial}\,\psi\|^2$. Thus if $\Delta_{\bar{\partial}}\psi = 0$, we have to get

$$\|\bar{\partial}^*\psi\| = 0 \quad \text{and} \quad \|\bar{\partial}\,\psi\| = 0,$$

hence $\bar{\partial}^*\Psi = 0$ and $\bar{\partial}\,\psi = 0$. The converse is trivial.

We may consider "Dirichlet integral"

$$D(\psi) = \tfrac{1}{2}(\|\bar{\partial}\,\psi\|^2 + \|\bar{\partial}^*\psi\|^2).$$

We observe that harmonic forms are the critical points of D. Indeed, for 1-parameter variation $t \to \psi + t\,\eta$, where $\psi, \eta \in A^{p,q}(M)$,

$$\begin{aligned} d/dt\,D(\psi + t\,\eta)_{|t=0} &= d/dt(D(\psi) + t\langle\bar{\partial}\psi, \bar{\partial}\eta\rangle + \\ &+ t\langle\bar{\partial}^*\psi, \bar{\partial}^*\eta\rangle + t^2 D(\eta))_{|t=0} = \langle\bar{\partial}\psi, \bar{\partial}\eta\rangle + \\ &+ \langle\bar{\partial}^*\psi, \bar{\partial}^*\eta\rangle = \langle\Delta_{\bar{\partial}}\psi, \eta\rangle. \end{aligned}$$

Thus ψ is a critical point of D if $\langle\Delta_{\bar{\partial}}\psi, \eta\rangle = 0$ for every $\eta \in A^{p,q}$, which is equivalent $\Delta_{\bar{\partial}}\psi = 0$. From the above considerations it follows that every critical point gives the absolute minimum of D.

Let us denote $\mathcal{H}^{p,q}_{\bar{\partial}}(M) = \{\psi \in A^{p,q}(M)\colon\ \Delta_{\bar{\partial}}\psi = 0\}$. Similarly as in the real case, we have

THEOREM 10.4 (The Hodge decomposition). *For each integer* p *and* q *with* $0 \le p \le m$, $0 \le q \le m$, $\mathcal{H}^{p,q}_{\bar{\partial}}$ *is finite dimensional and the following orthogonal direct decomposition holds:*

$$A^{p,q} = \mathcal{H}^{p,q}_{\bar{\partial}} \oplus \Delta_{\bar{\partial}}(A^{p,q}) = \mathcal{H}^{p,q}_{\bar{\partial}} \oplus \bar{\partial}(A^{p,q-1}) \oplus \partial(A^{p-1,q}).$$

Proof. Analogously as in the real case.

If we denote by $\pi_{(\mathcal{H}^{p,q}_{\bar{\partial}})^{\perp}}$ the orthogonal projection on $(\mathcal{H}^{p,q}_{\bar{\partial}})^{\perp}$ then by the Hodge decomposition we can define the operator $G = (\Delta_{\bar{\partial}}|(\mathcal{H}^{p,q}_{\bar{\partial}})^{\perp})^{-1} \circ \pi_{(\mathcal{H}^{p,q}_{\bar{\partial}})^{\perp}}\colon A^{p,q} \to (\mathcal{H}^{p,q}_{\bar{\partial}})^{\perp}$. This operator is bounded, selfadjoint and compact. Moreover, it has the following properties:

$$\bar\partial \circ G = G \circ \bar\partial \quad \text{and} \quad \bar\partial^* \circ G = \bar\partial^* \circ G \tag{1}$$

$$id = \pi_{\mathcal{H}^{p,q}_{\bar\partial}} + \Delta_{\bar\partial} \circ G \quad \text{on} \quad A^{p,q}. \tag{2}$$

This operator is called the Green's operator for $\Delta_{\bar\partial}$ and, by the definition, $G(\alpha)$ is the unique solution of $\Delta_{\bar\partial}\omega = \alpha - \pi_{\mathcal{H}^{p,q}_{\bar\partial}}\alpha$ in $(\mathcal{H}^{p,q}_{\bar\partial})^{\perp}$.

COROLLARY 10.5. *Each Dolbeault cohomology class contains the unique harmonic representative, i.e.:*

$$\mathcal{H}^{p,q}_{\bar\partial} \simeq H^{p,q}_{\bar\partial}.$$

Proof. Let α be an arbitrary (p,q)-form on M. From the definition of the Green's operator G, we have

$$\alpha = \bar\partial\bar\partial^* G\,\alpha + \bar\partial^*\bar\partial\, G\alpha + \pi_{\mathcal{H}^{p,q}}\alpha.$$

Since G commutes with $\bar\partial$ we get

$$\alpha = \bar\partial\bar\partial^* G\alpha + \bar\partial^* G\,\bar\partial\,\alpha + \pi_{\mathcal{H}^{p,q}}\alpha.$$

Thus, if α is $\bar\partial$-closed,

$$\alpha = \bar\partial\bar\partial^* G\,\alpha + \pi_{\mathcal{H}^{p,q}}\alpha,$$

so $\pi_{\mathcal{H}^{p,q}}\alpha$ is a harmonic (p,q)-form in the same Dolbeault cohomology class as α is. If two harmonic forms α_1 and α_2 differ by an exact form $\bar\partial\beta$, then

$$0 = \bar\partial\,\beta + (\alpha_1 - \alpha_2).$$

Yet $\bar\partial\,\beta$ and $(\alpha_1 - \alpha_2)$ are orthogonal since

$$\langle\bar\partial\,\beta, \alpha_1 - \alpha_2\rangle = \langle\beta, \bar\partial^*\alpha_1 - \bar\partial^*\alpha_2\rangle = \langle\beta, 0\rangle = 0.$$

Thus, $\bar\partial\,\beta = 0$ (direct sum decomposition) and $\alpha_1 = \alpha_2$, which proves that there is the unique harmonic form in each Dolbeault cohomology class.

On a compact Hermitian manifold we define a number of operators on the space A such as d, ∂, $\bar\partial$, their adjoints δ, ∂^*, $\bar\partial^*$ and the associated Laplacians $\Delta_d = d\,\delta + \delta\,d$, Δ_∂ and $\Delta_{\bar\partial}$, respectively.
We define three more operators:

(1) $d^c = i/4\pi(\bar\partial - \partial)$;

(2) $L\colon A^{p,q} \to A^{p+1,\,q+1}$ by the formula
$L(\eta) = \eta \wedge \omega$, where ω is the fundamental form;

(3) $\Lambda = L^*\colon A^{p,q} \to A^{p-1,\,q-1}$, formally adjoint to L.

Note that d^c (like d) is a real operator and

$$d\, d^c = - d^c\, d.$$

On a general Hermitian manifold there are no simple relations between these operators. In the Kähler case, however, we shall establish some of Hodge identities joining them together.

LEMMA 10.6. (i) $[\Lambda, d] = \Lambda \circ d - d \circ \Lambda = - 4\pi\, d^{c*}$,

(ii) $[L, d^*] = 4\pi\, d^c$,

(iii) $[\Lambda, \bar\partial] = - i\, \partial^*$,

(iv) $[\Lambda, \partial] = i\, \bar\partial^*$.

Proof. For the proof see [7].

From the above lemma we get easily the following

PROPOSITION 10.7. *On a compact Kähler manifold*

$$[L, \Delta_d] = 0 \quad \text{and} \quad [\Lambda, \Delta_d] = 0.$$

Proof. Since ω is closed,

$$d(\omega \wedge \eta) = \omega \wedge d\eta \quad \text{or} \quad [L, d] = 0 \quad \text{and}$$
$$[\Lambda, d^*] = 0.$$

Then $\Lambda(dd^* + d^*d) = (d\Lambda d^* - 4\pi\, d^{c*} d^*) + d^* \Lambda d = d\Lambda d^* + 4\pi\, d^* d^{c*} + d^* \Lambda d = (dd^* + d^*d)\Lambda$.

Lemma 10.6 and Proposition 10.7 enable us to prove the following fundamental fact about complex Laplacians:

THEOREM 10.8. *On a compact Kähler manifold*

$$\Delta_d = 2\,\Delta_{\bar\partial} = 2\,\Delta_\partial .$$

Proof. First, we shall show that $\partial\bar\partial^* + \bar\partial^*\partial = 0$. Since

$$\Lambda\partial - \partial\Lambda = i\,\bar\partial^*,$$

so

$$i(\partial\bar\partial^* + \bar\partial^*\partial) = \partial(\Lambda\partial - \partial\Lambda) + (\Lambda\partial - \partial\Lambda)\partial = \partial\Lambda\partial - \partial\Lambda\partial = 0.$$

Then

$$\Delta_d = (\partial + \bar\partial)(\partial^* + \bar\partial^*) + (\partial^* + \bar\partial^*)(\partial + \bar\partial) = \Delta_\partial + \Delta_{\bar\partial} + (\partial\bar\partial^* + \bar\partial^*\partial) = \Delta_\partial + \Delta_{\bar\partial} .$$

Finally, we show that $\Delta_\partial = \Delta_{\bar\partial}$:

$$- i\,\Delta_\partial = \partial(\Lambda\bar\partial - \bar\partial\Lambda) + (\Lambda\bar\partial - \bar\partial\Lambda)\partial = \partial\Lambda\bar\partial - \partial\bar\partial\Lambda + \Lambda\bar\partial\partial - \bar\partial\Lambda\partial$$

and

$$i\,\Delta_{\bar\partial} = \bar\partial(\Lambda\partial - \partial\Lambda) + (\Lambda\partial - \partial\Lambda)\bar\partial = \bar\partial\Lambda\partial - \bar\partial\partial\Lambda + \Lambda\partial\bar\partial + \partial\Lambda\bar\partial = i\,\Delta_{\partial},$$

which proves our theorem.

COROLLARY 10.9. *On a compact Kähler manifold* Δ_d *preserves the bidegree.*

Now let us see what this implies on cohomology. To avoid confusion, set

$$H_d^{p,q} = Z_d^{p,q}/B_d^{p,q}$$

$$\mathcal{H}_d^{p,q} = \{\eta \in A^{p,q}:\ \Delta_d\,\eta = 0\}$$

$$\mathcal{H}_d^{r} = \{\eta \in A^{r}:\ \Delta_d\,\eta = 0\}$$

and, similarly, for ∂ and $\bar\partial$. Since $\Delta_d = 2\,\Delta_{\bar\partial}$, we can immediately see that

$$\mathcal{H}_d^{p,q} = \mathcal{H}_{\bar\partial}^{p,q}.$$

We also have

$$\mathcal{H}_d^{r} = \bigoplus_{p+q=r} \mathcal{H}_d^{p,q}.$$

Indeed, all (p,q)-components of a harmonic form are harmonic since $[\Delta_d, \pi_{p,q}] = 0$ (Δ_d preserves the type). Since Δ_d is real we have also $\mathcal{H}_d^{p,q} = \overline{\mathcal{H}_d^{q,p}}$. If η is a closed form of the type (p,q) then

$$\eta = \mathcal{H}(\eta) + dd^*G(\eta),$$

where $\mathcal{H}(\eta)$ is the harmonic part of η which is also of the type (p,q). Hence $H_d^{p,q} \simeq \mathcal{H}_d^{p,q}$. When combining them with Hodge isomorphism $H^*_{Dr} \simeq \mathcal{H}^*$, we get

Hodge decomposition: For a compact Kähler manifold we have the following isomorphisms for complex cohomologies:

$$H^r(M, C) \simeq \bigoplus_{p+q=r} H_d^{p,q} \simeq \bigoplus_{p+q=r} H_{\bar\partial}^{p,q} \simeq \bigoplus_{p+q=r} H^q(M, \Omega^p)$$

and

$$H_d^{p,q} = \overline{H_d^{q,p}}.$$

As a special case of this decomposition we have

$$H^{p,0} \simeq H^0(M, \Omega^p) \quad \text{(the space of holomorphic } p\text{-forms).}$$

In fact, we have the following

PROPOSITION 10.10. *The holomorphic* p*-forms on a compact Kähler manifold are the harmonic* $(p,0)$*-forms for any Kähler metric.*

P r o o f. We have to prove the equality between spaces in question and not only isomorphism. Since $\Delta_d = 2\Delta_{\bar\partial}$, we have $\mathcal{H}_d^{p,0} = \mathcal{H}_{\bar\partial}^{p,0}$. In the Hodge decomposition we have, in general,

$$Z_{\bar\partial}^{p,q} = \mathcal{H}_{\bar\partial}^{p,q} \oplus B_{\bar\partial}^{p,q}$$

with $B_{\bar\partial}^{p,q} = \bar\partial A^{p,q-1}$.
For $q = 0$, we have, therefore, $Z_{\bar\partial}^{p,0} = \mathcal{H}_{\bar\partial}^{p,0}$. And $Z_{\bar\partial}^{p,0}$ is precisely the space of holomorphic p-forms.

Positive numbers $h^{p,q} = \dim_{\mathbb{C}} H_{\bar\partial}^{p,q}$ are called *Hodge numbers*.

These numbers on a Kähler manifold satisfy various conditions.

PROPOSITION 10.11. *On a compact Kähler manifold*

$$b_r = \sum_{p+q=r} h^{p,q}$$

$$h^{q,p} = h^{p,q}$$

$$h^{p,q} = h^{n-p\ n-q}$$

$$h^{p,p} \geq 1.$$

COROLLARY 10.12. *The odd Betti numbers of a compact Kähler manifold are even.*

P r o o f. $b_{2s+1} = \sum_{p=0}^{2s+1} h^{p,\,2s+1-p} = 2\sum_{p=0}^{s} h^{p,\,2s+1-p}$.

We now pause to consider a few examples. Since Betti numbers are independent of the metric, we can compute them by picking a simple metric and counting the number of harmonic forms with respect to this metric.

Example 1. Consider S^6 with the standard metric of the constant curvature. It is not difficult to show that with respect to this metric the only harmonic forms are the constant functions and multiples of the volume form. Thus, $b_o = b_6 = 1$, $b_i = 0$ for $i = 1, 2, \ldots, 5$, and the Euler number $\chi = 2$.

Example 2. Consider $S^3 \times S^3$ with the metric induced by the standard metric on S^3. In addition to $b_o = b_6 = 1$ we have $b_2 = 0$ and $b_3 = 2$. Since the volume form on S^3 is harmonic, so $\chi = 0$.

Example 3. Consider $S^2 \times S^2 \times S^2$. Now we have $b_o = b_6 = 1$, $b_2 = b_4 = 3$, $b_1 = b_3 = b_5 = 0$, and $\chi = 8$.

Example 4. Consider T^6. All the constant forms are harmonic. Hence $b_o = b_6 = 1$, $b_1 = b_5 = 6$, $b_2 = b_4 = 15$, $b_3 = 20$, and $\chi = 0$.

We also know that S^6 is not a complex manifold; $S_2^3 \times S_2^3$ is a complex manifold but does not admit any Kähler metric, $S^2 \times S^2 \times S^2$ admits Kähler metric but not Ricci flat one while T^6 admits Ricci flat Kähler metric.

References

[1] AHLFORS, L.V.: 'Quasiconformal deformations and mappings in $\mathbb{R}^n$', *J. Analyse Math.* 30 (1976), 74-97.

[2] ——: 'A singular integral equation connected with quasiconformal mappings in space', *Enseign. Math.* (2) 24 (1978), 225-236.

[3] BEALS, M., FEFFERMAN, C., GROSSMAN, R.: 'Strictly pseudoconvex domains in C^n', *Bull. Amer. Math. Soc.* (New Series), 8 (1983), 125-322.

[4] BRANSON, T.P.: 'Differential operators canonically associated to a conformal structure', *Math. Scand.* 57 (1985), 295-345.

[5] ——, ØRSTED, B.: Conformal deformation of the heat operator', preprint, *Kobenhavns Universität*, 1986.

[6] GILKEY, P.B.: *Invariance theory, the heat equation, and the Atiyah--Singer index theorem*, Publish or Perish 1984.

[7] GRIFFITHS, P., HARRIS, J.: *Principles of algebraic geometry*, John Wiley, New York, 1978.

[8] KOBAYASHI, S., NOMIZU, K.: *Foundations of differential geometry II*, Interscience Publ., New York-London-Sidney, 1969.

[9] NARASIMHAN, R.: *Analysis on real and complex manifolds*, North Holland Publ. Company, Amsterdam 1968.

[10] NEWLANDER, A., NIRENBERG, L.: 'Complex analytic coordinates in almost complex manifolds', *Ann. Math.* 65 (1954), 391-404.

[11] NIRENBERG, L.:'A complex Frobenius theorem', *In Seminar Analytic Functions*, 1, *Princeton Univ. Press*, Princeton 1957, 172-189.

[12] PALAIS, R.: *Seminar on the Atiyah-Singer index theorem*, Princeton NJ, Princeton Univ. Press 1965.

[13] PIERZCHALSKI, A.: 'On quasiconformal deformations of manifolds and hypersurfaces, *Ber. Univ. Jyväskylä*, Math. Inst. 28 (1984), 79-94.

[14] WARNER, F.: *Foundations of differential geometry and Lie groups*, Scott-Foresman, Glenview, Illinois, 1971.

EMBEDDING OF SOBOLEV SPACES INTO LIPSCHITZ SPACES

Ari Lehtonen
Department of Mathematics
University of Jyväskylä
SF-40100 Jyväskylä, Finland

ABSTRACT. The main result of the paper is that if Ω is a bounded uniform domain in $\mathbb{R}^n$ and $p > n$, then the Sobolev space $W^{1,p}(\Omega)$ embeds continously into $C^\alpha(\bar{\Omega})$, $\alpha = 1 - n/p$.

1. INTRODUCTION

It is well-known that for a smooth domain Ω in $\mathbb{R}^n$ (e.g. a bounded Lipschitz domain) each function u, which belongs to the Sobolev space $W^{1,p}(\Omega)$, is in fact Hölder-continuous in $\bar{\Omega}$, if p is greater than n (cf. Adams [1], Kufner et al. [4] or Nečas [10]). A similar embedding property holds for Orlicz-Sobolev spaces, too (cf. [1] or [4]).

Typically, the boundary behaviour of u is handled by straightening the boundary to a half space and deriving estimates for the Hölder-norm of u in terms of the Sobolev-norm (cf. [10, Ch. 2.3.5]). Instead of using estimates on the boundary we first show that if $p > n$ then for any domain Ω the Sobolev spaces $W^{1,p}(\Omega)$ can be embedded in a certain local Hölder class $\text{loc Lip}_\alpha(\Omega)$, $\alpha = 1 - n/p$. The embedding to $C^\alpha(\bar{\Omega})$ is then derived for a large class of domains by showing that $\text{loc Lip}_\alpha(\Omega)$ embeds to $C^\alpha(\bar{\Omega})$. As a result the following is obtained:

THEOREM 1.1. *If Ω is a bounded uniform domain and $p > n$, then $W^{1,p}(\Omega)$ embeds continuously into $C^\alpha(\bar{\Omega})$.*

The proof is based on classical Hölder continuity estimates together with Gehring and Martio's [2] and Lappalainen's [5] results on Lip_α-extension domains.

2. NOTATION

A domain $\Omega \subset \mathbb{R}^n$ is called *c-uniform*, $c \geq 1$, if each pair of points $x, y \in \Omega$ can be joined by a rectifiable curve γ in Ω such that

J. Ławrynowicz (ed.), Deformations of Mathematical Structures, 29–32.

$$\ell(\gamma) \leq c|x - y|, \tag{2.1}$$

where $\ell(\gamma)$ denotes the length of γ, and

$$\operatorname{dist}(\gamma(t), \partial\Omega) \geq 1/c \min(t, \ell(\gamma) - t). \tag{2.2}$$

A function $u: \Omega \to R$ belongs to the *local Lipschitz class* $\operatorname{loc Lip}_\alpha(\Omega)$, $0 < \alpha \leq 1$, if there exists constants $b \in]0,1[$ and $M = m_b$ such that

$$|u(x) - u(y)| \leq M|x - y|^\alpha. \tag{2.3}$$

holds for each $x \in \Omega$ and $y \in B_b(x) := B(x, b \operatorname{dist}(x, \partial\Omega))$. In [5, Thm. 2.17] it is shown that it is equivalent to require that the condition holds for $b = 1/2$; the smallest $m_{1/2}$ defines a seminorm of u. It should be remarked that this definition differs from the standard definitions of local Hölder spaces. In fact, the class $\operatorname{loc Lip}_\alpha(\Omega)$ is not a local space but in some sense semiglobal.

A function u belongs to the *Lipschitz class* $\operatorname{Lip}_\alpha(\Omega)$ if there exists a constant $M < \infty$ such that (2.3) holds for all $x, y \in \Omega$. For bounded domains $\operatorname{Lip}_\alpha(\Omega) = C^\alpha(\bar{\Omega})$.

A domain Ω is a $\operatorname{Lip}_\alpha$*-extension domain* if $\operatorname{loc Lip}_\alpha(\Omega)$ embeds continuously into $\operatorname{Lip}_\alpha(\Omega)$.

A useful result to find a large class of $\operatorname{Lip}_\alpha$-extension domains is the following

THEOREM 2.1. *Uniform domains are* $\operatorname{Lip}_\alpha$*-extension domains for all* $0 < \alpha \leq 1$.

Proof. See [2, Thm. 2.24], [5, Thm. 4.17] or [6]. ∎

3. EMBEDDING OF SOBOLEV SPACES

Let p be a real number, $p > n$ and $\alpha = 1 - n/p$.

PROPOSITION 3.1. *Let* Ω *be an arbitrary domain in* $\mathbb{R}^n$. *Then* $W^{1,p}(\Omega) \subset \operatorname{loc Lip}_\alpha(\Omega)$.

Proof. It follows from [1, Thm. 5.35] applied to balls contained in Ω that each function $u \in W^{1,p}(\Omega)$ is continuous. Now let $B_b(x_0)$ be a ball contained in Ω and $x_1 \in B_b(x_0)$. Let $s := |x_0 - x_1|$ and choose a ball B of radius s such that $x_0, x_1 \in B \subset B_b$. We denote by $|B|$ the Lebesgue measure of B and by

$$u_B := \frac{1}{|B|} \int_B u(z)\, dz \tag{3.1}$$

the mean value of u in B.

Now, for all $z \in \Omega$

$$|u(x) - u(z)| \leq \int_0^1 |Du(tz + (1-t)x)|\, |x - z|\, dt. \tag{3.2}$$

Therefore, the following estimate holds for $x \in B$:

$$|u(x) - u_B| \leq \frac{2s}{|B|} \int_0^1 t^{-n} \int_{B_t} |Du(z)|\, dz, \tag{3.3}$$

where B_t denotes a ball of radius ts contained in B. Since

$$\begin{aligned} \int_{B_t} |Du(y)|\, dy &\leq \|Du\|_{p,B_t} \|1\|_{p,B_t} \\ &= \omega_n^{-1/p} s^{-n/p} t^{n-n/p} \|Du\|_{p,B_t}, \end{aligned} \tag{3.4}$$

where $\omega_n := |B(0, 1)|$, we obtain

$$|u(x) - u_B| \leq \frac{2s^\alpha}{\omega_n^{1/p}\alpha} \|Du\|_{p,\Omega}. \tag{3.5}$$

Thus

$$|u(x_0) - u(x_1)| \leq \frac{4s^\alpha}{\omega_n^{1/p}\alpha} \|Du\|_{p,\Omega},$$

which yields the desired result. ∎

COROLLARY 3.2. *Let* $p > n$ *and* Ω *a* Lip_α*-extension domain. Then* $W^{1,p}(\Omega) \subset \mathrm{Lip}_\alpha(\Omega)$.

Proof. The result follows immediately from Proposition 3.1 and the definition of Lip_α-extension domains. ∎

Proof of Theorem 1.1. By Theorem 2.1 a uniform domain is a Lip_α-extension domain. Therefore, the theorem follows from the above corollary. ∎

Remark 3.3. The Koch curve described in Mandelbrot [8, p. 42] is an example of a uniform domain whose boundary is very irregular. Examplex of domains, which are Lip_α-extension domains but not uniform, can be found in [5, Lemma 4.28] and [2, Ex. 2.26]. Also, in [5] there are given examples of Lip_β-extension domains which are not Lip_α-extension domains for any $\beta > \alpha$.

References

[1] ADAMS, R.A.: *Sobolev spaces*, Pure and Applied Mathematics 65, Academic Press, New York - San Francisco - London, 1975.

[2] GEHRING, F.W. and O. MARTIO: 'Lipschitz classes and quasiconformal mappings', *Ann. Acad. Sci. Fenn. Ser. A I Math.* 10 (1985), 203-219.

[3] ——— and B.S. OSGOOD: 'Uniform domains and the quasihyperbolic metric', *J. Analyse Math.* 36 (1979), 50-74.

[4] KUFNER, A., O. JOHN, and S. FUCIK: *Function spaces*, Noordhoff International Publishing Leyden; Academia, Prague 1977.

[5] LAPPALAINEN, V.: 'Lip$_h$-extension domains', *Ann. Acad. Sci. Fenn. Ser. A I Math. Dissertationes* 56 (1985).

[6] ——: *Local and Global Lipschitz Classes*, Seminar on Deformations, Łódź-Lublin 1985/87, ed. by J. Ławrynowicz, D. Reidel Publishing Company, Dordrecht (to appear).

[7] —— and A. LEHTONEN: 'Embedding of Orlicz-Sobolew spaces in Hölder spaces', *Math. Scand.* (to appear).

[8] MANDELBROT, B.: *The fractal geometry of nature*, W.H. Freeman and Company, San Francisco 1982.

[9] MARTIO, O.: 'Definitions for uniform domains', *Ann. Acad. Sci. Fenn. Ser. A I Math.* 5 (1980), 179-205.

[10] NEČAS, J.: *Les méthodes directes en théorie des équations elliptiques*, Masson et Cie Editeurs, Paris; Academia, Editeurs, Prague 1967.

Keywords and phrases. Sobolev space, Lipschitz class, Hölder class, uniform domain, Lip$_\alpha$-extension domain
1980 Mathematics subject classifications (Amer. Math. Soc.): 26B35, 46E35

QUASIREGULAR MAPPINGS FROM R^n TO CLOSED ORIENTABLE n-MANIFOLDS

Jorma Jormakka
Department of Mathematics
University of Helsinki
SF - 00100 Helsinki, Finland

ABSTRACT. The growth of the fundamental group of a Riemannian manifold can be used to prove the nonexistence of quasiregular mappings from the Euclidean space to the manifold. Some constructive methods provide existence results. This paper is a preliminary report on a study of these problems.

Let M be a closed orientable Riemannian k-manifold, $k \geq 2$. A continuous function $f: R^k \to M$ is called quasiregular (qr) if f has locally L^k-integrable generalized first order partial derivatives and for some K, $1 \leq K < \infty$, at almost every point $x \in R^k$, $|df(x)|^k \leq K J_f(x)$, where $|\cdot|$ is the supremum norm of the formal derivative of f and J_f is the Jacobian. Since M is compact the existence of qr mappings from R^k to M does not depend on the Riemannian metric chosen for M.

A qr mapping $f: R^k \to M$ can be lifted to a qr mapping from R^k to the universal covering space $\tilde{M}$ of M. If the volume of balls in $\tilde{M}$ grows too rapidly with the radius then the isoperimetric inequality, preserved by qr mappings, can be used to prove the nonexistence of qr mappings $f: R^k \to M$. This is the case f. ex. M has a hyperbolic structure whence $\tilde{M} = H^k$, the hyperbolic space. Milnor [1] and Shvarc noticed that the growth of balls in $\tilde{M}$ can be studied by the growth of the fundamental group $\pi_1(M)$ in the word metric. The growth of $\pi_1(M)$ can also be used to show that a family of paths in $\tilde{M}$ going to infinity has positive modulus. Since the ideal boundary of R^k has zero capacity, this means that the lift of the path family has zero modulus. This gives a contradiction to the quasiregularity of f.

Consider the case where $\pi_1(M)$ contains an infinite subsemigroup W generated by two elements x and y of $\pi_1(M)$ using only positive exponents of x and y. For $n \in N$ denote by W_n the set of different expressions w of elements $|w| \in W$ in n generators where w cannot be expressed as a word in m generators for $m < n$, i.e. w has length n. Take small tubular neighbourhoods of some representatives α_1 and α_2 of x and y and let Γ be a family of closed

J. Ławrynowicz (ed.), Deformations of Mathematical Structures, 33–44.

curves nearly parallel to $\alpha_1 \circ \alpha_2$ foliating the neighbourhood. Then Γ can be lifted to a family Γ_n of paths in $\tilde{M}$ with every path corresponding to a word of W_n. If there were a qr mapping $f: R^k \to M$ then the lift $\tilde{f}: R^k \to \tilde{M}$ would be quasiregular and the modulus $M(\Gamma_n')$ of a family Γ_n' in R^k with $f\Gamma_n' = \Gamma_n$ and the modulus $M(\Gamma_n)$ would satisfy for some K, $1 \le K < \infty$, not depending on n the inequality

$$M(\Gamma_n) \le K M(\Gamma_n').$$

In the limit when n goes to infinity the right-hand side goes to zero, since the paths of Γ_∞' go to the ideal boundary of R^k. The number $N_n \ne W_n$ of elements in W_n is at most 2^n and the number of separate words $\# |W_n| = \{|w| \; | w \in W_n\}$ is at most $\# W_n$. If no n-word can be expressed as a shorter word, $\# W_n = 2^n$. Choose positive numbers p_i, $i = 1, \ldots, N_n$, with the sum $p_1 + \ldots + p_{N_n} = 1$ and $p_i = p_i'$ if $|w_i| = |w_i'|$. Let $F(\Gamma_n)$ be

$$F(\Gamma_n) = \{\phi \in C^\infty \; | \int_\beta \phi \, ds \ge 1 \quad \text{for all} \quad \beta \in \Gamma_n\}$$

and choose an arbitrary $\phi \in F(\Gamma_n)$. Define

$$\psi = \sum_{i=1}^{N_n} p_i \chi_{w_i},$$

where χ_w is the characteristic function of all paths in Γ_n in the tubular neighbourhood of the lift of $\alpha_1 \circ \alpha_2$ to representatives of w_i. It follows that $\psi \in F(\Gamma)$ since for each $\beta \in \Gamma$

$$\int_\beta \psi \, ds = \sum_{i=1}^{N_n} p_i \int_{\beta_{n,i}} \phi \, ds \ge \sum_{i=1}^{N_n} p_i = 1,$$

where $\beta_{n,i}$ is the lift of β to the path in Γ_n corresponding to w_i. All paths of Γ are of finite length so the modulus $M(\Gamma)$ is positive. Then for some $\varepsilon > 0$

$$\int_M \chi_\Gamma \psi^k \, ds \ge M(\Gamma) > \varepsilon.$$

The weighted characteristic function of Γ_n can be written as a sum over words of unequal value:

$$\chi_p = \sum_{i=1}^{N_n} p_i \chi_{w_i} = \sum_{j=1}^{n} \sum_{i=1}^{\# |W_j|} p_{ij} \chi_{ij},$$

where χ_{ij} is the characteristic function of those paths corresponding to the word $|w_i|$ of length j. By Hölder's inequality the kth power ψ is estimated from above:

$$\psi^k = \Big(\sum_{j=1}^{n} \sum_{i=1}^{\# |W_j|} p_{ij} \chi_{ij} \phi \Big)^k \le$$

$$\Big(\sum_{j=1}^{n}\sum_{i=1}^{\#|W_j|} p_{ij}^{\frac{k}{k-1}}\Big)^{k-1} \cdot \sum_{j=1}^{n}\sum_{i=1}^{\#|W_j|} \chi_{ij}\,\phi^k = S_n^{k-1}\,\chi_{\Gamma_n}\,\phi^k,$$

where χ_Γ is the characteristic function of Γ_n and S_n is the sum of the series. Integrating over M

$$\int_M \chi_\Gamma\,\psi^k\,dV = \int_{\tilde{M}} \Big(\sum_{i=1}^{N_n} p_i\,\chi_{w_i}\,\phi\Big)^k dV \le S_n^{k-1} \int_{\tilde{M}} \chi_{\Gamma_n}\,\phi^k\,dV.$$

This holds for all $\phi \in F(\Gamma_n)$. So

$$\varepsilon \le \int_M \chi_\Gamma\,\psi^k\,dV \le S_n^{k-1} \inf_{\phi\in F(\Gamma_n)} \int_{\tilde{M}} \chi_{\Gamma_n}\,\phi^k\,dV.$$

The contradiction is obtained if the values S_n have un upper boundary S,

$$\varepsilon \le S^{k-1} \lim_{n\to\infty} M(\Gamma_n) \le K\,S^{k-1} \lim_{n\to\infty} M(\Gamma_n') = 0.$$

This gives the following theorem.

THEOREM 1. *If there exists a finite upper boundary* S *to the sequence*

$$S_n = \sum_{j=1}^{n}\sum_{i=1}^{\#|W_j|} p_{ij}^{\frac{k}{k-1}}$$

then there exist no quasiregular mappings from R^k *to* M.

COROLLARY 1. *If* $\pi_1(M)$ *has a free subsemigroup in two generators then there exist no qr mappings from* R^k *to* M.

Proof. In this case $N_n = W_n = 2^n$ and each $p_{ij} = 2^{-j}$ so

$$S_n = \sum_{j=1}^{n}\sum_{i=1}^{2^j} 2^{-\frac{k}{k-1}j} < \sum_{j=1}^{\infty} 2^{-\frac{j}{k-1}} = (1-2^{-\frac{1}{k-1}})^{-1}.$$

This Corollary is proved by other methods in [4] and there credited to Alhfors and Picard. The use of the modulus method closely follows the ideas of Väisälä and Rickman.

Next we reprove Pansu's result that there does not exist any qr mapping from R^3 to the three-dimensional Heisenberg group G_3, i.e. the group of 3×3 upper triangular matrices on reals with ones on the diagonal. The group G_3 has cocompact subgroups of the form

$$H_{3,k} = \left\{ \begin{pmatrix} 1 & i & m+n/k \\ 0 & 1 & j \\ 0 & 0 & 1 \end{pmatrix} \;\middle|\; i,j,m,n \in Z \right\},\quad k = 1,2,\ldots$$

Closed 3-manifolds $G_3/H_{3,k}$ are all finitely covered by $M = G_3/H_{3,1}$

so it is sufficient to establish the convergence of the sequence S_n for $\pi_1(M) = H_{3,1}$. The discrete Heisenberg group $H_3 = H_{3,1}$ is generated by two elements $x = \mathbb{1} + \delta_{12}$ and $y = \mathbb{1} + \delta_{23}$ where δ_{ij} has 1 in the (ij)-entry and zeros elsewhere. The commutator $z = [x, y] = x y x^{-1} y^{-1} = \mathbb{1} + \delta_{1,3}$. All elements of H_3 have an expression in the form $w = y^a x^b z^c = \mathbb{1} + a\,\delta_{2,3} + b\,\delta_{1,2} + c\,\delta_{1,3}$. This follows f.ex. from the central series

$$H_3 \supset [H_3, H_3] \supset 1,$$

which has sets $\{x, y\}$ and $\{z\}$ as generators of the factor groups. This is a convenient expression for calculating the growth powers of the group since the order of the generators is fixed and is used f.ex. by Milnor [1], Bass [7], and Wolf [6], but it is not sufficient for the modulus method. For that reason and for further applications to a study of qr mappings from G_3 to closed 3-manifolds we diverge to a closer look on the growth of H_3. Let $s_{n,q,r}$ be the number of words w in W_n which have the same value $|w|$ i.e. $s_{n,q,r}$ is the number of different expressions as a word of length n in x and y in positive generators for the element

$$|w| = |w|_{n,q,r} = \begin{pmatrix} 1 & n-q & r \\ 0 & 1 & q \\ 0 & 0 & 1 \end{pmatrix} \quad \text{of } H_3.$$

The number r takes all integer values from 0 to $q(n-q)$ depending on the order of elements x and y in the word w. It is easy to calculate the number of unequal n-words:

$$\#|W_n| = \sum_{q=0}^{n} \sum_{r=0}^{q(n-q)} 1 = n + 1 + \tfrac{1}{2} n^2 (n+1) - \tfrac{1}{6} n(n+1)(2n+1)$$
$$= \tfrac{1}{6} n^3 + \tfrac{5}{6} n + 1.$$

The integral of $\#|W_n|$ is usually called the growth function (in positive exponents)

$$\gamma(n) = \sum_{j=1}^{n} \# |W_j| = \tfrac{1}{24}(n^4 + 2n^3 + 11n^2 + 34n).$$

The formal power series

$$p_s(t) = \sum_{n=1}^{\infty} \#|W_n| t^n = \tfrac{1}{6}(t + 4t^2 + t^3)/(1-t)^4 + \tfrac{5}{6} t/(1-t)^2 + t/(1-t)$$
$$= t(2 - 4t + 4t^2 - t^3)/(1-t)^4$$

is called the growth power series. There are some interesting results concerning the growth function and the growth power series but we proceed to calculate the numbers $s_{n,q,r}$ which constitute what might be called the growth distribution of H_3. The numbers $s_{n,q,r}$ have the symmetries

$$s_{n,q,r} = s_{n,\, n+1-q,\, r} = s_{n,\, q,\, q(n-q)-r},$$

$$s_{n,q,r} + s_{n,\, q-1,\, r+q-n-1} = s_{n,\, q-1,\, r} + s_{n,\, q,\, r-q}.$$

The element $|w|_{n,q,r}$ is obtained either by multiplying $|w|_{n-1,q-1,r+q-n}$ from right by y or by multiplying $|w|_{n-1,q,r}$ from right by x. This gives a recursion formula to the numbers $s_{n,q,r}$:

$$s_{1,0,0} = s_{1,1,0} = 1,$$

$$s_{n,q,r} = s_{n-1,\, q,\, r} + s_{n-1,\, q-1,\, r+q-n}, \qquad q \le n/2.$$

A similar recursion is obtained by multiplying from left:

$$s_{n,q,r} = s_{n-1,\, q,\, r-q} + s_{n-1,\, q-1,\, r}.$$

In these formulas we define $s_{n,q,r} = 0$ if $q < 0$, $r < 0$, $q > n$ or $r > q(n-q)$. There is also an inversion formula:

$$s_{n,q,r} = \sum_{j=0}^{n} \sum_{i=0}^{j(m-j)} s_{m,j,i}\, s_{n-m,\, q-j,\, r+j(q-n)+i},$$

and by simple properties of permutating q x:s among n letters

$$\binom{n}{q} = \sum_{r=0}^{q(n-q)} s_{n,q,r}, \qquad 2^n = \sum_{q=0}^{n} \sum_{r=0}^{q(n-q)} s_{n,q,r}.$$

The maximal element is $s_{n,\lfloor\frac{n}{2}\rfloor,\lfloor\frac{n^2}{8}\rfloor}$ in the center of the distribution (here $\lfloor x\rfloor$ is the largest integer $\le x$). Let $G_n(u,v)$ be the generating function

$$G_n(u,v) = \sum_{q,r=0}^{\substack{q=n\\ r=q(n-q)}} s_{n,q,r}\, u^q\, v^{r+\frac{1}{2}q^2}.$$

The recursion formula for $s_{n,q,r}$ can be written with $G_n(u,v)$ as

$$G_n(u,v) = (1 - u\, v^{n-\frac{1}{2}})\, G_{n-1}(u,v),$$

which can be solved:

$$G_n(u,v) = \prod_{k=1}^{n} (1 + uv^{k-\frac{1}{2}}).$$

Then

$$s_{n,q,r} = \frac{1}{q!\,(r+\frac{1}{2}q^2)!}\;\frac{\partial^{r+\frac{1}{2}q^2+q}}{\partial v^{r+\frac{1}{2}q^2}\,\partial u^q}\,G_n(u,v)\Big|_{u=v=0}$$

$$= \frac{1}{q!}\sum_{\substack{k_1,\ldots,k_q \\ k_i \neq k_j}}^{n} 1 = 1 \;\Big|\; \sum_{i=1}^{q} k_i = r + \tfrac{1}{2}q + \tfrac{1}{2}q^2$$

$$= \text{Numbers of partitions}\{n \geq k_1 > \ldots > k_q \geq 1 \;\Big|\; \sum_{i=1}^{q} k_i = r + \tfrac{q}{2} + \tfrac{q^2}{2}\}.$$

Let $(m,q,\leq j)$ be the number of partitions of m to q non-empty classes which are in strictly increasing order with the largest $\leq j$. The numbers $(m,q,<\infty) = (m,<\infty,q)$ are well known, probably also the numbers $(m,q,\leq j)$ have been calculated, perhaps not. By the Ferrer diagram, $(m,q,\leq j) = (m,\leq r,j)$. The recursion formulas for $s_{n,q,r}$ can be written as

$$(m,\, q,\, \leq j) = (m-q,\, q,\, \leq j-1) + (m-q,\, q-1,\, \leq j-1)$$

and

$$(m,\, q,\, \leq j) = (m,\, q,\, \leq j-1) + (m-j,\, q-1,\, \leq j-1),$$

where as definition $(0,\, 0,\, \leq 0) = 1$.

These formulas can also be seen by taking away one from each part in the first equation and separating cases where the largest part has size j and less than j in the second equation. These identities reduce j to a parameter. Write

$$\nu_s(k) = \nu_s^{(j)}(k) = (k,\, s,\, \leq j-1).$$

Elimination of $(m,q,\leq j)$ gives

$$\nu_q(m) - \nu_q(m-q) = \nu_{q-1}(m-q) - \nu_{q-1}(m-j).$$

Let $F_q(t)$ be the generating function

$$F_q(t) = F_q^{(j)}(t) = \sum_{m=0}^{\infty} \nu_q(m)\, t^m.$$

The equation for $\nu_q(m)$ is expressed and solved with $F_q(t)$ as

$$(1 - t^q)\, F_q(t) = (t^q - t^j)\, F_{q-1}(t).$$

Then

$$F_q(t) = \prod_{k=1}^{q} \frac{t^k - t^j}{1 - t^k} = t^{\frac{1}{2}q(q+1)} \prod_{k=1}^{q} (1 - t^{j-k}) \sum_{k=0}^{\infty} P_q^{k+q}\, t^k,$$

where P_k^i is the number of partitions of i into k nonempty classes:

$i = a_1 + \ldots + a_k$ with $a_1 \geq \ldots \geq a_k > 0$. Expand the Gaussian polynomial as

$$\prod_{k=1}^{q} (1 - t^{j-k}) = \sum_{i=0}^{q\,j-\frac{1}{2}q(q+1)} a_{qi}\, t^i .$$

The generating function is

$$F_q(t) = \sum_{s=0}^{\infty} \Big(\sum_{i=0}^{s} a_{qi}^{(j)}\, p_q^{q+s-i} \Big)\, t^{s+\frac{1}{2}q(q+1)} .$$

The upper index j in a_{qi} simply indicates dependence on j. This gives

$$(m, q, \leq j) = \sum_{i=1}^{m-\frac{1}{2}q(q+1)} a_{qi}^{(j+1)}\, p_q^{q-i+m-\frac{1}{2}q(q+1)}$$

and

$$s_{n,q,r} = (r + \tfrac{1}{2}q + \tfrac{1}{2}q^2, q, \leq n) = \sum_{i=0}^{r} a_{qi}^{(n+1)}\, p_q^{q-i+r}, \qquad q \geq 1,$$

$$s_{n,0,0} = 1.$$

The coefficients p_k^i and $a_{qi} = a_{qi}^{(j)}$ satisfy recursion formulas

$$p_k^i = p_{k-1}^{i-1} + p_k^{i-k} \qquad \text{with } p_1^i = 1,\ i \geq 1, \quad p_k^i = 0,\ i < k$$

and

$$a_{qi} = a_{q-1,i} - a_{q-1,\, i-j+q}, \quad \text{where } a_{q-1,j} = 0 \text{ if } i < 0 \text{ or } i > (q-1)j - \frac{q(q-1)}{2},$$

$$a_{10} = 1, \quad a_{1,j-1} = -1, \quad a_{1i} = 0, \quad 1 \leq i < j-1.$$

It appears that the numbers $s_{n,q,r}$ do not admit any expression in a closed form. The asymptotic behaviour of the numbers is that n^2 largest near the center $s_{n,\left|\frac{n}{2}\right|,\left|\frac{n^2}{8}\right|}$ will be of size $\simeq 2^n n^{-2}$ and the rest will be negligible. Then

$$\sum_{q=0}^{n} \sum_{r=0}^{q(n-q)} s_{n,q,r}^{\alpha} = 2^{\alpha n} n^{2-2\alpha} .$$

We return to the problem of qr-mappings from R^3 to $M = G_3/H_3$. Recall the construction in the Theorem 1. Choose a function $\phi \in F(\Gamma_n)$ and let

$$\psi = \sum_{i=1}^{2^n} p_i \chi_{w_i} \phi ,$$

where the weights sum up to 1. Also choose $p_i = p_{i'}$ if the w_i and $w_{i'}$ have the same value as elements of H_3. Denote by q_{ij} and r_{ij} the integers for which the restriction $w_{i|j}$ of w_i to j letters has the value $|w|_{j,q_{ij},r_{ij}}$.

Let $\chi_{j,q_{ij},r_{ij}}$ be the part of χ_{w_i} corresponding to $w_{i|j}$. Then

$$\chi_{w_i} = \sum_{j=1}^{2^n} \chi_{j,q_{ij},r_{ij}} .$$

The total characteristic function can be written as a sum of parts $\chi_{j,q,r}$

$$\sum_{i=1}^{2^n} p_i \chi_{w_i} = \sum_{j=1}^{n} \sum_{i=1}^{2^n} p_i \chi_{j,q_{ij},r_{ij}} = \sum_{j=1}^{n} \sum_{q=0}^{j} \sum_{r=0}^{q(j-q)} p_{j,q,r}\chi_{j,q,r},$$

where the summation is changed to go over words of unequivalent value. The weights $p_{j,q,r}$ can be calculated from the following recursion formula

$$p_{n,q,r} = s_{n,q,r}\, p_i, \qquad |w_i| = |w|_{n,q,r},$$

$$p_{j-1,q,r} = \frac{s_{j-1,q,r}}{s_{j,q,r}} p_{j,q,r} + \frac{s_{j-j,q,r}}{s_{j,q-1,r-q-1+j}} p_{j,q-1,r-q-1+j} .$$

To explain this formula notice that there are $p_{j,q,r}$ paths in the part $\chi_{j,q,r}$ and $\frac{s_{j-1,q,r}}{s_{j,q,r}}$ of them go to $\chi_{j-1,q,r}$ and the other $\frac{s_{j-1,q-1,r+q-j}}{s_{j,q,r}}$ go to $\chi_{j-1,q-1,r+q-j}$. Also there are $p_{j,q+1,r-q-1+j}$ paths through $\chi_{j,q+1,r-q-1+j}$ and $\frac{s_{j-1,q,r}}{s_{j,q+1,r-q-1+j}}$ of them go to $\chi_{j-1,q,r}$ and the rest, $\frac{s_{j-1,q+1,r-q-1+j}}{s_{j,q+1,r-q-1+j}}$ paths go to $\chi_{j-1,q+1,r-q-1+j}$.

The sum of weights is constant in each level j as it of course must be

$$\sum_{q=0}^{j-1} \sum_{r=0}^{q(j-1-q)} p_{j-1,q,r} = \sum_{q=0}^{j} \sum_{r=0}^{q(j-q)} \frac{s_{j-1,q,r} + s_{j-1,q-1,r+q-1}}{s_{j,q,r}} p_{j,q,r}$$

$$\sum_{q=0}^{j} \sum_{r=0}^{q(j-q)} p_{j,q,r} = \sum_{q=0}^{n} \sum_{r=0}^{q(n-q)} p_{n,q,r} = 1.$$

The sum which should have a finite upper boundary independent of n is

$$S_n = \sum_{j=1}^{n} \sum_{q=0}^{j} \sum_{r=0}^{q(j-q)} p^{\frac{3}{2}}_{j,q,r} .$$

The optimal choise of the weights is

$$p_{n,q,r} = M_n\, s^{\frac{2}{3}}_{n,q,r} ,$$

where

$$M_n = \Big(\sum_{q=0}^{n} \sum_{r=0}^{q(n-q)} s^{\frac{2}{3}}_{n,q,r} \Big)^{-1} .$$

We check the convergence of the sequence S_n by comparing the difference $S_n - S_{n-1}$ to a term in a superharmonic series.

It turns out that the convergence of (S_n) is not essentially improved by the choice of the weights $p_{n,q,r}$, so for simplicity we take

$$p_{n,q,r} = 2^{-n} s_{n,q,r}.$$

In the difference $S_n - S_{n-1}$ all other terms cancel but one sum:

$$S_n - S_{n-1} = \sum_{q=0}^{n} \sum_{r=0}^{q(n-q)} s_{n,q,r}^{\frac{3}{2}} 2^{-\frac{3}{2}n}.$$

The growth of the numbers $s_{n,q,r}$ is such, that the difference has only a harmonic upper bound. However, the choice of paths can be modified to give a convergent sequence. For this purpose the asymptotic behaviour of $s_{n,q,r}$ is not needed, merely the facts that the numbers $s_{n,q,r}$ grow fastest near the center $q = \frac{n}{2}$, $r = \frac{n^2}{8}$, and slowest near the sides and the average growth of $s_{n,q,r}$ is sufficiently slow - $2^n(\# |W_n|)^{-1} \cong 2^n n^{-3}$ - to make the sequence converge if the values of the $s_{n,q,r}$ were more evenly distributed. The length of the paths of Γ_n is n in every direction. Yet the manifold G_3 is more euclidean in those direction where the numbers $s_{n,q,r}$ are large and more hyperbolic in the directions where they are small. We divide the paths of Γ_∞ in levels L_n corresponding to the levels Γ_n so that in L_n the length of the paths near the center is $n^{1-\alpha}$ for a small $\alpha > 0$ which is chosen so that the largest term grows to power $2^n n^{-2-\delta}$ for some $\delta > 0$. The paths near the sides will be chosen to grow faster than n so that the total sum of the corresponding numbers s_{n_i,q_i,r_i} will be 2^n. The weights will be chosen so that all other terms in the difference $S_n - S_{n-1}$ cancel but the n-level. The difference $S_n - S_{n-1}$ is just as above with the exeption that it can be approximated from above by the square root of the largest $p_{n,q,r}$:

$$S_n - S_{n-1} \leq n^{-1-\delta/2}.$$

The sequence will have a superharmonic majorant and the conclusion of the Theorem 1 completes the proof of a theorem Pansu [5] has obtained by an isometric inequality.

COROLLARY 2. *No qr mappings from* R^3 *to* G_3 *or to* $G_3/H_{3,1}$ *exist.*

In the general case of closed orientable manifolds it seems plausible, that unless $\pi_1(M)$ has polynomial growth, i.e. $\gamma(n) \cong n^d$ for some d, the sequence (S_n) in Theorem 1 should converge. However Corollary 1 applies only to the case where $\pi_1(M)$ has a free subgroup on two generators. This need not be the case even if $\gamma(n)$ grows as exponent or, which is a related but not equivalent condition, π, M is nonamenable. By isoperimetric inequality Gromov [4] has shown that if $\pi_1(M)$ is nonamenable the universal covering space of M is open at infinity and no qr mappings from R^k to M exist. If $\pi_1(M)$ is amenable it may still have exponential growth, see Milnor [1], also a group may have subexpotential growth and hence be amenable but grow faster than any polynomial, cf. Grigorchuk [8] for any example. Since M is compact it has a finitely generated fundamental group. If $\pi_1(M)$ is solvable then it either has polynomial growth or it contains a free subsemigroup on two generators by a theorem of Rosenblatt in [9]. For a nonsolvable amenable group there is a possibility for exponential or subexponential but nonpolynomial growth without contradicting the Corollary 1 or Gromov's theorem but no examples are known. In the sequel let M be 3-dimensional. Modulo three old conjectures a fairly complete solution for the qr mapping problem is obtained.

If $\pi_1(M)$ is nonsolvable, amenable and not of polynomial growth with no free subsemigroup on two generators, then the Kneser-Milnor decomposition of M is $M = M' \# K(\pi,1)$, where M' is a homotopy sphere since otherwise $\pi_1(M)$ contains a free subgroup of rank 2. Also by a theorem of Evans and Moser [10] the factor $K(\pi,1)$ cannot be sufficiently large for a sufficiently large $K(\pi,1)$ either has a solvable fundamental group π or π contains a free subgroup of rank 2. If a conjecture of Waldhausen 1977 is true and every nonsufficiently large closed 3-manifold with an infinite fundamental group has a finite covering by a Haken manifold then applying the Corollary 1 to the covering it could be shown that no qr mappings to the manifold exist. Notice that a manifold M with $\pi_1(M)$ of growth between the exponential and polynomial types would be a counterexample to Waldhausen's conjecture.

In all other remaining cases $\pi_1(M)$ is a finitely generated group of polynomial growth. By a theorem of Gromov [3] $\pi_1(M)$ is almost nilpotent and so polycyclic. Bass in [7] gives a formula to the growth power of the nilpotent subgroup G of finite index in $\pi_1(M)$. Let the lower central series of G be

$$G = H_1 \geq H_2 \geq \dots H_{s-1} \geq H_s = 1, \qquad H_i = [H_{i-1}, H].$$

The factor groups H_i/H_{i-1} are finitely generated abelian groups and since H is a 3-manifold group it is torsion free or finite. Bass states that $\gamma(n) \simeq n^d$, where

$$d = \sum_{i \leq 1} i \operatorname{rank}(H_i/H_{i-1}).$$

It follows that either $G = 1, Z, Z \oplus Z, Z \oplus Z \oplus Z$ or $d \geq 4$.

If G is trivial then $\pi_1(M)$ is finite. The only known examples of closed 3-manifolds with finite fundamental groups are Seifert fibered spaces (including S^3 as a lens space). If the Poincaré conjecture is true and if all smooth free actions on S^3 are ortogonal these are the only possibilities. Nothing can be said of qr mappings of R^3 to exotic homotopy spheres so we ignore them waiting how the late rumors of the Poincaré conjecture turn out. The Seifert fibered spaces with finite fundamental groups have all been classified and they can all be taken as having orientable fibers and orientable Seifert surface S of genus 0 (those with genus 1 being homeomorphic to prism manifolds). Divide R^2 to squares and map every second to the lower resp. upper hemisphere of $S = S^2$. Map the fibers R in $R^3 = R^2 \times R$ to the fibers S^1 of M with the exponential map. The mapping obtained is quasiregular with exeptional fibers in the image of the branch set.

If $G = Z$ then by the Kneser-Milnor decomposition $M = M' \# S^2 \times S^1$, where M' is a homotopy sphere. If $M = S^2 \times S^1$ then the same construction as above gives a qr mapping $R^3 \to M$, actually the sphere bundle is a lens space and included in the previous case.

If $G = Z \oplus Z$ or $Z \oplus Z \oplus Z$ then only the latter can occur for if $G = Z \oplus Z$, M would be a I-bundle (ignoring homotopy spheres) and the boundary ∂M could not be empty. If $G = Z \oplus Z \oplus Z$ then M is finitely covered by the 3-torus T^3 and the covering $R^3 \to T^3 \to M$ of the flat space form M is conformal.

In all other cases G contains a subgroup H generated by two non-commuting elements. The formula of Bass shows that the growth power of H is at least 4. The situation is similar to that of the Heisenberg group H_3.

As a 3-manifold group on two generators H is finitely presented and so has a growth distribution where some directions are more euclidean and some more hyperbolic. By choosing paths with length growing faster in the more hyperbolic directions the sequence (S_n) can be made to converge and by the Theorem 1 no qr mappings $f: R^3 \to M$ exist.

The advantage of the modulus method over the use of the isoperimetric inequality is that the latter must be established for all large sets whether as it is sufficient to find one path family for a contradiction.
I thank prof. S. Rickman, who supervised this work, for many valuable discussions. For details of the proofs see [11].

Bibliography

[1] MILNOR, J.: 'A note on curvature and fundamental group', *J. Diff. Geom.* 2, (1968), 1-7.

[2] ——: 'Growth of finitely generated solvable groups', *ibid.*, 447-49.

[3] GROMOV, M.: 'Groups of polynomial growth and expanding maps', *Publ. Math. I.H.E.S.* 53 (1981), 53-73.

[4] ——: 'Hyperbolic manifolds, groups and actions', *Ann. Math. Studies* 97 Conf. Stony Brook 1978, 183-213.

[5] PANSU, P.: 'An isoperimetric inequality on the Heisenberg group', *C.R. Acad. Sc. Paris* 295 (1982), 127-130.

[6] WOLF, J.: 'Growth of finitely generated solvable groups and curvature of Riemannian manifolds', *J. Diff. Geom.* 2 (1968), 421-446.

[7] BASS, H.: ' The degree of polynomial growth of finitely generated

nilpotent groups', *Proc. Lond. M.S.* (3) **25** (1972), 602-614.

[8] GRIGORCHUK, R.I.: 'Degrees of growth of finitely generated groups and the theory of invariant means', *Math. USSR Iz.* **25** (1985), 259-99.

[9] ROSENBLATT, J.: 'Invariant measures and growth conditions', *Trans. Amer. M.S.* **193** (1974), 33-53.

[10] EVANS, B. and MOSER, L.: 'Solvable fundamental groups of compact 3-manifolds', *Trans. Amer. M.S.* **168** (1972), 189-210.

[11] JORMAKKA, J.: 'Existence of quasiregular mappings from $\mathbb{R}^3$ to closed orientable 3-manifolds', to appear.

SOME UPPER BOUNDS FOR THE SPHERICAL DERIVATIVE

Sakari Toppila
Department of Mathematics
University of Helsinki
SF-00100 Helsinki 10, Finland

ABSTRACT. The author answers the question: what can be said on functions with a Nevanlinna deficient value and of positive lower order λ less than 1/2 ? It is given in the following three theorems: Thm. 1. Let f be a meromorphic function of lower order λ, $0<\lambda<1/2$, such that $\delta(\infty, f) > 1-\cos \pi\lambda$. Then

$$\liminf_{r\to\infty} \frac{\log \mu(r, f)}{T(r, f)} \leq \frac{-\pi\lambda}{\sin \pi\lambda}(\cos \pi\lambda + \delta(\infty, f) - 1). \tag{1}$$

Thm. 2. Given λ, $0<\lambda<1/2$, and d, $1-\cos\pi\lambda<d<1$, there exists a meromorphic function f of order λ and of lower order λ such that $\delta(\infty, f) = d$ and

$$\liminf_{r\to\infty} \frac{\log \mu(r, f)}{T(r, f)} = \frac{-\pi\lambda}{\sin \pi\lambda}(\cos \pi\lambda + d - 1). \tag{2}$$

Thm. 3. Given λ, $0<\lambda<1/2$, and d, $0<d<1-\cos\pi\lambda$, there exists a meromorphic function f of order λ and of lower order λ such that $\delta(\infty, f) = d$ and

$$\liminf_{r\to\infty} \frac{r\mu(r, f)}{T(r, f)} \geq \frac{\pi\lambda^2}{2}. \tag{3}$$

1. INTRODUCTION AND STATEMENT OF RESULTS

Let f be a meromorphic func... in the complex plane. We write

$$\rho(f(z)) = |f'(z)|/(1+|f(z)|^2)$$

and

$$\mu(r, f) = \max\{\rho(f(z)) : |z| = r\}.$$

The following results have been proved by Anderson and me in [1]:

J. Ławrynowicz (ed.), Deformations of Mathematical Structures, 45–49.

THEOREM A. *Suppose that* f *is a meromorphic function of lower order* b *for some positive* b. *Then*

$$\liminf_{r\to\infty} [r\,\mu(r, f)/T(r, f)] \leq 126\sqrt{2}\, b^2. \tag{4}$$

THEOREM B. *Suppose that* f *is a meromorphic function such that*

$$\liminf_{r\to\infty} [T(r, f)/r^b] \leq K$$

for some $b > 0$ *and* $0 < K < \infty$. *Then*

$$\liminf_{r\to\infty} [r\,\mu(r, f)/r^b] \leq 24\sqrt{2}\, K b^2. \tag{5}$$

The following result shows that Theorems A and B give sharp results.

THEOREM C. *Given any* $b > 0$ *and* $K > 0$ *there is a meromorphic function* f *with*

$$T(r, f) = (K + o(1))\, r^b \qquad (r \to \infty)$$

and such that

$$\liminf_{r\to\infty} [r\,\mu(r, f)/r^b] \geq K b^2.$$

I have proved in [7] that for entire functions of order k less than 1/2 an estimate essentially stronger than (4) and (5) holds.

THEOREM D. *Let* f *be a transcendental entire function of order* k, $0 \leq k < 1/2$. *Then*

$$\liminf_{r\to\infty} [\log \mu(r, f)/\log M(r, f)] \leq -\cos \pi k.$$

The following result of [7] shows that Theorem D is sharp.

THEOREM E. *Suppose that* $0 < k < 1/2$. *There exists an entire function* f *of order* k *such that*

$$\liminf_{r\to\infty} [\log \mu(r, f)/\log M(r, f)] = -\cos \pi k.$$

The following theorem of [7] shows that Theorems A and B remain essentially sharp for entire functions whose order is greater than 1/2.

THEOREM F. *Given any* $k > 1/2$ *and* $K > 0$ *there is an entire function* g *with*

$$T(r, g) = (K + o(1))\, r^k \qquad (r \to \infty)$$

and such that

$$\liminf_{r\to\infty} [r\,\mu(r, g)/r^k] \geq \tfrac{1}{2}\pi K k^2.$$

Furthermore, I have proved in [6] the following results:

THEOREM G. *Let* f *be a transcendental meromorphic function. Then*

$$\liminf_{r\to\infty} [\log \mu(r, f)/T(r, f)] \geq -1.$$

THEOREM H. *Let* f *be a transcendental meromorphic function of lower order zero. Then*

$$\liminf_{r\to\infty} [\log \mu(r, f)/T(r, f)] \leq -\delta(\infty, f).$$

These results imply that

$$\liminf_{r\to\infty} [\log \mu(r, f)/T(r, f)] = -1$$

for transcendental entire functions of lower order zero.

In view of these results and Theorem D there arises the following question: What can be said on functions with a Nevanlinna deficient value and of positive lower order λ less than 1/2 ? I can give the following answer to this question:

THEOREM 1. *Let* f *be a meromorphic function of lower order* λ, $0 < \lambda < 1/2$, *such that* $\delta(\infty, f) > 1 - \cos\pi\lambda$. *Then* (1) *holds.*

The following theorem shows that Theorem 1 is sharp.

THEOREM 2. *Given* λ, $0 < \lambda < 1/2$, *and* d, $1 - \cos\pi\lambda < d < 1$, *there exists a meromorphic function* f *of order* λ *and of lower order* λ *such that* $\delta(\infty, f) = d$ *and* (2) *holds.*

The following theorem shows that Theorem A remains sharp for functions which do not satisfy the inequality $\delta(\infty, f) > 1 - \cos\pi\lambda$.

THEOREM 3. *Given* λ, $0 < \lambda < 1/2$, *and* d, $0 < d < 1 - \cos\pi\lambda$, *there exists a meromorphic function* f *of order* λ *and of lower order* λ *such that* $\delta(\infty, f) = d$ *and* (3) *holds.*

2. PROOFS

Let f be as in Theorem 1. We write

$$L(r, f) = \min\{|f(z)| : |z| = r\}.$$

Since

$$\log \rho(f(z)) \le \log|f'(z)/f(z)| + \log|1/f(z)|,$$

we deduce, using the method by Edrei [2], by Fuchs [4], or by Essen, Rossi and Shea [3], as well as the standard methods of the Nevanlinna theory for the logarithmic derivative, that there exists a sequence (r_n), $r_n \to \infty$ as $n \to \infty$, such that

$$\log L(r_n, f) \ge \frac{\pi\lambda}{\sin \pi\lambda} (\cos \pi\lambda - 1 + \delta(\infty, f) + o(1)) T(r_n, f)$$

and

$$\log M(r_n, f'/f) \le o(T(r_n, f))$$

as $n \to \infty$. These estimates prove Theorem 1.

The following lemma is a consequence of a result of Nevanlinna [5, p. 229].

LEMMA 1. *Let* λ *satisfy* $0 < \lambda < 1/2$ *and let*

$$g(z) = \prod_{n=1}^{\infty} (1 + z n^{-1/\lambda}).$$

Then for any ε, $0 < \varepsilon < \pi$,

$$\log g(z) = [\pi(\sin \pi\lambda)^{-1} + o(1)] z^{\lambda}$$

as $z \to \infty$ *in the angle* $|\arg z| \le \pi - \varepsilon$.

Let λ and g be as above and let a be chosen so that $0 < a^{\lambda} < 1/\cos \pi\lambda$. We set $f(z) = g(z)/g(-az)$.

If $0 < a^{\lambda} < \cos \pi\lambda$, then an explicit calculation shows that $\delta(\infty, f) = I - a^{\lambda}$ and f satisfies (2) with $d = \delta(\infty, f)$. This proves Theorem 2.

If $\cos \pi\lambda < a^{\lambda} < 1/\cos \pi\lambda$, then, using Lemma 1, we deduce that f satisfies (3) and

$$\delta(\infty, f) = 1 - |1 - a^{\lambda} e^{i\pi\lambda}|^{-1} a^{\lambda} \sin \pi\lambda.$$

This implies that if a^{λ} increases from $\cos \pi\lambda$ to $1/\cos \pi\lambda$ then $\delta(\infty, f)$ decreases from $1 - \cos \pi\lambda$ to zero. This shows that a suitable choice of a^{λ} gives us the function desired for Theorem 3.

References

[1] ANDERSON, J.M. and S. TOPPILA: 'The growth of the spherical derivative of a meromorphic function of finite lower order', *J. London Math. Soc.* (2), 27 (1983), 289-305.

[2] EDREI, A.: 'A local form of the Phragmén-Londelöf indicator', *Mathematika* 17 (1970), 149-172.

[3] ESSEN, M.R., J.F. ROSSI and D.F. SHEA: 'A convolution inequality with applications in function theory', *Contemporary Mathematics* 25 (1983), 141-147.

[4] FUCHS, W.H.J.: 'A theorem on $\min_{|z|} \log|f(z)|/T(r, f)$', Proc. of the Symposium on Complex Analysis in Canterbury 1973. *London Math. Soc. Lecture Note Series* 12 (1974), 69-72.

[5] NEVANLINNA, R.: *Analytic functions*. Springer-Verlag, Berlin - Heidelberg - New York, 1970.

[6] TOPPILA, S.: 'On the growth of the spherical derivative of a meromorphic function', *Ann. Acad. Scient. Fenn. Ser. A I Math.* 7 (1982), 119-139.

[7] ———: 'On the spherical derivative of an entire function', *Manuscripta Math.* 53 (1985), 225-234.

ON THE CONNECTION BETWEEN THE NEVANLINNA CHARACTERISTICS OF AN ENTIRE FUNCTION AND OF ITS DERIVATIVE

Julian Ławrynowicz
Institute of Mathematics
Polish Academy of Sciences
PL - 90-136 Łódź, Poland

and Sakari Toppila
Department of Mathematics
University of Helsinki
SF - 00100 Helsinki 10, Finland

ABSTRACT. We use the usual notation of the Nevanlinna theory. The following result is mentioned but not proved in [2]: There exists an absolute constant $Q > 1$ such that

$$\limsup_{r\to\infty} [T(Qr, f')/T(r, f)] \geq 1 \tag{1}$$

for any transcendental entire function f. Now we give a numerical estimate for the constant Q. We prove that if f is a transcendental entire function then

$$\limsup_{r\to\infty} [T(10r, f')/T(r, f)] \geq 1. \tag{2}$$

1. INTRODUCTION AND STATEMENT OF RESULTS

We aim at giving a numerical estimate for the constant Q in

THEOREM 1. *There exists an absolute constant* $Q > 1$ *such that the inequality* (1) *holds for any transcendental function* f.

Namely, we prove

THEOREM 2. *Let* f *be a transcendental entire function. Then we have the estimate* (2).

Remark. A direct calculation, analogous to that given below, with $Q = 9$ instead of $Q = 10$, does not lead to the positive conclusion. The best possible Q is certainly within $(1; 10)$, but its precise determining seems to be very involved.

2. PROOF OF THEOREM 2

Let f be a transcendental entire function. Let λ be the order of f. We suppose first that

J. Ławrynowicz (ed.), Deformations of Mathematical Structures, 51–53.

$$\lambda < 1 - (\log 10)^{-1} \sim 0.5657. \tag{3}$$

Then we may write

$$f'(z) = a z^m \prod_{p=1}^{\infty} (1 - z/a_p)$$

and it follows from Theorem 1.11 of Hayman [1], p. 27, that

$$\begin{aligned} \log M(r, f') &\le \int_0^r (n(t, 0, f') - n(0, 0, f'))t^{-1}\,dt + \\ &\quad + r \int_r^{\infty} n(t, 0, f')t^{-2}\,dt + O(\log r) \\ &= N(r, 0, f') + r \int_r^{\infty} n(t, 0, f')t^{-2}\,dt + \\ &\quad + O(\log r) \qquad \text{as} \quad r \to \infty. \end{aligned} \tag{4}$$

We write $b = 1 - (\log 10)^{-1}$. Since $\lambda < b$, there exists a sequence r_n, $r_n \to \infty$ as $n \to \infty$, such that

$$n(t, 0, f') \le (t/r_n)^b\, n(r_n, 0, f') \qquad \text{for all} \quad t > r_n.$$

This together with (4) implies

$$\begin{aligned} \log M(r_n, f') &\le N(r_n, 0, f') + \\ &\quad + r_n \int_{r_n}^{\infty} n(r_n, 0, f') r_n^{-b}\, t^{b-2}\,dt + O(\log r_n) \\ &= N(r_n, 0, f') + n(r_n, 0, f') \log 10 + \\ &\quad + O(\log r_n) \qquad \text{as} \quad n \to \infty. \end{aligned} \tag{5}$$

From the first main theorem of the Nevanlinna theory it follows that

$$\begin{aligned} T(10\, r_n, f') &\ge N(10\, r_n, 0, f') + O(1) \\ &= N(r_n, 0, f') + \int_{r_n}^{10 r_n} n(t, 0, f')t^{-1}\,dt + O(1) \\ &\ge N(r_n, 0, f') + n(r_n, 0, f') \log 10 + O(1) \qquad \text{as} \quad n \to \infty. \end{aligned}$$

This together with (6) implies that

$$\log M(r_n, f') \le (1 + o(1))\, T(10\, r_n, f') \qquad \text{as} \quad n \to \infty. \tag{6}$$

Since

$$M(r, f) \le |f(0)| + r\, M(r, f') \tag{7}$$

and

$$T(r, f) \le \log M(r, f) \tag{8}$$

for all large r, we deduce from (6) that (2) holds.

Suppose now that

$$\lambda > 0.565. \tag{9}$$

We write $d = 0.565$. Since $\lambda > d$, there exists a sequence r_n, $r_n \to \infty$ as $n \to \infty$, such that

$$T(10\, r_n, f')/T(4\, r_n, f') \geq (5/2)^d \tag{10}$$

for any n. This implies that

$$\log M(r_n, f') \leq (4\, r_n + r_n)/(4\, r_n - r_n)\, T(4\, r_n, f')$$
$$\leq (5/3)(2/5)^d\, T(10\, r_n, f') < T(10\, r_n, f')$$

for all large n, which together with (7) and (8) shows that (4) holds also in this case. Theorem 2 is proved.

References

[1] HAYMAN, W.K.: *Meromorphic functions*, Clarendon Press, Oxford 1964.

[2] TOPPILA, S.: 'On Nevanlinna's proximity function', Commentationes in memoriam Rolf Nevanlinna, *Ann. Acad. Sci. Fenn. Ser.* A I *Math.* 7 (1982), 59-64.

CHARACTERISTIC HOMOMORPHISM FOR TRANSVERSELY HOLOMORPHIC FOLIATIONS VIA THE CAUCHY-RIEMANN EQUATIONS

Grzegorz Andrzejczak
Institute of Mathematics
Polish Academy of Sciences
PL-90-136 Łódź, Poland

ABSTRACT. In our earlier paper [1], we have presented a detailed exposition of the Bott characteristic homomorphism for transversely holomorphic foliations. The original Bott construction [4] arises from a comparison between two sets of connections in the normal bundle. Our intention is to give an alternative construction which is slightly more intrinsic and comes from integrating cocycles associated with the normal bundle. The existence of such cocycles allows us to modify the construction of [6] in order to make it applicable to transversely holomorphic foliations (and even to a wide class of foliations with an integrable transverse G-structure [2]).

1. THE CONSTRUCTION

1.1. For any smooth foliation F of a (real) manifold M its principal normal bundle is the manifold of 1-jets $P_F = \{j^1_x\phi;\ \phi(x) = 0\}$ of submersions (to $\mathbb{R}^N$) defining F locally. A 2q-codimensional foliation F is called *transversely holomorphic* (TH-) if it is endowed with an integrable $Gl^{\mathbb{C}}_q$-subbundle $P \subset P_F$, where one identifies the complex linear group with the subgroup

$$Gl^{\mathbb{C}}_q = \{ \begin{bmatrix} A & B \\ -B & A \end{bmatrix};\ \det(A + iB) \neq 0\} \subset Gl(2q, \mathbb{R}).$$

Integrability means that the distinguished subbundle P admits (sufficiently many) local sections of the form

$$U \ni x \to j^1_x(t_{-\phi(x)}\ \phi) \in P, \tag{1.2}$$

where t_a stands for the translation $z \to z + a$ and $\phi\colon M \supset U \to \mathbb{R}^{2q}$ is a submersion defining $F|U$.

Analogously to complex structures on real 2q-manifolds, P introduces a complex structure in the direction transverse to the foliation.

J. Ławrynowicz (ed.), Deformations of Mathematical Structures, 55–63.

1.3. Let (F,P) be any TH-foliation of M of complex codimension q. Any collection $\{s_a;\ a \in A\}$ of (smooth) local sections $s_a\colon U_a \to P$ over a covering $\mathcal{U} = \{U_a;\ a \in A\}$ of M determines a transition $Gl_q^{\mathbb{C}}$-cocycle $\{g_{ab}\}$ such that

$$s_b(x) = s_a(x)\, g_{ab}(x) \quad \text{for} \quad x \in U_a \cap U_b.$$

When interpreted as a homomorphism of groupoids,

$$M_{\mathcal{U}} \xrightarrow{\ g\ } Gl_q^{\mathbb{C}}$$

the cocycle gives rise to a homomorphism of the de Rham algebras

$$A^{**}(N\, Gl_q^{\mathbb{C}}) \xrightarrow{\ g^*\ } A^{**}(N\, M_{\mathcal{U}}) = \check{C}^*(\mathcal{U}, A^*)$$

(we pass to semi-simplicial nerves of the groupoids, see [5], [7]). In cohomology, this yields a classical characteristic homomorphism

$$H^*(B\, Gl_q^{\mathbb{C}}) \to H^*(M)$$

for the complex normal bundle P (we recall that $B\, Gl_q^{\mathbb{C}}$ is the classifying space; its real cohomology algebra is freely generated by the universal Chern classes).

1.4. Taking advantage of integrability of the transverse complex structure P, we may restrict ourselves to *integrating* $Gl_q^{\mathbb{C}}$-cocycles, i.e. those cocycles which come from local sections of the form (1.2) only. Then each map $g_{ab}\colon U_a \cap U_b \to Gl_q^{\mathbb{C}}$ is locally the *derivative* of a biholomorphic automorphism of $\mathbb{C}^q \cong \mathbb{R}^{2q}$. Indeed, if s_a comes from a submersion $\phi_a\colon U_a \to \mathbb{R}^{2q}$, $a \in A$, then

$$g_{ab}(x) = D\,\gamma_{ab}^{(x)}\,(\phi_b(x)),$$

where

$$\phi_a = \gamma_{ab}^{(x)} \circ \phi_b \quad \text{around} \quad x \in U_a \cap U_b;$$

the condition $g_{ab}(x) \in Gl_q^{\mathbb{C}}$ is nothing but the Cauchy-Riemann equations on $\mathbb{C}^q$ (we identify $(x_1 + i y_1, \dots, x_q + i y_q) \in \mathbb{C}^q$ with $(x_1, \dots, x_q, y_1, \dots, y_q) \in \mathbb{R}^{2q}$).

Let $\Gamma_q^{\mathbb{C}}$ be the groupoid of germs of biholomorphic local automorphisms of $\mathbb{C}^q$. The homomorphism of groupoids associated with any integrating cocycle factorizes clearly as follows

$$\begin{array}{ccc} M_{\mathcal{U}} & \xrightarrow{\ g\ } & Gl_q^{\mathbb{C}} \\ & {\scriptstyle\gamma}\searrow \quad \nearrow{\scriptstyle D} & \\ & \Gamma_q^{\mathbb{C}} & \end{array},$$

where γ is given by the $\Gamma_q^{\mathbb{C}}$-cocycle

$$U_a \cap U_b \ni x \xrightarrow{\ \gamma_{ab}\ } [\gamma_{ab}^{(x)}, \phi_b(x)] \in \Gamma_q^{\mathbb{C}}$$

and D stands for the derivative. The induced factorization at the d.g. algebras level,

$$\begin{array}{ccc} A^{**}(N\,Gl^{\mathbb{C}}_q) & \xrightarrow{\;g^*\;} & \check{C}^*(U, A^*) \\ & {\scriptstyle D^*}\searrow \quad \nearrow{\scriptstyle \gamma^*} & \\ & A^{**}(N\,\Gamma^{\mathbb{C}}_q) & \end{array}$$

is of particular interest, for the cochain complex $A^{**}(N\,\Gamma^{\mathbb{C}}_q)$ computes the real cohomology of $B\Gamma^{\mathbb{C}}_q$, the classifying space of 2q-codimensional transversely holomorphic foliations, and D^* (and so every g^*) has a *nontrivial kernel*. Whatever the kernel of D^* is, one immediately obtains:

<u>THEOREM 1.5.</u> *Given an arbitrary* 2q-*codimensional* TH-*foliation* (F, P) *of a manifold* M, *for any integrating cocycle* g *(over a covering* U *of* M) *associated with the transverse complex structure* P *there is a well-defined d.g.a. homomorphism*

$$A^{**}(N\,Gl^{\mathbb{C}}_q)/\ker D^* \xrightarrow{\;g^*\;} \check{C}^*(U, A^*).$$

In cohomology, the induced characteristic homomorphism

$$\chi = \chi_{(F,P)} : H(A^{**}(N\,Gl^{\mathbb{C}}_q)/\ker D^*) \to H^*(M)$$

depends on the TH-*foliation only.*

We shall not compute the domain of χ. Instead, we shall characterize the ideal $\ker D^*$ and relate χ to the original Bott characteristic homomorphism.

1.6. Let $(x^a)_{a\le 2q}$ be the canonical coordinates in $\mathbb{R}^{2q}$, and $[x^a_b]_{a,b\le 2q}$ the coordinates in $Gl(2q,\mathbb{R})$ restricted to $Gl^{\mathbb{C}}_q$, so that $x^{q+i}_{q+j} = x^i_j$ and $x^i_{q+j} = -x^{q+i}_j$ for $i, j \le q$.

The groupoid $\Gamma^{\mathbb{C}}_q$ as well as the Lie group $Gl^{\mathbb{C}}_q$ carry canonical complex structures - one induced by the source projection $\alpha: \Gamma^{\mathbb{C}}_q \to \mathbb{R}^{2q} = \mathbb{C}^q$, and the other given by the global chart $Gl^{\mathbb{C}}_q \ni [x^a_b]_{a,b\le 2q} \to [x^k_l + i\,x^{q+k}_l]_{k,l\le q} \in \mathbb{C}^{q\times q}$. It is an obvious second order consequence of the Cauchy-Riemann equations that the derivative $D: \Gamma^{\mathbb{C}}_q \to Gl^{\mathbb{C}}_q$ is a *holomorphic* map. Consequently, the tangent homomorphism D_* preserves the decomposition of the complex tangent bundles

$$T_{\mathbb{C}}\Gamma^{\mathbb{C}}_q = T^{10}\Gamma^{\mathbb{C}}_q \oplus T^{01}\Gamma^{\mathbb{C}}_q \to T^{10} \oplus T^{01} = T_{\mathbb{C}}\,Gl^{\mathbb{C}}_q.$$

Being involutive, the subbundles can be interpreted as foliations; this point of view leads to the following observation:

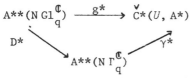

Let J^+_1 (J^-_1) be an ideal of the complex-valued forms on $Gl^{\mathbb{C}}_q$ whose restriction to T^{01} (resp., to T^{10}) is 0. D pulls J^{+}_1 back to forms on $\Gamma^{\mathbb{C}}_q$ that annihilate $T^{01}\Gamma^{\mathbb{C}}_q$, i.e. to an ideal generated by q 1-forms $\alpha^*(dx^k + i\,dx^{q+k})$, $k = 1, \ldots, q$. Thus $D^*(J_1^+)^{q+1} = 0$; more generally,

$$(J_1^+)^{q+1} + (J_1^-)^{q+1} \subset \ker D^*.$$

An analogous reasoning holds for the maps

$$D_n\colon N_n\Gamma_q^{\mathbb{C}} \to N_n Gl_q^{\mathbb{C}}$$

$n = 1, 2, \ldots$, induced from D. As the complex manifolds $N_n\Gamma_q^{\mathbb{C}}$ are all q-dimensional, we obtain

$$(J_n^+)^{q+1} + (J_n^-)^{q+1} \subset \ker D_n^*,$$

where J_n^+ (J_n^-) is an ideal of the forms on $N_n Gl_q^{\mathbb{C}}$ annihilating $T^{01} \subset T_{\mathbb{C}} N_n Gl_q^{\mathbb{C}}$ (resp., T^{10}).

Let $J^{\pm} := \bigoplus_{n \geq 1} J_n^{\pm} \subset A_{\mathbb{C}}^{**}(N\,Gl_q^{\mathbb{C}})$. One can easily see that both J^+ and J^- are differential ideals such that

$$(J^+)^{q+1} + (J^-)^{q+1} \subset \ker(D^*\colon A_{\mathbb{C}}^{**}(N\,Gl_q^{\mathbb{C}}) \to A_{\mathbb{C}}^{**}(N\,\Gamma_q^{\mathbb{C}}))$$

(note that the face operators of the semi-simplicial manifolds are holomorphic maps). According to Theorem 1.5, this implies

COROLLARY 1.7. *Any* TH-*foliation* (F, P) *of complex codimension* q *of a manifold* M *gives rise to natural characteristic homomorphisms*

$$\chi^{\mathbb{C}}\colon H(A_{\mathbb{C}}^{**}(N\,Gl_q^{\mathbb{C}})/(J^+)^{q+1} + (J^-)^{q+1}) \to H^*(M, \mathbb{C})$$

and

$$\chi^{\mathbb{R}}\colon H(A^{**}(N\,Gl_q^{\mathbb{C}})/\mathrm{Re}(J^+)^{q+1} + \mathrm{Im}(J^+)^{q+1}) \to H^*(M, \mathbb{R}).$$

P r o o f. For any two real forms ϕ and ψ, $\phi + i\psi \in (J^+)^{q+1}$ iff $\phi - i\psi \in (J^-)^{q+1}$. If this is the case, then both ϕ and ψ are in $(J^+)^{q+1} + (J^-)^{q+1}$.

Clearly, $\chi^{\mathbb{C}}$ is the complexification of $\chi^{\mathbb{R}}$. We prefer the complex homomorphism for it is easier to handle.

2. A REDUCTION

The above construction provides no information about geometric significance of the characteristic homomorphism - partly because the universal algebra $H(A_{\mathbb{C}}^{**}(N\,Gl_q^{C})/J)$, $J = (J^+)^{q+1} + (J^-)^{q+1}$, has not been computed. In the sequel we shall show how to reduce the homomorphism to a more computable one (see below Theorem 2.7). As in [10] the reduction consists of two steps. Namely,

1° we blow up the ss-manifold to some natural $Gl_q^{\mathbb{C}}$-bundle (Lemma 2.3), and then

2° apply the semi-simplicial Weil homomorphism ([7], [8]; Lemma 2.6).

2.1. For shortness, let G denote the Lie group $Gl_q^{\mathbb{C}}$. The ss-manifold NG admits a canonical *universal* G-bundle $\bar{N}G = (\bar{N}_n G)$ such that

$$\bar{N}_n G = G \times \ldots \times G \quad (n+1 \ \text{factors}),$$

the action of G (on the right) is the diagonal one, and the projection is

$$\bar{N}_n G \ni (g_o, \ldots, g_n) \xrightarrow{\pi} (g_o g_1^{-1}, \ldots, g_{n-1} g_n^{-1}) \in N_n G.$$

The bundle projection induces a homomorphism of differential algebras,

$$A_{\mathbb{C}}^{**}(\bar{N}G/U(q))/(\tilde{J}^+)^{q+1} + (\tilde{J}^-)^{q+1} \xleftarrow{\pi^*}$$

$$\xleftarrow{\pi^*} A_{\mathbb{C}}^{**}(NG)/(J^+)^{q+1} + (J^-)^{q+1}, \quad (2.2)$$

where $U(q)$ is the unitary group, whereas $\tilde{J}^+$ $(\tilde{J}^-)$ is the ideal generated by $\pi^* J^+$ $(\pi^* J^-)$ and so consists of the forms annihilsting $\pi_*^{-1} T^{01}$ (resp., $\pi_*^{-1} T^{10}$).

LEMMA 2.3. *Homomorphism* (2.2) *induces an isomorphism in cohomology.*

Proof. It sufficies to observe that the π^* induces an isomorphism in d-cohomology. This is evident, for $U(q)$ is a maximal compact subgroup of $G = Gl_q^{\mathbb{C}}$ and any ideal lifted by π^* is invariant under homotopies along the fibers.

2.4. Let g be the Lie algebra of $G = Gl_q^{\mathbb{C}}$ and $g_{\mathbb{C}}$ its complexification. We consider the canonical g-valued 1-form θ on G an element of $A_{\mathbb{C}}^1(G) \otimes_{\mathbb{C}} g_{\mathbb{C}}$. θ gives rise to a semi-simplicial Weil homomorphism

$$W_{\mathbb{C}}(g) \xrightarrow{k_{\Delta}} A_{\mathbb{C}}^{**}(\bar{N}G)$$

defined briefly as follows.

For $n \geq 0$, the $n+1$ canonical projections of $\bar{N}_n G = G^{n+1}$ onto G lift θ to connections $\theta_o, \ldots, \theta_n$ in the G-bundle $\bar{N}_n G \to N_n G$. If $\Delta^n \subset \mathbb{R}^{n+1}$ stands for the geometric n-simplex, the assignment

$$\Delta^n \ni t \to \theta_t^{(n)} := t^o \theta_o + \ldots + t^n \theta_n \quad (2.5)$$

yields a connection $\theta^{(n)}$ in $\bar{N}_n G \times \Delta^n$. Consequently, there is a linear map

$$k_n \colon W_{\mathbb{C}}(g) \xrightarrow{k(\theta^{(n)})} A_{\mathbb{C}}^*(\bar{N}_n G \times \Delta^n) \xrightarrow{\int_{\Delta^n}} A_{\mathbb{C}}^*(\bar{N}_n G)$$

of degree $-n$ ($k(\theta^{(n)})$ - the classical Weil homomorphism). The semi-

-simplicial Weil homomorphism associated with θ is the sum

$$k_\Delta = k_o + k_1 + \dots .$$

The complexified Lie algebra $g_{\mathbb{C}}$ admits a canonical decomposition $g_{\mathbb{C}} = g_+ \oplus g_-$ according to the two eigenvalues of the complex structure inherited from G. Consequently,

$$g^*_{\mathbb{C}} = g^*_+ \oplus g^*_- ,$$

where the elements of g^*_+ (g^*_-) annihilate g_- (resp., g_+).

LEMMA 2.6. *The semi-simplicial Weil homomorphism* k_Δ *induces a homomorphism*

$$(W_{\mathbb{C}}(g)/(S^{>q}\, g^*_+) + (S^{>q}\, g^*_-))_{U(q)} \to A^{**}_{\mathbb{C}}(N\bar{G}/U(q))/(\tilde{J}^+)^{q+1} + (\tilde{J}^-)^{q+1}.$$

Note. The truncated Weil algebra on the left is usually denoted by $(W_{\mathbb{C}}(g_+ \oplus g_-)_{q,q})_{U(q)}$ (cf. [9]).

Proof. For any $n \geq 0$, the complex structure of $G = Gl^{\mathbb{C}}_q$ induces a canonical decomposition of the tangent bundles

$$\begin{array}{ccccc} T_{\mathbb{C}}\bar{N}_n G & = & \bar{T}^{10} & \oplus & \bar{T}^{01} \\ \downarrow \pi_* & & & & \downarrow \\ T_{\mathbb{C}} N_n G & = & T^{10} & \oplus & T^{10}. \end{array}$$

Both $\bar{T}^{10}$ and $\bar{T}^{01}$ are G-invariant involutive subbundles of $T_{\mathbb{C}}\bar{N}_n G$ and project to the involutive subbundles T^{10} and T^{01}, respectively. This fact has a clear interpretation in terms of foliations and partial connections over them. Namely, we may call a connection ω in the G-bundle *adapted* (to the complex structure) if it is g_+-valued on $\bar{T}^{10}$ (and g_--valued on $\bar{T}^{01}$). Alternatively, the adaptedness of ω means

$$\omega(J\,v) = J\,\omega(v), \quad \text{for } v \in T\bar{N}_n G,$$

where J stands for the complex structure (in the tangent space on the left, and in g on the right). If this is the case and Ω is the curvature of ω (a horizontal form), then its g_+-component annihilates the subbundle $\pi_*^{-1} T^{01} \subset T_{\mathbb{C}}\bar{N}_n G$, while its g_--component annihilates $\pi_*^{-1} T^{10}$. Clearly, the same property characterizes the difference form $\omega - \omega'$ of any two adapted connections ω and ω'.

Turning back now to the canonical form θ and the induced connections in the G-bundles $\bar{N}_n G \to N_n G$, let us observe that all these connections are adapted to the complex structure. Actually, it sufficies to check adaptedness of θ, for the property is inherited by the pull-backs (the structure operators are holomorphic). One has $\theta = X^{-1}\,dX$ in the matrix notation, and thus adaptedness of the connection follows immediately from the fact that the complex structures (in TG and in g) are just products with a matrix $J_o = \begin{bmatrix} 0 & I \\ -I & 0 \end{bmatrix}$ (note that $Gl_q^{\mathbb{C}} = \{A \in Gl(2q, \mathbb{R})\,;\, A J_o = J_o A\}$).

Thus, we have come to the point that for any $n \geq 0$ the connections $\theta_o, \ldots, \theta_n$ on $\bar{N}_n G$ - and so $\theta_t^{(n)}$ for $t \in \Delta^n$ (cf. (2.5)) - are all adapted. If we denote by $\Omega_t^{(n)}$ the curvature of $\theta_t^{(n)}$, then the curvature of $\theta^{(n)}$ is

$$\Omega^{(n)} = \Omega_+^{(n)} + dt^1 \wedge (\theta_1 - \theta_o) + \ldots + dt^n \wedge (\theta_n - \theta_o).$$

As we have observed, both $\Omega_t^{(n)}$ and the differences $\theta_i - \theta_o$ verify the property that their g_+-components (g_--components) annihilate $\pi_*^{-1} T^{01}$ (resp., $\pi_*^{-1} T^{10}$). This implies that $k_n : W_{\mathbb{C}}(g) \to A_{\mathbb{C}}^*(\bar{N}_n G)$ transfers the ideal $\Lambda g_{\mathbb{C}}^* \otimes_{\mathbb{C}} S^{>q} g_+^* \subset W_{\mathbb{C}}(g)$ to the q-th power of the ideal of those forms which annihilate $\pi_*^{-1} T^{01}$ (and an analogous assertion holds for $\Lambda g_{\mathbb{C}}^* \otimes_{\mathbb{C}} S^{>q} g_-^*$). In conclusion, the $U(q)$-basic homomorphism

$$k_\Delta : W_{\mathbb{C}}(g)_{U(q)} \to A_{\mathbb{C}}^{**}(\bar{N} G/U(q))$$

sends elements of $\Lambda g_{\mathbb{C}}^* \otimes_{\mathbb{C}} (S^{>q} g_+^* + S^{>q} g_-^*)$ to $(\tilde{J}^+)^{q+1} + (\tilde{J}^-)^{q+1}$, as was to be shown.

Lemmas 2.3 & 2.6 imply that there is a well-defined homomorphism of algebras

$$\kappa = (\pi^*)^{-1} k_{\Delta *} : H(W_{\mathbb{C}}(g_+ \oplus g_-)_{q,q})_{U(q)} \to$$
$$\to H(A_{\mathbb{C}}^{**}(N G)/(J^+)^{q+1} + (J^-)^{q+1}).$$

<u>THEOREM 2.7</u>. *For any* $2q$*-codimensional transversely holomorphic foliation* (F, P) *of a manifold* M*, the composition*

$$\begin{array}{ccc} H(W_{\mathbb{C}}(g_+ \oplus g_-)_{q,q})_{U(q)} & \xrightarrow{\ \kappa\ } & H(A_{\mathbb{C}}^{**}(N Gl_q^{\mathbb{C}})/(J^+)^{q+1} + (J^-)^{q+1}) \\ & \searrow & \downarrow \chi^{\mathbb{C}} \\ & & H^*(M, \mathbb{C}) \end{array}$$

is the Bott-Kamber-Tondeur characteristic homomorphism ([3], [9]).

<u>Question</u>. Is κ an isomorphism? It is reasonable to expect the answer is YES, but we cannot prove this at the moment.

<u>Proof</u> of the theorem. We adopt the notation of 1.3-1.4. Let $g = (g_{ab})$ be any integrating $Gl_q^{\mathbb{C}}$-cocycle for the complex normal bundle P, associated with a collection of submersions $\phi_a : U_a \to \mathbb{R}^{2q}$, $a \in A$, over a covering $\mathcal{U} = \{U_a\,;\, a \in A\}$ of M. The corresponding ss-map $g : N M_{\mathcal{U}} \to N Gl_q^{\mathbb{C}}$ admits a canonical lift to a homomorphism of ss-$Gl_q^{\mathbb{C}}$-bundles

$$\bar{g} : N P_{\pi^{-1}\mathcal{U}} \to \bar{N} Gl_q^{\mathbb{C}},$$

$\pi^{-1}(U_{a_0}\cap\dots\cap U_{a_n}) \ni j^1_x\phi \mapsto (D(\phi_{a_0}\phi^{-1})(0),\dots,D(\phi_{a_n}\phi^{-1})(0)) \in \bar N_n Gl_q^{\mathbb{C}}$.

Furthermore, $\bar g_0$ (0-th stage of $\bar g$) pulls back the canonical form θ to a collection of local connections ω_a in $P|U_a$, $a\in A$, which are adapted in the sense that the subbundles

$$\omega_a^{-1}(g_+) = (\bar g_0|\pi^{-1}U_a)^{-1}T^{10} \subset T_{\mathbb{C}}P|\pi^{-1}U_a$$

are restrictions of a global involutive subbundle $T^+\subset T_{\mathbb{C}}P$. If T^- is the complex conjugate to T^+, then clearly T^+ and T^- span the whole $T_{\mathbb{C}}P$, whereas their intersection $T^+\cap T^-$ is exactly the complexification of the real Bott partial connection in $P\subset P_F$. By definition, the connection forms ω_a are g_+-valued on T^+ (and g_--valued on T^-). Consequently, there is a diagram

$$\begin{array}{ccccc}
 & & A^{**}_{\mathbb{C}}(\bar N G/U(q))/\dots & \xleftarrow{\pi^*} & A^{**}_{\mathbb{C}}(NG)/\dots \\
 & {\scriptstyle k_\Delta}\nearrow & \downarrow{\scriptstyle \bar g^*} & & \downarrow{\scriptstyle g^*} \\
(W_{\mathbb{C}}(g_+\oplus g_-)_{qq})_{U(q)} & \to & A^{**}_{\mathbb{C}}(NP_{\pi^{-1}U}/U(q)) & \xleftarrow{\pi^*} & A^{**}_{\mathbb{C}}(NM_U), \\
 & {\scriptstyle k}\searrow & \uparrow{\scriptstyle i} & & \uparrow{\scriptstyle i} \\
 & & A^*_{\mathbb{C}}(P/U(q)) & \xleftarrow{\pi^*} & A^*_{\mathbb{C}}(M)
\end{array}$$

where the squares commute, all the π^*'s as well as the i's induce isomorphisms in homology, and k is a classical Weil homomorphism for any adopted connection in P (i.e. $g_\pm$-valued on $T^\pm$). Since the Bott-Kamber-Tondeur characteristic homomorphism does not depend on the particular adapted connection (nor a collection of such connections), the diagram commutes in homology, as was to be shown.

References

[1] ANDRZEJCZAK, G.: 'Transversely holomorphic foliations and characteristic classes', *Proceedings of the Second Finnish-Polish Summer School in Complex Analysis at Jyväskylä*, pp. 5-14. Univ. of Jyväskylä, Rep. 28, 1984.

[2] ——: 'On foliations of semi-simplicial manifolds and characteristic homomorphisms', preprint MPI Bonn/SFB 85 - 62

[3] BAUM, P. and R. BOTT,: 'Singularities of holomorphic foliations', *J. Diff. Geometry* 7 (1972), 279-342.

[4] BOTT, R.: 'On the Lefschetz formula and exotic characteristic classes', *Symp. Math.* 10 (1972), 95-105.

[5] ——: 'Lectures on characteristic classes and foliations', *Lectures on Algebraic and Differential Topology*, pp. 1-94. Lecture Notes in Math., vol. 279, Springer-Verlag, Berlin-Heidelberg-New York, 1972.

[6] ——, H. SHULMAN and J. STASHEFF, 'On the de Rham Theory of Certain Classifying Spaces', *Advances in Math.* 20 (1976), 43-56.

[7] DUPPONT, J.L.: *Curvature and Characteristic Classes.* Lecture Notes

in Math., vol. **640**, Springer-Verlag, Berlin-Heidelberg-New York, 1978.

[8] KAMBER, F.W. and Ph. TONDEUR, 'Semi-simplicial Weil algebras and characteristic classes for foliated bundles in Cech cohomology', *Proc. Symposia Pure Math.*, Vol. **27** (1975), 283-294.

[9] —— and ——: 'G-foliations and their characteristic classes', *Bull. Amer. Math. Soc.*, **84** (1978) 6, 1086-1124.

[10] SHULMAN, H. and J. STASHEFF: 'De Rham theory for $B\Gamma$', *Differential Topology, Foliations and Gelfand-Fuks Cohomology* (Proc. Rio de Janeiro 1976), pp. 62-74. Lecture Notes in Math., vol. **652**, Springer-Verlag, Berlin-Heidelberg-New York, 1978.

COMPLEX PREMANIFOLDS AND FOLIATIONS

Włodzimierz Waliszewski
Institute of Mathematics
Polish Academy of Sciences
PL-90-136 Łódź, Poland

ABSTRACT. The concept of a complex premanifold (c.p.) as well as the concept of an analytical mapping of c.p. are introduced in the paper. In the category of c.p. for any set of complex functions there exist the smallest c.p. containing this set. Any complex manifold is a c.p. Some characterization of a complex submanifold of C^n is given and it is shown that if Cartesian product of two c.p. is a complex manifold, then these c.p. are complex manifolds as well. The paper concludes with considerations concerning foliations on c.p. The concept of such objects is defined and it is proved that if c.p. is a complex manifold, then on this manifold any foliation in the new sense is the foliation in the classical sense.

INTRODUCTION. During the last twenty years the works with concepts being some generalizations of the concept of a smooth manifold have appeared. These concepts allow us to study the mathematical facts which do not require that the considered space is diffeomorphically Euclidean and, simultaneously, allow us to take an advantage of lots of methods acting in the theory of differentiable manifolds. Some generalizations as, for example, the concept of R. Sikorski's differential space ([3] and [4]), M.M. Postnikov's concept of premanifold ([2], see also [6]) as well as the concept of a complex premanifold presented below allow us to construct new objects by algebraic generating. The proposed concept is especially simple. We simply deal with a set of complex valued functions as the only basic concept. The topology and the differential structure are defined by this set of functions. In the present paper we propose some generalization of the concept of foliation to the category of complex premanifolds.

0. THE CONCEPTS OF A COMPLEX PREMANIFOLD AND AN ANALYTICAL MAPPING

Let M be a set of complex valued functions. Denote the union of all domains D_α of functions $\alpha \in M$ by $\underline{M}$. On the set $\underline{M}$ we consider the smallest topology, $\operatorname{top} M$, such that all the sets $\alpha^{-1} B$, where $\alpha \in M$

J. Ławrynowicz (ed.), Deformations of Mathematical Structures, 65–78.

and B is open in $\mathbb{C}$, are open. Let $S \subset \underline{M}$. We define M_S as the set of all functions β such that for any $p \in D_\beta$ there exist $\alpha \in M$ and $U \in \operatorname{top} M$ fulfilling the condition: $U \subset D_\alpha$, $p \in U \cap S \subset D_\beta$ and $\beta | U \cap S = \alpha | U \cap S$. It is easy to check that $\operatorname{top} M_S$ coincides with the topology on S induced from $\operatorname{top} M$. In particular, $\operatorname{top} M_{\underline{M}} = \operatorname{top} M$. It is also easy to check that $(M_T)_S = M_S$ whenever $S \subset T \subset \underline{M}$.

For any complex function ω analytical on the set D_ω open in $\mathbb{C}^m$ and for any complex functions $\alpha_1, \ldots, \alpha_m$ we define the function $\omega(\alpha_1, \ldots, \alpha_m)$ on the set of all points $p \in D_{\alpha_1} \cap \ldots \cap D_{\alpha_m}$ such that $(\alpha_1(p), \ldots, \alpha_m(p)) \in D_\omega$ by the formula $\omega(\alpha_1, \ldots, \alpha_m)(p) = \omega(\alpha_1(p), \ldots, \alpha_m(p))$. The set of all functions $\omega(\alpha_1, \ldots, \alpha_m)$, where $\alpha_1, \ldots, \alpha_m \in M$, ω is analytical on an open subset of C^m, $m = 1, 2, \ldots$, will be denoted by $\operatorname{an} M$. It is easy to see that $\operatorname{top} \operatorname{an} M = \operatorname{top} M$.

The set M of complex functions satisfying the condition $M = M_{\underline{M}} = \operatorname{an} M$ will be called a complex premanifold (c.p.). The concept of c.p. is related to the one of R. Sikorski's differential space [3], yet there are some essential differences. The differential structure of a differential space is a set of real functions defined on the same set regarded as the set of all points of the differential space. The topology of the differential space is meant to be the weakest topology such that all the functions of the differential structure are continuous. Hence it follows that if the topology of a differential structure is T_0-topology, in P. S. Alexandrov's terminology, then it is $T_{3\,1/2}$-topology. This fact does not transfer to the category of c.p.

Example. Let us set $\alpha_0(0) = \alpha_0(1) = 0$, $\alpha_1(1) = 1$, $D_{\alpha_0} = \{0, 1\}$, $D_{\alpha_1} = \{1\}$, $M = \{\alpha_0, \alpha_1\}$. Here we have $\operatorname{top} M = \{\emptyset, \{0,1\}, \{1\}\}$. It is easy to see that $\operatorname{top} M$ is T_0-topology but it is not T_1-topology, and then $\operatorname{top} M$ is not a Hausdorff topology.

For any c.p. M and any $S \subset \underline{M}$ we have a c.p. of the form M_S. Any c.p. N satisfying the condition $N = M_{\underline{N}}$ will be called a complex subpremanifold of M.

Let M and N be c.p. and f be a mapping from $\underline{M}$ into $\underline{N}$. We will say that f maps analytically M into N, what we write down in the form

$$f : M \to N \tag{0.1}$$

iff for any $\beta \in N$ we have $\beta \circ f \in M$. Here for any functions g and f the composition $g \circ f$ is meant to be the function given on the f-preimage $f^{-1} D_g$ of the domain D_g of g by the formula $(g \circ f)(x) = g(f(x))$. The analytical mapping (0.1) such that f is one-to-one and we have the analytical mapping $f^{-1} : N \to M$ is said to be a diffeomorphism. By an easy verification we have the following

0.1. PROPOSITION. *For* (0.1), *any* $S \subset \underline{M}$ *and any* $T \subset \underline{N}$ *such that* $f\, S \subset T$, *we have the analytical mapping* $f|S : M_S \to N_T$. *If* $g : N \to P$, *then* $g \circ f : M \to P$.

1. THE GENERATING OF C.P.

The following proposition allows us to define some c.p. whenever we have an arbitrary set of complex functions.

1.1. PROPOSITION. *For any set* G *of complex functions the set* $(\mathrm{an}\, G)_{\underline{G}}$ *is the smallest one among all c.p. containing* G.

Proof. It is easy to check that for any set G of complex functions we have $\mathrm{an}(G_{\underline{G}}) \subset (\mathrm{an}\, G)_{\underline{G}}$. Applying Lemma 1.1 from [5], p. 264, we end the proof.

For any $S \subset \underline{M}$ we will denote by $M|S$ the set of all functions $\alpha|D_\alpha \cap S$, where $\alpha \in M$. We have, of course, $M|S \subset M_S$.

1.2. PROPOSITION. *If* $f : \underline{M} \to \underline{N}$ *and* N *is a c.p. generated by* G, *then we have* (0.1) *if and only if* $\beta \circ f \in M$ *for* $\beta \in G$.

Proof. Assume that $\beta \circ f \in M$ for $\beta \in G$. Let $\gamma \in G_G$ and $p \in D_{\gamma \circ f}$. Then there exist $\beta \in G$ and $V \in \mathrm{top}\, G$ such that $f(p) \in V \subset D_\beta \cap \cap D_\gamma$ and $\gamma|D = \beta|V$. Thus, $\gamma \circ f|f^{-1}V = \beta \circ f|f^{-1}V$ and $\beta \circ f \in M$. Taking any $q \in f^{-1}V$ we have $f(q) \in V$. Then for some $\beta_1, \ldots, \beta_s \in N$ and sets $B_1, \ldots, B_s$ open in $\mathbb{C}$, we have

$$f(q) \in \bigcap_j \beta_j^{-1} B_j \subset V.$$

Thus

$$q \in \bigcap_j (\beta_j \circ f)^{-1} B_j \subset f^{-1} V.$$

Hence it follows that $f^{-1} V \in \mathrm{top}\, M$ and, therefore, $\gamma \circ f \in M_{\underline{M}} = M$. Now, let γ an G. Then $\gamma = \omega(\gamma_1, \ldots, \gamma_m)$, where ω is analytical on the set D_ω open in $\mathbb{C}^m$ and $\gamma_1, \ldots, \gamma_m \in G$. Thus $\gamma \circ f = \omega(\gamma_1 \circ f, \ldots, \gamma_m \circ f) \in \mathrm{an}\, M = M$. Hence it follows (0.1). Q.E.D.

For any function f we have a pull-back functor f^* defined by the equality $f^*(\beta) = \beta \circ f$ for any function β. The domain of the function $f^*(\beta)$ is equal to $f^{-1} D_\beta$.

Let us take an endexed set $(M_i;\ i \in I)$ of c.p. We define the functions

$$\mathrm{pr}_i : \times_{j \in I} \underline{M_j} \to \underline{M_i}$$

by the equalities $\mathrm{pr}_i(x) = x_i$ for

$$x = (x_j;\ j \in I) \in \times_{j \in I} \underline{M_j},$$

where $i \in I$. Then we have the Cartesian product $\times_{j \in I} M_j$ meant to be the smallest c.p. containing the set $\bigcup_{i \in I} \mathrm{pr}_i^* M_i$. In the particular case, for two c.p. M_1 and M_2 the Cartesian product $\times_{i \in \{1,2\}} M_i$

is denoted by $M_1 \times M_2$. 0.1. Proposition yields

1.3. PROPOSITION. *For analytical mappings* (0.1) *and* $g : P \to Q$ *we have the analytical mapping* $f \times g : M \times P \to N \times Q$. *If* $A \subset \underline{M}$ *and* $B \subset \underline{N}$, *then* $M_A \times N_B = (M \times N)_{A \times B}$.

Every point set $\{j\}$ may be treated as the set of all points of a c.p. (j) such that $\underline{(j)} = \{j\}$. This c.p. consists of all complex functions on the set $\{j\}$, where j is its only point. We have then a well defined c.p. $M_j \times (j)$ for $j \in I$, diffeomorphic in the natural way with M_j. Taking the set of all unions $\bigcup_{j\in I} \alpha_j$ of functions α_j belonging to $M_j \times (j)$ for $j \in I$, we get a c.p. called the disjoined union of c.p. of the indexed set $(M_i;\ i \in I)$ and denoted by $\bigoplus_{i\in I} M_i$. Then $\operatorname{top} \bigoplus_{i\in I} M_i$ coincides with the disjoined union $\bigoplus_{i\in I} \operatorname{top} M_i$ of topologies $\operatorname{top} M_i$ for $i \in I$.

2. THE TANGENT SPACE TO C.P.

Let M be a c.p. and $p \in \underline{M}$. We set

$$M(p) = \{\alpha;\ \alpha \in M \text{ and } p \in D_\alpha\}$$

and

$$M(p;\ 0) = \{\alpha;\ \alpha \in M(p) \text{ and } \alpha(p) = 0\}.$$

Let us consider the equivalence $\equiv$ on $\bigcup\{M(p) \times \{p\};\ p \in \underline{M}\}$ defined as follows: $(\alpha,p) \equiv (\beta,q)$ iff $p = q \in \underline{M}$, $\alpha,\beta \in M(p)$ and there exists some $U \in \operatorname{top} M$, $p \in U \subset D_\alpha \cap D_\beta$ satisfying the condition $\alpha|U = \beta|U$. The quotient set $(\bigcup\{M(p) \times \{p\};\ p \in \underline{M}\,) / \equiv$ will be denoted by $[M]$ and its elements will be called M-germs. For any $\xi \in [M]$ we have the only point p such that $(\alpha,p) \in \xi$. This point will be denoted by $a\xi$ and the number $\alpha(p)$ by $b\xi$; $a\xi$ and $b\xi$ are called the source and the target of ξ, respectively. For any $p \in \underline{M}$ and any $\alpha \in M(p)$ let $[\alpha,p]_M$ denote the M-germ such that $(\alpha,p) \in [\alpha,p]_M$. We define natural operations by the equalities:

$$[\alpha,p]_M + [\beta,p]_M = [\alpha + \beta,\ p]_M, \quad [\alpha,p]_M \cdot [\beta,p]_M = [\alpha\beta,\ p]_M,$$

$$c[\alpha,p]_M = [c\alpha,\ p]_M$$

for $\alpha,\beta \in M(p)$, $c \in \mathbb{C}$. Define also the sets

$$[M,p] = \{[\alpha,p]_M;\ \alpha \in M(p)\}$$

and

$$[M,p;\ 0] = \{\xi;\ \xi \in [M,p] \text{ and } b\xi = 0\}.$$

The set $[M,p]$ may be treated as a $\mathbb{C}$-algebra. Denote it by $\underline{[M,p]}$. The set $[M,p;\ 0]$ is a maximal ideal in this algebra. When denoting the square of this ideal by $[M,p;\ 0]^2$ we get a well defined vector space $[M,p;\ 0]/[M,p;\ 0]^2$ denoted by $T^*_p M$. The $\mathbb{C}$-dual to $T^*_p M$ vector space will be denoted by $T_p M$. On the other hand, we define the $\mathbb{C}$-vector space M_p. The vectors of M_p are meant to be such $v : M(p) \to \mathbb{C}$ that $v(\alpha+\beta) = v(\alpha) + v(\beta)$, $v(c \cdot \alpha) = cv(\alpha)$ and $v(\alpha \cdot \beta) = \alpha(p)v(\beta) + \beta(p)v(\alpha)$ for $\alpha, \beta \in M(p)$ and $c \in \mathbb{C}$. By setting $(v+w)(\alpha) = v(\alpha) + w(\alpha)$ and $(cv)(\alpha) = cv(\alpha)$ for $\alpha \in M(p)$, where c is any complex number, we define the vector space M_p. By a standard proof we get the following

2.1. PROPOSITION. *If* N *is a complex subpremanifold of* M, *i.e.* $\underline{N} \subset \underline{M}$ *and* $N = M_{\underline{M}}$, *then we have the natural mappings with commutative diagram*

$$\begin{array}{ccccc} M(p;\,0) & \to & [M,\ p;\ 0] & \to & T^*_p M \\ \downarrow & & \downarrow & & \downarrow \\ N(p;\,0) & \to & [N,\ p;\ 0] & \to & T^*_p N , \end{array} \tag{2.1}$$

where the horizontal arrows as well as the second and third vertical ones are epimorphisms. These mappings are induced by the correspondence $M(p;\,0) \ni \alpha \mapsto \alpha|D_\alpha \in \underline{N}$ *and by the equivalence relations. The linear mapping*

$$T_p N \to T_p M \tag{2.2}$$

is a monomorphism and we have the natural isomorphism

$$v \mapsto \overset{o}{v} : T_p M \to M_p, \tag{2.3}$$

where $\overset{o}{v}(\alpha) = v([\alpha - \alpha(p),\ p]_M + [M,\ p;\ 0]^2)$ *for any* v *in* $T_p M$. *Here* $\xi + [M,\ p;\ 0]^2$ *stands for the set of all germs* $\xi + \eta$, η *being in* $[M,\ p;\ 0]^2$. *By setting for* (1) $f_{*p}(v)(\beta) = v(\beta \circ f)$, *when* $\beta \in N(f(p))$ *and* v *is in* M_p, *we get the linear mapping* $f_{*p} : M_p \to N_{f(p)}$ *called the tangent mapping to the* (1) *at* p. *If* N *is a complex subpremanifold of* M, $p \in \underline{N}$, *then the square*

$$\begin{array}{ccc} T_p N & \to & T_p M \\ \downarrow & & \downarrow \\ N_p & \to & M_p \end{array}$$

is commutative, where we have the tangent mapping $id_{*p} : N_p \to M_p$ *to the inclusion* $id : N \to M$.

Let $\mathbb{C}_n$ be the set of all complex analytic functions on all the open sets of the complex space $\mathbb{C}^n$. $\mathbb{C}_n$ will be called the natural c.p. of $\mathbb{C}^n$. We have the following consequence of 2.1.:

2.2. PROPOSITION. *If* M *is a complex subpremanifold of* $\mathbb{C}_n$, *then we have the natural monomorphism*

$$T_p M \to T_p \mathbb{C}_n \tag{2.4}$$

and the natural isomorphisms

$$T_p M \to M_p, \tag{2.5}$$

$$T_p\mathbb{C}_n \to \mathbb{C}_{np} \quad \textit{and} \quad n_p : \mathbb{C}_{np} \to \mathbb{C}^n, \tag{2.6}$$

where $n_p(\partial_{hp}) = \delta_h$, $\delta_h = (\delta_h^1,\ldots,\delta_h^n)$, δ_h^j *being the so-called Kronecker's delta and* $\partial_{hp}\beta = (\partial_h\beta)(p)$ *for* $\beta \in C_n(p)$, *(*$\partial_h\beta$ *denotes the partial derivative of* β *with respect to* h-*th variable),* $h,j = 1,\ldots,n$. *The following diagram with horizontal arrows being isomorphisms and the vertical ones being monomorphisms*

$$\begin{array}{ccccc} T_p\mathbb{C}_n & \to & \mathbb{C}_{np} & \to & \mathbb{C}^n \\ \uparrow & & \uparrow & & \uparrow \\ T_pM & \to & M_p & \to & M'_p \end{array}$$

is commutative. Here M'_p *is the image of* T_pM *given by the composition of mappings* (2.4) *and* (2.6), *and* $M'_p \to \mathbb{C}^n$ *is an identity mapping.*

3. ANALYTIC COMPLEX MANIFOLD AS C.P.

Let M be an m-dimensional analytic complex manifold (with Hausdorff topology and a countable basis of open sets). Then this manifold is uniquely determined by the set of all complex analytic functions defined on the open sets of M. This set of functions is a c.p. It will be denoted by the same letter M. The analytic complex manifold may then be treated as a c.p. (if it has Hausdorff topology with a countable basis of open sets) locally diffeomorphic to $\mathbb{C}_m$, i.e. such a c.p. that for any $p \in \underline{M}$ there exist $U \in \operatorname{top} M$, $V \in \operatorname{top}\mathbb{C}_m$, and a diffeomorphism

$$\mu : M_U \to \mathbb{C}_{mV}, \tag{3.1}$$

where $p \in U$.

<u>3.1. LEMMA</u>. *If* M *is an* m-*dimensional analytic complex submanifold of* $\mathbb{C}_n$, *then the set* $p + M'_p$ *of all points* $p + u$, *where* u *is in* M'_p *(see 2.2. Proposition), is the tangent plane (in the classical sense) to* M *at the point* p.

<u>Proof</u>. Let us take a diffeomorphism, $p \in U \in \operatorname{top} M$, $V \in \operatorname{top}\mathbb{C}_m$, and $\mu(p) = 0$. We have a vector base $\mu_{1p},\ldots,\mu_{mp}$ of M_p defined by the equalities:

$$\mu_{jp}(\alpha) = \partial_j(\alpha\circ\mu^{-1})(\mu(p)) \quad \text{for} \quad \alpha \in M(p), \quad j = 1,\ldots,m.$$

Thus the vectors $n_p(\mathrm{id}_{*p}(\mu_{1p})),\ldots,n_p(\mathrm{id}_{*p}(\mu_{mp}))$, where $\mathrm{id} : M \to \mathbb{C}_n$ is the identity mapping, constitute a base of the vector space M_p. For any $\beta \in \mathbb{C}_n(p)$ we have

$$\mathrm{id}_{*p}(\mu_{jp})(\beta) = \mu_{jp}(\beta\circ\mathrm{id}) = \partial_j(\beta\circ\mathrm{id}\circ\mu^{-1})(\mu(p)) = \partial_j(\beta\circ\mu^{-1})(0)$$

$$= (\partial_h\beta)(\mu^{-1}(0))\partial_j(\mu^{-1})^h(0),$$

where $\mu^{-1}(z) = ((\mu^{-1})^1(z), \ldots, (\mu^{-1})^n(z))$, $z = (z^1, \ldots, z^m)$. This yields

$$id_{*p}(\mu_{jp})(\beta) = \partial_j(\mu^{-1})^h(0)\, \partial_{hp}\, \beta.$$

Hence it follows that $id_{*p}(\mu_{jp}) = \partial_j(\mu^{-1})^h(0)\, \partial_{hp}$. Applying n_p we get

$$n_p(id_{*p}(\mu_{jp})) = \partial_j(\mu^{-1})^h(0) n_p(\partial_{hp}) = \partial_j(\mu^{-1})^h(0)\delta_h = \partial_j\, \mu^{-1}(0).$$

Then we have the base $\partial_1\mu^{-1}(0), \ldots, \partial_m\mu^{-1}(0)$ of the vector space M_p. Therefore $p + M'_p$ is the tangent space to M at p in the classical sense. Q.E.D.

From the above lemma there immediately follows

3.2. PROPOSITION. *If* M *is an* m*-dimensional analytic complex submanifold of* $\mathbb{C}_n$, *then for any* $p \in \underline{M}$ *the orthogonal projection of the set* $\underline{M}$ *into* $p + M'_p$ *is locally open at the point* p, *i.e. there exists a set* $U \in \text{top}\, M$ *such that* $p \in U$ *and for any* $V \in \text{top}\, M$, $p \in V \subset U$, *the orthogonal projection* $V_p^{\perp}$ *of* V *into* $p + M'_p$ *is open in* $p + M'_p$.

4. THE VECTOR SPACE SPANNED BY DIRECTIONS OF A SET IN $\mathbb{C}^n$.

Let S be any subset of $\mathbb{C}^n$ and $p \in S$. We say that the sequences

$$p_1, p_2, \ldots \quad \text{and} \quad q_1, q_2, \ldots \tag{4.1}$$

of points belonging to S define a direction w of S at p iff the sequences (4.1) tend to p, $p_k \neq q_k$ for all k, and

$$(p_k - q_k)/|p_k - q_k| \to w \quad \text{as} \quad k \to \infty.$$

Here $|z| = (z^1\bar{z}^1 + \ldots + z^n\bar{z}^n)^{\frac{1}{2}}$ for $z = (z^1, \ldots, z^n) \in \mathbb{C}^n$. The smallest vector subspace of $\mathbb{C}^n$ including all the directions of S at p will be called the direction tangent space of S at p and denoted by $\text{dir}_p S$.

4.1. PROPOSITION. *For any* $p \in S$ *the space* $\text{dir}_p S$ *is a subspace of* $(\mathbb{C}_{nS})'_p$.

Proof. Let us set $M = \mathbb{C}_{nS}$. According to 2.1. and 2.2. we are to prove that $\text{dir}_p S$ is included in the image of M_p given by the linear mapping $n_p \circ id_{*p}$, where $id : M \to \mathbb{C}^n$ is the inclusion. Take any direction S at p. For any α $M(p)$ there exists $\beta \in \mathbb{C}_n(p)$ such that $\alpha|U \cap S = \beta|U \cap S$, where $p \in U \in \text{top}\, \mathbb{C}_n$, $U \cap S \subset D_\alpha$, $U \subset D_\beta$. We have sequences (4.1) defining the direction w of S at p. Thus,

$$(\alpha(q_k) - \alpha(p_k))/|q_k - p_k| \underset{k\to\infty}{\to} (\partial_w \beta)(p),$$

where $(\partial_w \beta)(p)$ stands for the directional derivative of the function β at the point p in the direction w. The complex number $(\partial_w \beta)(p)$ depends only on α, p and w, and we have $(\partial_w \beta)(p) = w^h \partial_{hp} \beta$, where $w = (w^1, \dots, w^n) \in \mathbb{C}^n$. Then, by setting

$$w_p(\alpha) = w^h \partial_{hp} \beta \tag{4.2}$$

we get the vector w_p of the space M_p such that $w_p(\gamma | D_\gamma \cap S) = w^h \partial_{hp} \gamma$ for $\gamma \in C_n$. This yields

$$(\mathrm{id}_{*p} w_p)(\gamma) = w_p(\gamma \circ \mathrm{id}) = w_p(\gamma | D_\gamma \cap S) = w^h \partial_{hp} \gamma.$$

Hence it follows that $\mathrm{id}_{*p}(w_p) = w^h \partial_{hp}$. Therefore

$$n_p(\mathrm{id}_{*p}(w_p)) = w^h n_p(\partial_{hp}) = w^h \delta_h = w.$$

Thus, $\mathrm{dir}_p S$ is included in $n_p \, \mathrm{id}_{*p} M_p = M'_p$. Q.E.D.

For any $S \subset \mathbb{C}^n$ and $p \in S$ the set $p + (\mathbb{C}_{nS})'_p$ is denoted by S_p.

4.2. PROPOSITION. *If* $p \in S \subset \mathbb{C}^n$ *and* $p \in V \in \mathrm{top}\, \mathbb{C}_n$ *then there exist* $U \in \mathrm{top}\, \mathbb{C}_n$, $U \subset V$, *and a continuous mapping* $g : D_g \to S$ *such that* $p \in D_g \subset S_p \cap U$, $g D_g = S \cap U$ *and for any* $q \in D_g$ *the orthogonal projection* $\mathbb{C}^n$ *onto* S_p *of* $g(q)$ *is equal to* q.

P r o o f. Let p_S denote the orthogonal projection of $\mathbb{C}^n$ onto S_p. By 4.1, for any neighbourhood V of p there exists some $U \in \mathrm{top}\, \mathbb{C}_n$ such that $p \in U \subset V$, $p_S U \subset U$, and for any $q \in S_p \cap U$ the set $U \cap S \cap p_S^{-1}\{q\}$ has at most one point. Denote by A the set of all points $q \in S_p \cap U$ for which the set $U \cap S \cap p_S^{-1}\{q\}$ is non-empty, and then for any $q \in A$ let $g(q)$ be the only element of this set. Thus we have defined the mapping g with the domain A such that $p_S(g(q)) = q$ for $q \in A$, and $g A \subset S \cap U$. Taking any $u \in S \cap U$ and setting $q = p_S(u)$ we get $q \in S_p \cap U$ and $u \in U \cap S \cap p_S^{-1}\{q\}$. Then $q \in A$ and $u = g(q)$. Therefore $g A = S \cap U$.

Now. let us presume that there exists a neighbourhood V of p open in $\mathbb{C}^n$ such that for any neighbourhood U of p included in V and any g such that

$$g : D_g \to S \quad (p \in D_g \subset S_p \cap U, \quad g D_g = S \cap U \quad \text{and} \quad p_S \circ g = \mathrm{id})$$

has a discontinuity point in D_g. Then, taking the ball $U_l \subset V$ with the center p and the radius $r_l > 0$ such that $r_l \to 0$ as $l \to \infty$, we find mappings:

$$g_l : D_{g_l} \to S \quad (p \in D_{g_l} \subset S_p \quad U_l \quad \text{and} \quad g_l D_{g_l} = S \cap U_l$$

as well as points x_l, y_l in D_{g_l} such that

$$|g_l(x_l) - g_l(y_l)|/|x_l - y_l| > l, \quad l = 1,2,\ldots . \tag{4.3}$$

Notice that $g_l(x_l) - g_l(y_l) = x_l - y_l + z_l$, where the vector $z_l = g_l(x_l) - x_l + y_l - g_l(y_l)$ is orthogonal to S_p. By setting $\varepsilon_l = g_l(x_l) - g_l(y_l)$ and $w_l = (g_l(x_l) - g_l(y_l))/\varepsilon_l$ we get

$$w_l - (x_l - y_l)/\varepsilon_l = z_l/\varepsilon_l, \quad l = 1,2,\ldots . \tag{4.4}$$

We may assume that $w_l \to w$ as $l \to \infty$. By (4.3) we have $|x_l - y_l|/\varepsilon_l < 1/l$. Hence it follows that $z_l/\varepsilon_l \to w$ as $l \to \infty$. Thus w is orthogonal to S_p. We have $g_l(x_l)$, $g_l(y_l) \in U_l$. This yields $w \in \mathrm{dir}_p S$. By 4.1 $w \in (\mathbb{C}_{nS})'_p$. Hence it would follow that $p + w$ S_p, $|w| = 1$ and w is orthogonal to S_p, which is impossible. Q.E.D.

5. A NECESSARY AND SUFFICIENT CONDITION FOR A SUBSET OF $\mathbb{C}^n$ TO BE A SUBMANIFOLD.

We will prove the following

5.1. THEOREM. *Let* m *be a natural number,* $m \leqq n$, *and* $S \subset \mathbb{C}^n$. *The c.p.* $\mathbb{C}_{nS}$ *is an* m*-dimensional submanifold of* $\mathbb{C}_n$ *if and only if at any* $p \in S$ *we have* $\dim S_p = m$ *and the orthogonal projection of the set* S *into the hyperplane* S_p *is open at* p, *i.e. there exists* $V \in \mathrm{top}\,\mathbb{C}_{nS}$ *such that* $p \in V$ *and for any* $U \subset V$, $p \in U \in \mathrm{top}\,\mathbb{C}_{nS}$, *the orthogonal projection of* U *is open in* S_p.

Proof. Necessity of the condition is an immediate consequence of 3.2. To prove its sufficiency we take any $p \in S$ and an orthonormal base $e_1,\ldots,e_n$ such that $e_1,\ldots,e_m$ is a base of M'_p, where $M = \mathbb{C}_{nS}$. Let us set

$$e^j(z) = (z - p)e_j \quad \text{for } z \in \mathbb{C}^n, \quad j = 1,\ldots,n. \tag{5.1}$$

Here $(z-p)e_j = (z^h - p^h)\bar{e}_{hj}$, where $z = (z^1,\ldots,z^n)$ and $e_j = (e_{1j},\ldots, e_{nj})$, $j = 1,\ldots,n$. We have then the analytic functions $e^1,\ldots,e^n$. When setting

$$e^j_p = [e^j|S,\ p]_M + [M,\ p;\ 0]^2, \quad j = 1,\ldots,n, \tag{5.2}$$

we get the vectors of T^*_pM. We will prove that $e^1_p,\ldots,e^m_p$ are linearly independent. Let

$$\phi : T^*_p\mathbb{C}_n \to T^*_pM \quad \text{and} \quad \psi : T_p\mathbb{C}_n \to \mathbb{C}_{np} \tag{5.3}$$

be the natural mappings as in 2.1. and 2.2., respectively. Then we have

$$\phi([\alpha,p]_{\mathbb{C}_n} + [\mathbb{C}_n,\ p;\ 0]^2) = [\alpha|S,\ p]_M + [M,\ p;\ 0]^2$$

and

$$\psi(t)(\alpha) = t([\alpha,p]_{\mathbb{C}_n} + [\mathbb{C}_n,\ p;\ 0]^2)$$

for $\alpha \in \mathbb{C}_n(p;0)$. From (5.3) we get, according to 2.2., the natural mapping

$$\phi^* : T_p M \to T_p \mathbb{C}_n, \quad \phi^*(w) = w \circ \phi \quad \text{for} \quad w \quad \text{in} \quad T_p^* \mathbb{C}_n. \tag{5.4}$$

Thus, M_p' is the image of $T_p M$ given by $n_p \circ \psi \circ \phi^*$. Then there exist e_h' in $T_p M$ such that $n_p(b_h) = e_h$, where $b_h = \psi(\phi^*(e_h'))$, $h=1,\dots,m$. This yields

$$\begin{aligned} b_h(e^j) &= \psi(\phi^*(e_h'))(e^j) = \phi^*(e_h')([e^j, p]_{\mathbb{C}_n} + [\mathbb{C}_n, p; 0]^2) \\ &= e_h'(\phi([e^j, p]_{\mathbb{C}_n} + [\mathbb{C}_n, p; 0]^2)) = e_h'([e^j | S, p]_M + [M, p; 0]^2) \\ &= e_h'(e_p^j) \end{aligned}$$

and

$$e_h'(e_p^j) = b_h(e^j). \tag{5.5}$$

On the other hand, b_h are vectors in $\mathbb{C}_{np}$. Thus, we have $b_h = b_h^k \partial_{kp}$, where $b_h^k \in \mathbb{C}$, $h = 1,\dots,m$, $k = 1,\dots,n$. Hence it follows that

$$b_h(e^j) = b_h^k \partial_{kp} e^j = b_h^k (\partial_k e^j)(p) = b_h^k \bar{e}_{kj}.$$

We also have

$$e_h = b_h^k n_p(\partial_{kp}) = b_h^k \delta_k = (b_h^1, \dots, b_h^n).$$

By (5.5) and (5.1) we get $e_h'(e_p^j) = b_h(e^j) = e_h e_j = \delta_{hj}$. Therefore $e_p^1,\dots,e_p^m$ are linearly independent. From the assumption that M_p' is m-dimensional, by 2.2., we get $\dim T_p^* M = m$. Thus, $e_p^1,\dots,e_p^m$ defined by (5.2) constitute a vector base for $T_p^* M$. The elements $e_p^{m+1},\dots,e_p^n$ as vectors of $T_p^* M$ may linearly be expressed by $e_p^1,\dots,e_p^m$. Then there exist complex numbers a_h^j such that $e_p^j = a_h^j e_p^h$. Hence it follows that there are analytic functions

$$\alpha_r^j, \ \beta_r^j \in \mathbb{C}_n(p;0), \quad j = m+1,\dots,n, \quad r = 1,\dots,s$$

and a neighbourhood V of the point p, open in $\mathbb{C}^n$ and contained in domains of all the functions α_r^j and β_r^j, satisfying

$$e^j | S \cap V = (a_h^j e^h + \sum_{r=1}^{s} \alpha_r^j \beta_r^j) | S \cap V, \quad j = m+1,\dots,n. \tag{5.6}$$

According to 4.2. there exist a neighbourhood U of p open in $\mathbb{C}^n$ and included in V as well as a mapping $g : A \to S$ such that $p \in A \subset S_p \cap \cap U$, $gA = S \cap U$ and $p_S(g(q)) = q$ for $q \in A$, where p_S is the orthogonal projection of $\mathbb{C}^n$ onto S_p. Notice that

$$w = p_S(w) + e^j(w) e_j \quad \text{and} \quad p_S(w) = p + e^h(w) e_h$$

for $w \in \mathbb{C}^n$. Thus,

$$q - p = e^h(g(q))e_h \quad \text{and} \quad q - p = e^h(q)e_h$$

for $q \in A$. This yields $e^h(g(q)) = e^h(q)$, $h = 1,\dots,m$. Then, from (5.6) we get

$$e^j(g(q)) = a^j_h e^h(q) + \sum_{r=1}^{s} \alpha^j_r(g(q))\beta^j_r(g(q)) \quad \text{for} \quad q \in A. \tag{5.7}$$

Denote by e the isometry of $\mathbb{C}^n$ such that $e(0) = p$ and $e(\delta_k) = e_k$, $k = 1,\dots,n$, and set $f = g \circ e \circ i_m$, where $i_m : \mathbb{C}^m \to \mathbb{C}^n$ is the standard inclusion. Then $D_f = i_m^{-1} e^{-1} A$ and, by (5.7),

$$e^j(f(z)) = a^j_h z^h + \sum_r \alpha^j_r(f(z))\, \beta^j_r(f(z)) \quad \text{for} \quad z \in D_f,$$

where $z = (z^1,\dots,z^n)$. Hence it follows that for $z \in D_f$

$$f(z) = z^h e_h + (a^j_h z^h + \sum_r \alpha^j_r(f(z))\, \beta^j_r(f(z)))e_j. \tag{5.8}$$

When setting

$$F(z,w) = w - z^h e_h - (a^j_h z^h + \sum_r \alpha^j_r(w)\beta^j_r(w)) \quad \text{for} \quad (z,w) \in \mathbb{C}^m \times V, \tag{5.9}$$

by (5.8), we get

$$F(z,\ f(z)) = 0 \quad \text{for} \quad z \in D_f. \tag{5.10}$$

The assumption that the mapping p_S is open at the point p yields the existence of $U_o \in \operatorname{top} \mathbb{C}_m$ such that $0 \in U_o \subset D_f$. An implicit function theorem for analytic mappings gives analyticity of f and, according to (5.9), regularity of f in some neighbourhood of 0 in the space $\mathbb{C}^m$. Therefore we may assume that U_o is a polydisc with center 0, $U_o = D_f$, and f is an analytic homeomorphism U_o onto a neighbourhood of p open in M. When taking the mapping f^{-1} as a chart around p, we conclude that M is an m-dimensional complex analytic manifold. Q.E.D.

As an application of 5.1. we will prove the following

5.2. THEOREM. *If the Cartesian product* $M \times N$ *of c.p.* M *and* N *is a complex analytic manifold, then* M *and* N *are complex analytic manifolds as well.*

Proof. Let $x \in M$ and $y \in N$. An assumption that $M \times N$ is a complex analytic manifold yields the existence of $U \in \operatorname{top} M$, $V \in \operatorname{top} N$ and a diffeomorphism

$$f : M_U \times N_V \to \mathbb{C}_{nW} \tag{5.11}$$

such that $x \in U$, $y \in V$ and $W = f(U \times V) \in \operatorname{top} \mathbb{C}_n$. By setting $a(u) = (u,\ y)$ for $u \in U$ and $b(v) = (x,\ v)$ for $v \in V$, we have diffeomorphisms

$$a : M_U \to (M\times N)_{U\times\{y\}} \quad \text{and} \quad b : N_V \to (M\times N)_{\{x\}\times V}. \tag{5.12}$$

Let

$$P = \mathbb{C}_{nf(U\times\{y\})}, \quad Q = \mathbb{C}_{nf(\{x\}\times V)}$$

and

$$h(p,q) = f(a^{-1}(f^{-1}(p)),\ b^{-1}(f^{-1}(q))) \quad \text{for} \quad (p,q)\in \underline{P}\times\underline{Q}.$$

By (5.11) and (5.12) h is one-to-one and we get the diffeomorphism

$$h : P\times Q \to \mathbb{C}_{nW}. \tag{5.13}$$

This yields for any p in P and q in Q the isomorphism

$$h_{*(p,q)} : (P\times Q)_{(p,q)} \to (\mathbb{C}_{nW})_{g(p,q)}. \tag{5.14}$$

On the other hand, $(P\times Q)_{(p,q)}$ is isomorphic to $P_p + Q_q$, and W is open in C_n. Hence it follows that $\dim P_p + \dim Q_q = n$ for $p\in \underline{P}$ and $q\in \underline{Q}$. Then we get the constant function $p \mapsto \dim P_p$. Let $m = \dim P_p$ and

$$p_{\underline{P}} : \mathbb{C}^n \to \underline{P}_p \tag{5.15}$$

be the orthogonal projection. By 4.2. there exist $U_1 \in \operatorname{top}\mathbb{C}_n$, $U_1 \subset U$, and a continuous mapping $g_1 : A_1 \to \underline{P}$ such that $p\in A_1 \subset \underline{P}_p \cap U_1$, $g_1 A_1 = \underline{P}\cap U_1$ and $p_{\underline{P}}(g_1(s)) = s$ for $s\in \bar{A}_1$. By the same argument there exist $U_2 \subset \operatorname{top}\mathbb{C}_n$, $U_2 \subset U$, and a continuous mapping $g_2 : A_2 \to \underline{Q}$ such that

$$q\in A_2 \subset \underline{Q}_q \cap U_2, \quad g_2 A_2 = \underline{Q}\cap U_2 \quad \text{and} \quad q_{\underline{Q}}(g_2(t)) = t \quad \text{for} \quad t\in A_2,$$

where

$$q_{\underline{Q}} : \mathbb{C}^n \to \underline{Q}_q \tag{5.16}$$

is the orthogonal projection. Let us set

$$l(s,t) = h(g_1(s), g_2(t)) \quad \text{for} \quad (s,t)\in A_1\times A_2. \tag{5.17}$$

Then we have

$$l : A_1\times A_2 \to W_o, \quad \text{where} \quad W_o = h((\underline{P}\cap U_1)\times(\underline{Q}\cap U_2)). \tag{5.18}$$

When applying (5.15) and (5.16) we can state that l is one-to-one and

$$l^{-1}(z) = (p_{\underline{P}}(pr_1 h^{-1}(z)),\ q_{\underline{Q}}(pr_2 h^{-1}(z))) \quad \text{for} \quad z\in W_o. \tag{5.19}$$

From the continuity of (5.15), (5.16), g_1, g_2, h and h^{-1}, by (5.17) and (5.19), we say that (5.18) is a homeomorphism. The mapping (5.13) as a homeomorphism transforms any open set in $P\times Q$ onto an open set in C_{nW}. Therefore W_o is open in $\mathbb{C}_n$. We have

$$A_1\times A_2 \subset \underline{P}_p\times\underline{Q}_q \quad \text{and} \quad l^{-1} : W_o \to A_1\times A_2.$$

From the equality $\dim(\underline{P}_p\times\underline{Q}_q) = n$, by Brouwer's theorem on open sets, we state that $A_1\times A_2$ is open in $\underline{P}_p\times\underline{Q}_q$. Thus, A_1 is open in $\underline{P}_p$.

From the equality

$$p_{\underline{P}} \circ g_1 = \mathrm{id}_{A_1}$$

it follows that for any neighbourhood G of p open in C^n and included in U_1 we have

$$p_{\underline{P}}(\underline{P} \cap G) = g_1^{-1} G.$$

By the continuity of g_1 we conclude that $p_{\underline{P}}(\underline{P} \cap G)$ is open in A_1, and then in $\underline{P}_p$. In other words, the orthogonal projection $p_{\underline{P}} : \underline{P} \to \underline{P}_p$ is open at p. Therefore, by 5.1., P is an m-dimensional complex analytic submanifold of $\mathbb{C}_n$. From (5.11), by 0.1., we get the diffeomorphism

$$f|U \times \{y\} : (M \times N)_{U \times \{y\}} \to P. \tag{5.20}$$

The expressions (5.12) and (5.20) give the diffeomorphism

$$f|U \times \{y\} \circ a : M_U \to P.$$

Therefore M is a complex analytic manifold. Q.E.D.

6. FOLIATIONS ON C.P.

For any set S let $\operatorname{discr} S$ denote the discreet c.p. with the set S of all points, i.e. $\operatorname{discr} S$ is the set of all complex functions α such that $D_\alpha \subset S$. Let M and F be c.p. A diffeomorphism

$$\phi : M_U \to F_V \times M_W, \tag{6.1}$$

where $U \in \operatorname{top} M$, $V \in \operatorname{top} F$, $p \in U \cap V \cap W$ and $W \subset \underline{M}$ are such that there exists an induced diffeomorphism $\phi|U : F_U \to F_V \times \operatorname{discr} W$, will be called a local diffeomorphism on M adapted by F at the point p. The set of all local diffeomorphisms on M adapted by F at p will be denoted by $\operatorname{adap}(M, F, p)$. A c.p. F will be called a prefoliation on M iff $\underline{M} = \underline{F}$ and for any $p \in \underline{M}$ we have $\operatorname{adap}(M, F, p) \neq \emptyset$. Any connected component of F, i.e. a c.p. F_C, where C is a connected component of $\operatorname{top} F$, is said to be a leaf of F. The c.p. F is said to be locally homogeneous iff for any p, $q \in \underline{F}$ there exists a diffeomorphism $f : F_V \to F_W$, where $p \in V$, $q \in W$ and $V, W \in \operatorname{top} F$.

For any diffeomorphism (6.1) the set W is well defined by ϕ. The c.p. M_W in (6.1) will be called the transversal piece of M given by ϕ. A prefoliation F on M such that the c.p.

$$\bigoplus_{\phi \in I} \phi_2 M, \quad \text{where} \quad I = \bigcup_{p \in M} \operatorname{adap}(M, F, p)$$

and $\phi_2 M$ is the piece of M given by ϕ, is locally homogeneous will be called transversally homogeneous.

<u>6.1. THEOREM</u>. *If a c.p.* M *is a complex manifold and a c.p.* F *is a locally homogeneous or transversally homogeneous prefoliation on* M, *then* F *is a complex manifold being a foliation on* M *in the classical*

sense.

<u>Proof</u>. To prove that F is a complex manifold take any point p of $\underline{F}$. Then there exists $\phi \in \operatorname{adap}(M, F, p)$, ϕ being of the form (6.1). According to 5.2. the c.p. F_V and M_W are complex manifolds. Denote by d_p the dimension of F_V at the point p. Both the homogeneity and transversal homogeneity of F yield $d_p = d_q$ for $p, q \in \underline{F}$. When setting $k = d_p$, $p \in \underline{F}$, we state that F is a k-diemnsional complex manifold. By taking around p charts μ and ν of F_V and M_W, respectively, and setting for $x \in U$

$$\rho(x) = (\mu(\phi_1(x)),\ \nu(\phi_2(x))), \quad \text{where} \quad \phi(x) = (\phi_1(x),\ \phi_2(x)), \tag{6.2}$$

we get a chart $\rho : U \to \mu D_\mu \times \nu D_\nu$, where μD_μ and νD_ν are open subsets of $\mathbb{C}^k$ and $\mathbb{C}^{m-k}$, respectively. Here $m = \dim M$. It is easy to conclude that the induced mapping

$$\rho : \operatorname{top} F | U \to (\operatorname{top} \mathbb{C}_k) | \mu D_\mu \times \operatorname{top\,discr} \nu D_\nu$$

is a homeomorphism. This (cf. [1]) completes the proof that F is a foliation (in the classical sense) on a complex manifold M.

The above theorem allows us to regard a prefoliation F which is locally homogeneous or transversally homogeneous as a generalization of the concept of foliation from the category of complex manifolds to the category of complex premanifolds.

<u>References</u>

[1] ANDRZEJCZAK, G.: 'On foliations of semi-simplicial manifolds and their holonomy', preprint MPI Bonn/SFB 84-55.

[2] POSTNIKOV, M.M.: *Introduction to Morse Theory* (Russian), Moscow 1971.

[3] SIKORSKI, R.: 'Abstract covariant derivative', *Colloq. Math.* 18 (1967), 251-272.

[4] ——: 'Differential modules', *ibid.* 24 (1871), 45-79.

[5] WALISZEWSKI, W.: 'Regular and coregular mappings of differential spaces', *Ann. Polon. Math.* 30 (1975), 263-281.

[6] ——: 'Analytical premanifolds', *Zeszyty Naukowe Politechniki Śląskiej*, Seria: Matematyka-Fizyka 48 (1986), 217-226.

MÖBIUS TRANSFORMATIONS AND CLIFFORD ALGEBRAS OF EUCLIDEAN AND ANTI-EUCLIDEAN SPACES

Pertti Lounesto
Institute of Mathematics
Helsinki University of Technology
SF-02150 Espoo, Finland

and

Arthur Springer
San Diego State University
San Diego, CA 92182, U.S.A.

ABSTRACT. L. Ahlfors studied Möbius transformations employing Clifford algebras of anti-euclidean spaces with negative definite quadratic forms. This paper gives a passage from the euclidean space (positive definite) to the anti- euclidean space (negative definite). The computations are realized without any extra dimensions or projective representations of the Möbius transformations, and so no book-keeping of superfluous parameters is needed. Conformal transformations in higher dimensions are applied to Cauchy-Riemann equations and Dirac spinors.

INTRODUCTION. Classical complex analysis can be generalized from the complex plane to higher dimensions in three different ways: function theory of several complex variables (commutative), higher-dimensional one-variable hypercomplex analysis (anti-commutative), and (quasi-)conformal or Möbius transformations (geometric).

In this paper the Möbius group in $\mathbf{R}^n$ is scrutinized following a method used by Ahlfors [1] and previously by Vahlen [15] and Maass [11]. The Möbius transformations are represented by 2×2-matrices with entries in the Lipschitz group. The Lipschitz group belongs to a geometric algebra, called the Clifford algebra, connecting algebraic anti-commutativity and geometric orthogonality [8], [12], [13]. This approach to Möbius transformations has very important advantages. It is not only a natural generalization of the use of 2×2-matrices in the classical case, thus replacing non-linear conformal transformations by linear transformations, but also the computations of conformal transformations are realized without any extra dimensions or projective representations of conformal groups, and so no book-keeping of superfluous components or coefficients is needed. Compare Ref. [1] to Refs. [2], [5], [10].

The essential content of this paper is that both the negative and positive definite vector spaces are considered together with a passage from one case to the other. Both cases have their own

J. Ławrynowicz (ed.), Deformations of Mathematical Structures, 79–90.

advantages: vectors with negative squares are closer to the classical complex case while vectors with positive squares are closer to the present vector algebra formalism of physics. For physical applications of Clifford algebras, see Refs. [7], [8].

The Möbius transformations are employed to hypercomplex analysis. Hypercomplex analysis uses the Clifford algebras to study the Cauchy-Riemann equations, Cauchy's theorem, Cauchy's integral formula and Laurent series expansion [3], [6], [9]. The Cauchy-Riemann equations are invariant under the Möbius transformations, a result due to J. Ryan [14]. Finally, the conformal transformations of the Minkowski space and Dirac spinors are discussed. For more details on spinors, see Ref. [4], and for more details on conformal groups of indefinite forms, see Refs. [2], [5], [10].

1. PROPERTIES OF THE CLIFFORD ALGEBRAS

If one wants to multiply vectors of a euclidean space $\mathbf{R}^n$, it seems natural to require that a vector $\mathbf{x}$ in $\mathbf{R}^n$ multiplied by itself $\mathbf{xx} = \mathbf{x}^2$ gives the length $|\mathbf{x}|$ of $\mathbf{x}$ squared: $\mathbf{x}^2 = |\mathbf{x}|^2$.

EXAMPLE. In the plane $\mathbf{R}^2$ one requires $(x\mathbf{e}_1 + y\mathbf{e}_2)^2 = x^2 + y^2$, where $\mathbf{e}_1$ and $\mathbf{e}_2$ are perpendicular unit vectors. Use the distributive rule to obtain $x^2\mathbf{e}_1{}^2 + y^2\mathbf{e}_2{}^2 + xy(\mathbf{e}_1\mathbf{e}_2 + \mathbf{e}_2\mathbf{e}_1) = x^2 + y^2$. This is satisfied if

$$\mathbf{e}_1{}^2 = \mathbf{e}_2{}^2 = 1, \qquad \mathbf{e}_1\mathbf{e}_2 = -\mathbf{e}_2\mathbf{e}_1.$$

Use associativity to obtain $(\mathbf{e}_1\mathbf{e}_2)^2 = -\mathbf{e}_1{}^2\mathbf{e}_2{}^2 = -1$. Here $\mathbf{e}_{12} = \mathbf{e}_1\mathbf{e}_2$ is a new kind of unit *bivector*, representing an oriented plane area. □

DEFINITION. *For a euclidean space* $\mathbf{R}^n$ *there is an associative algebra* $\mathbf{A}$ *over* $\mathbf{R}$ *containing copies of* $\mathbf{R}$ *and* $\mathbf{R}^n$ *in such a way that:*

(a) *for all* $\mathbf{x}$ *in* $\mathbf{R}^n$, $\mathbf{x}^2 = |\mathbf{x}|^2$

(b) $\mathbf{A}$ *is generated as an algebra by* $\mathbf{R}^n$

(c) $\mathbf{A}$ *is not generated by any proper subspace of* $\mathbf{R}^n$.

$\mathbf{A}$ *is unique and it is called* <u>*the Clifford algebra*</u> $C\ell_n$ *of* $\mathbf{R}^n$. □

Using an orthonormal basis $\mathbf{e}_1, \mathbf{e}_2, \ldots, \mathbf{e}_n$ of $\mathbf{R}^n$ the condition (a) can be replaced by the relations $\mathbf{e}_\nu^2 = 1$ and $\mathbf{e}_\mu \mathbf{e}_\nu = -\mathbf{e}_\nu \mathbf{e}_\mu$, $\mu \neq \nu$, and the condition (c) by the test $\mathbf{e}_1 \mathbf{e}_2 \ldots \mathbf{e}_n \neq \pm 1$, necessary only for $n \bmod 4 = 1$.

EXAMPLE. In $\mathbf{R}^4$ the orthogonal unit vectors $\mathbf{e}_1, \mathbf{e}_2, \mathbf{e}_3, \mathbf{e}_4$ and their product $\mathbf{e}_{1234} = \mathbf{e}_1 \mathbf{e}_2 \mathbf{e}_3 \mathbf{e}_4$ can be represented by the 2×2-matices

$$\begin{pmatrix} 0 & -\mathbf{i} \\ \mathbf{i} & 0 \end{pmatrix}, \begin{pmatrix} 0 & -\mathbf{j} \\ \mathbf{j} & 0 \end{pmatrix}, \begin{pmatrix} 0 & -\mathbf{k} \\ \mathbf{k} & 0 \end{pmatrix}, \begin{pmatrix} 1 & 0 \\ 0 & -1 \end{pmatrix} \quad \text{and} \quad \begin{pmatrix} 0 & 1 \\ 1 & 0 \end{pmatrix}$$

with entries in the division ring of quaternions $\mathbf{H}$. Therefore $C\ell_4$ is isomorphic to $\mathbf{H}(2)$, the real algebra of 2×2-matrices with entries in $\mathbf{H}$.

Since the vectors $\mathbf{e}_1, \mathbf{e}_2, \mathbf{e}_3, \mathbf{e}_4$ and their product $\mathbf{e}_{1234}$ have square 1 and anticommute with each other, these 5 elements might serve as candidates for an orthonormal basis of $\mathbf{R}^5$. However, $\mathbf{e}_1 \mathbf{e}_2 \mathbf{e}_3 \mathbf{e}_4 \mathbf{e}_{1234} = 1$ and so the condition (c) is not fulfilled. Notwithstanding, one can choose the matrices corresponding to the elements

$$(\mathbf{e}_1, -\mathbf{e}_1),\ (\mathbf{e}_2, -\mathbf{e}_2),\ (\mathbf{e}_3, -\mathbf{e}_3),\ (\mathbf{e}_4, -\mathbf{e}_4),\ (\mathbf{e}_{1234}, -\mathbf{e}_{1234})$$

to represent an orthonormal basis of $\mathbf{R}^5$. These matrices belong to the direct product ${}^2\mathbf{H}(2) = \mathbf{H}(2) \times \mathbf{H}(2)$, where addition and multiplication are defined componentwise. In fact, $C\ell_5 \simeq {}^2\mathbf{H}(2)$. □

$C\ell_n$ is a linear space of dimension 2^n. It is a sum of the spaces $C\ell_n^k$ with basis elements $\mathbf{e}_{\underline{\nu}} = \mathbf{e}_{\nu_1} \mathbf{e}_{\nu_2} \ldots \mathbf{e}_{\nu_k}$, $1 \leq \nu_1 < \ldots < \nu_k \leq n$ for fixed $k = 0, 1, \ldots, n$. More precisely, the basis elements are:

k	$\mathbf{e}_{\underline{\nu}}$	
0	1	
1	$\mathbf{e}_\nu$	$1 \leq \nu \leq n$
2	$\mathbf{e}_{\mu\nu} = \mathbf{e}_\mu \mathbf{e}_\nu$	$1 \leq \mu < \nu \leq n$
$\vdots$		
n	$\mathbf{e}_{12\ldots n}$	

$C\ell_n^1$ shall be identified with $\mathbf{R}^n$. The sum of the $C\ell_n^k$ with even k will be denoted by $C\ell_n^{(0)}$, while $C\ell_n^{(1)}$ refers to odd k. $C\ell_n^{(0)}$ is a subalgebra of $C\ell_n$.

1.1. Involutions

The Clifford algebra has three important involutions, similar to complex conjugation. The first, called the *main involution*, is the isomorphism $a \to a^{\iota}$ obtained by replacing each $\mathbf{e}_{\nu}$ by $-\mathbf{e}_{\nu}$, thereby replacing each $\mathbf{a}$ in $C\ell_n^k$ by $\mathbf{a}^{\iota} = (-1)^k\mathbf{a}$. By definition $(ab)^{\iota} = a^{\iota}b^{\iota}$.

The second involution, called the *reversion*, is an anti-isomorphism $a \to a^*$ obtained by reversing the order of factors $\mathbf{e}_{\nu_h}$ in each $\mathbf{e}_{\underline{\nu}}$, thereby replacing each $\mathbf{a}$ in $C\ell_n^k$ by $\mathbf{a}^* = (-1)^{[k/2]}\mathbf{a}$. By definition $(ab)^* = b^*a^*$. The third involution, called the *conjugation*, is a combination of the two others $\bar{a} = a^{*\iota} = a^{\iota *}$.

EXAMPLE. For $a = 1 + \mathbf{e}_1 + \mathbf{e}_{12} + \mathbf{e}_{123}$ in $C\ell_3$ on gets:
$a^{\iota} = 1 - \mathbf{e}_1 + \mathbf{e}_{12} - \mathbf{e}_{123}$, $a^* = 1 + \mathbf{e}_1 - \mathbf{e}_{12} - \mathbf{e}_{123}$, $\bar{a} = 1 - \mathbf{e}_1 - \mathbf{e}_{12} + \mathbf{e}_{123}$. □

1.2. The absolute value

The euclidean square norm on $\mathbf{R}^n$ extends to the whole Clifford algebra $C\ell_n$ by defining

$$|a|^2 = \sum a_{\underline{\nu}}^2 \quad \text{for} \quad a = \sum a_{\underline{\nu}}\mathbf{e}_{\underline{\nu}}$$

where the sum ranges over all ordered multi-indices $\underline{\nu} = \nu_1\nu_2...\nu_k$, $1 \leq \nu_1 < \nu_2 < \ldots < \nu_k \leq n$. This gives the *absolute value* $|a|$ of a, also obtained from $|a|^2 = \mathrm{Re}(a^*a)$.

EXAMPLE. If $a = 6 + 3\mathbf{e}_1 + 2\mathbf{e}_{12}$, then $a^*a = 49 + 36\mathbf{e}_1 + 12\mathbf{e}_2$ and $|a| = 7$.

1.3. The Lipschitz group

The products of non-zero vectors in $\mathbf{R}^n$ form the *Lipschitz group* $\mathbf{\Gamma}_n$ of $\mathbf{R}^n$. If a is in $\mathbf{\Gamma}_n$ then a^*a is real and so $|a|^2 = a^*a$, from which it follows that $|ab| = |a|\,|b|$.

If $\mathbf{x}$ is in $\mathbf{R}^n$ and a is in $\mathbf{\Gamma}_n$ then $a^{\iota -1}\mathbf{x}a$ is again a vector in $\mathbf{R}^n$. Furthermore, the transformation $\mathbf{x} \to a^{\iota -1}\mathbf{x}a$ is a euclidean isometry. In other words, for every a in $\mathbf{\Gamma}_n$ there is a matrix U_a in $\mathbf{O}(n)$ such that $a^{\iota -1}\mathbf{x}a = U_a(\mathbf{x})$. Conversely, every orthogonal matrix can be represented in this way.

The Lipschitz group Γ_n splits in even and odd parts $\Gamma_n = \Gamma_n^{(0)} \cup \Gamma_n^{(1)}$, where $\Gamma_n^{(i)} = \Gamma_n \cap C\ell_n^{(i)}$. The even part $\Gamma_n^{(0)}$ covers the rotation group $\mathbf{SO}(n)$ so that the unit elements a, $|a| = 1$, in $\Gamma_n^{(0)}$ form a two-fold covering group $\mathbf{Spin}(n)$ of $\mathbf{SO}(n)$.

1.4. Anti-euclidean spaces and their Clifford algebras

The following approach is closer to the algebra of complex numbers. Consider an associative algebra generated by the elements $\mathbf{i}_1, \mathbf{i}_2, \ldots, \mathbf{i}_n$ subject to the relations $\mathbf{i}_\nu^2 = -1$, and $\mathbf{i}_\mu \mathbf{i}_\nu = -\mathbf{i}_\nu \mathbf{i}_\mu$, $\mu \neq \nu$, and also $\mathbf{i}_1 \mathbf{i}_2 \ldots \mathbf{i}_n \neq \pm 1$, necessary only when $n \bmod 4 = 3$. Such an algebra is unique and will be called <u>the Clifford algebra</u> $C\ell_{0,n}$ of the anti-euclidean space $\mathbf{R}^{0,n}$ spanned by $\mathbf{i}_1, \mathbf{i}_2, \ldots, \mathbf{i}_n$. The absolute value $|a|$ of $a = \sum a_\nu \mathbf{i}_\nu$ in $C\ell_{0,n}$ is the square root of $\sum a_\nu^2 = \mathrm{Re}(\bar{a}a)$.

The following table gives isomorphic images of Clifford algebras $C\ell_n$ and $C\ell_{0,n}$, with $n < 8$, as matrix algebras over $\mathbf{R}$, $\mathbf{C}$, $\mathbf{H}$, ${}^2\mathbf{R}$ and ${}^2\mathbf{H}$.

n	0	1	2	3	4	5	6	7
$C\ell_n$	$\mathbf{R}$	${}^2\mathbf{R}$	$\mathbf{R}(2)$	$\mathbf{C}(2)$	$\mathbf{H}(2)$	${}^2\mathbf{H}(2)$	$\mathbf{H}(4)$	$\mathbf{C}(8)$
$C\ell_{0,n}$	$\mathbf{R}$	$\mathbf{C}$	$\mathbf{H}$	${}^2\mathbf{H}$	$\mathbf{H}(2)$	$\mathbf{C}(4)$	$\mathbf{R}(8)$	${}^2\mathbf{R}(8)$

1.5. Passage from euclidean to anti-euclidean

Consider $\mathbf{R}^{0,n-1}$ as a subspace of $\mathbf{R}^{0,n}$ formed by deleting the vector $\mathbf{i}_n$, $\mathbf{i}_n^2 = -1$, from the orthonormal basis $\mathbf{i}_1, \mathbf{i}_2, \ldots, \mathbf{i}_{n-1}, \mathbf{i}_n$ of $\mathbf{R}^{0,n}$. Then map $u = u_0 + u_1$ in $C\ell_{0,n-1} = C\ell_{0,n-1}^{(0)} + C\ell_{0,n-1}^{(1)}$ to $u_0 + u_1 \mathbf{i}_n$ in $C\ell_{0,n}^{(0)}$. Next compute $(u_0 + u_1 \mathbf{i}_n)(v_0 + v_1 \mathbf{i}_n) = (u_0 v_0 + u_1 v_1) + (u_0 v_1 + u_1 v_0)\mathbf{i}_n$ to get an isomorphism $C\ell_{0,n-1} \simeq C\ell_{0,n}^{(0)}$. Obviously, $C\ell_{0,n}^{(0)} \simeq C\ell_n^{(0)}$. Therefore $C\ell_{0,n-1} \simeq C\ell_n^{(0)}$ and one may identify

$$\mathbf{i}_\nu \simeq \mathbf{e}_\nu \mathbf{e}_n \quad \text{for} \quad \nu = 1, 2, \ldots, n-1.$$

The vector $\mathbf{x} = x_1 \mathbf{e}_1 + x_2 \mathbf{e}_2 + \ldots + x_{n-1} \mathbf{e}_{n-1} + x_n \mathbf{e}_n$ in $\mathbf{R}^n$ is replaced by the element $x = x_0 + x_1 \mathbf{i}_1 + x_2 \mathbf{i}_2 + \ldots + x_{n-1} \mathbf{i}_{n-1}$, $x_0 = x_n$, in the direct sum $R^n = \mathbf{R} + \mathbf{R}^{0,n-1}$. In other words, one has the correspondences $x \simeq \mathbf{x}\mathbf{e}_n$ and $R^n \simeq \mathbf{R}^n \mathbf{e}_n$. The length $|x|$ of x in R^n is the square root of $\bar{x}x = x_0^2 + x_1^2 + \ldots + x_{n-1}^2$.

In case of the anti-euclidean space $\mathbf{R}^{0,n-1}$ it is more convenient to consider the group formed by the products of non-zero elements in $R^n = \mathbf{R} + \mathbf{R}^{0,n-1}$. This group is called the Lipschitz group Γ_n of R^n. Products of unit elements a, $|a| = 1$, in R^n form the spin group $Spin(n)$. Obviously, $\Gamma_n \simeq \mathbf{\Gamma}_n^{(0)}$ and $Spin(n) \simeq \mathbf{Spin}(n)$.

If $\mathbf{a}$ is a non-zero vector in $\mathbf{R}^n$ then $\mathbf{x} \to \mathbf{a}^{\iota-1}\mathbf{x}\mathbf{a}$ is a reflection of $\mathbf{R}^n$ with respect to the mirror through the origin perpendicular to $\mathbf{a}$. If we replace $\mathbf{a}$ and $\mathbf{x}$ in $\mathbf{R}^n$ by $a \simeq \mathbf{a}\mathbf{e}_n$ and $x \simeq \mathbf{x}\mathbf{e}_n$ in R^n, then $x \to a^{\iota-1}xa$ rotates the plane through the real axis and a.

Remark. The main involution of $C\ell_n$ restricted to $C\ell_n^{(0)}$ corresponds to the identity automorphism of $C\ell_{0,n-1}^{(0)}$ since $\mathbf{e}_\nu\mathbf{e}_n \to (\mathbf{e}_\nu\mathbf{e}_n)^\iota = \mathbf{e}_\nu\mathbf{e}_n$. Both the reversion and the conjugation of $C\ell_n^{(0)}$ correspond to the conjugation $\mathbf{i}_\nu \to \bar{\mathbf{i}}_\nu$ of $C\ell_{0,n-1}$. The main involution $\mathbf{i}_\nu \to \mathbf{i}_\nu{}^\iota$ of $C\ell_{0,n-1}$ is represented by an inner automorphism $\mathbf{e}_\nu\mathbf{e}_n \to \mathbf{e}_n(\mathbf{e}_\nu\mathbf{e}_n)\mathbf{e}_n{}^{-1}$ of $C\ell_n^{(0)}$. □

2. MÖBIUS TRANSFORMATIONS

A conformal mapping preserves angles between intersecting curves. It is well known that a conformal transformation sending a region in $\mathbf{R}^n$, $n > 2$, into $\mathbf{R}^n$ is a restriction of a Möbius transformation. Möbius transformations send spheres (and hyperplanes) to spheres (or hyperplanes). The Möbius group is generated by translations, reflections and the inversion $\mathbf{x} \to \mathbf{x}^{-1} = \mathbf{x}/\mathbf{x}^2$ or, on the other hand, by affine reflections and inversions in spheres.

A sense-preserving Möbius transformation can be written as a combination of the following four different kinds of mappings:

a rotation	$\mathbf{x} \to U(\mathbf{x})$	U in $\mathbf{SO}(n)$
a translation	$\mathbf{x} \to \mathbf{x} + \mathbf{b}$	$\mathbf{b}$ in $\mathbf{R}^n$
a dilatation	$\mathbf{x} \to \mathbf{x}\delta$	$\delta > 0$

and a *transversion* defined by

$$\mathbf{x} \to \frac{\mathbf{x} + \mathbf{x}^2\mathbf{c}}{1 + 2\mathbf{x}\cdot\mathbf{c} + \mathbf{x}^2\mathbf{c}^2} \qquad \mathbf{c} \text{ in } \mathbf{R}^n.$$

Rewriting the transversion in the form $\mathbf{x} \to (\mathbf{x}^{-1} + \mathbf{c})^{-1}$ one sees that it is a combination of the inversion, a translation and the inversion.

Using the multiplicative notation of Clifford algebra, the transversion can still be written in the form $\mathbf{x} \to \mathbf{x}(\mathbf{c}\mathbf{x}+1)^{-1}$. This might suggest the following: In case a,b,c,d are in $C\ell_n$ and $(a\mathbf{x}+b)(c\mathbf{x}+d)^{-1}$ is in $\mathbf{R}^n$ for all vectors $\mathbf{x}$ such that $c\mathbf{x}+d$ is invertible, the mapping $\mathbf{x} \to (a\mathbf{x}+b)(c\mathbf{x}+d)^{-1}$ is a Möbius transformation of $\mathbf{R}^n$. Even though this is true, the group obtained in this way is too large to be a good covering group for the Möbius group. Therefore introduce

DEFINITION. *A matrix* $g = (\begin{smallmatrix} a & b \\ c & d \end{smallmatrix})$ *fulfilling the conditions*

(1) $a, b, c, d \in \Gamma_n \cup \{0\}$

(2) $ad^* - bc^* = \pm 1$

(3) $a^{-1}b,\ ca^{-1} \in \mathbf{R}^n$ if $a \neq 0$ or $ac^{-1},\ c^{-1}d \in \mathbf{R}^n$ if $c \neq 0$

is called a <u>*Vahlen matrix*</u> *of* $\mathbf{R}^n$. □

The Vahlen matrices form a group under matrix multiplication. For instance, to see that a product $g_1 g_2$ inherits the property (1) take as a sample $a_1a_2 + b_1c_2 = b_1(b_1^{-1}a_1 + c_2a_2^{-1})a_2$ which clearly is in $\Gamma_n \cup \{0\}$.

The condition (2) makes it impossible for two elements of g in the same row or column to be simultaneously zero. If both a and c are non-zero, then the two alternatives in the condition (3) can be shown to be equivalent: $b = ad^*c^{*-1} \pm c^{*-1} = ac^{-1}d \pm c^{*-1}$ and hence $a^{-1}b = c^{-1}d \pm a^{-1}c^{*-1} = c^{-1}d \pm (c^*a)^{-1} \in \mathbf{R}^n$. The last step follows from $c^*a \in \mathbf{R}^n$ being equivalent to $ac^{-1} \in \mathbf{R}^n$, since $c^*a = c^*(ac^{-1})c = \pm c^*(ac^{-1})\bar{c}^{-1}|c|^2$. Similarly, $ba^*,\ dc^* \in \mathbf{R}^n$.

The condition (2) together with $ab^* = ba^*$ and $cd^* = dc^*$ results in

$$\begin{pmatrix} a & b \\ c & d \end{pmatrix}\begin{pmatrix} d^* & -b^* \\ -c^* & a^* \end{pmatrix} = \pm\begin{pmatrix} 1 & 0 \\ 0 & 1 \end{pmatrix}$$

which further gives the inverse of a Vahlen matrix.

The matrix $g = (\begin{smallmatrix} a & b \\ c & d \end{smallmatrix})$ acts on $\mathbf{x}$ in $\mathbf{R}^n$ according to the rule:

$$g\mathbf{x} = (a\mathbf{x}+b)(c\mathbf{x}+d)^{-1}.$$

In order for this to make sense, it is necessary to pass from $\mathbf{R}^n$ to the compactification $\mathbf{R}^n \cup \{\infty\}$. If $c\mathbf{x}+d \neq 0$, then it is in Γ_n since $c\mathbf{x}+d = c(\mathbf{x}+c^{-1}d)$.

EXAMPLE. To compute the inverse map of $\mathbf{x} \to \mathbf{y} = (a\mathbf{x} + b)(c\mathbf{x} + d)^{-1}$ multiply from the right by $c\mathbf{x} + d$ to get $\mathbf{y}(c\mathbf{x} + d) = a\mathbf{x} + b$. Taking reversion gives $(\mathbf{x}c^* + d^*)\mathbf{y} = \mathbf{x}a^* + b^*$ and $\mathbf{x} = (d^*\mathbf{y} - b^*)(-c^*\mathbf{y} + a^*)^{-1}$. □

In a Vahlen matrix $g = \left(\begin{smallmatrix} a & b \\ c & d \end{smallmatrix}\right)$ both the diagonal elements a, d are either even or odd (both the off-diagonal elements b, c are odd or even, respectively). The group of the Vahlen matrices has four components labelled by the parity of the diagonal and the sign of the *pseudo-determinant* $ad^* - bc^*$. It is sufficient to pick up two components out of the group of the Vahlen matrices to cover all of the two-component Möbius group of $\mathbf{R}^n$.

THEOREM. *The Vahlen matrices of* $\mathbf{R}^n$, *with even diagonal and pseudo-determinant* 1, *form a two-fold covering group of the sense-preserving Möbius group of* $\mathbf{R}^n$. *The remaining Möbius transformations in* $\mathbf{R}^n$, *which swap the sense of orientation opposite, can be represented by the Vahlen matrices of* $\mathbf{R}^n$ *with odd diagonal and pseudo-determinant* -1. □

2.1. Step in anti-euclidean

A matrix $g = \left(\begin{smallmatrix} a & b \\ c & d \end{smallmatrix}\right)$ fulfilling the conditions

(1) $a, b, c, d \in \Gamma_n \cup \{0\}$

(2) $ad^* - bc^* = 1$

(3) $a^{-1}b,\ ca^{-1} \in R^n$ if $a \neq 0$ or $ac^{-1},\ c^{-1}d \in R^n$ if $c \neq 0$

is a Vahlen matrix of R^n. For x in R^n define $gx = (ax + b)(cx + d)^{-1}$. Observe the factorization:

$$\begin{pmatrix} a & b \\ c & d \end{pmatrix} = \begin{pmatrix} 1 & ac^{-1} \\ 0 & 1 \end{pmatrix}\begin{pmatrix} c^{*-1} & 0 \\ 0 & c \end{pmatrix}\begin{pmatrix} 0 & -1 \\ 1 & 0 \end{pmatrix}\begin{pmatrix} 1 & c^{-1}d \\ 0 & 1 \end{pmatrix}.$$

The first and last factors represent parallel translations by ac^{-1} and $c^{-1}d$ in R^n. The second factor represents a combined rotation and multiplication by a positive scalar. The third factor, representing an inversion $x \to -x^{-1} = -\bar{x}/|x|^2$, is also sense-preserving.

The Vahlen group of R^n is connected. It is a two-fold covering group of the sense-preserving Möbius group of R^n.

2.2. The derivative of a Möbius transformation

The difference of the Möbius transformations of x and y in R^n is given by:

$$gx - gy = (cy + d)^{*-1}(x - y)\,(cx + d)^{-1}.$$

The derivative of $x \to gx$ is the linear transformation $|cx + d|^{-2}U_{(cx+d)'}$ of R^n.

2.3. The Lie algebra of the Vahlen group

If a matrix $\begin{pmatrix} a & \mathbf{b} \\ \mathbf{c} & d \end{pmatrix}$ is in the Lie algebra of the Vahlen group then the element $\mathbf{y} = a\mathbf{x} + \mathbf{b} - \mathbf{xcx} - \mathbf{x}d$ must be a vector in $\mathbf{R}^n$ for all $\mathbf{x}$ in $\mathbf{R}^n$. It follows that $\mathbf{b}$ and $\mathbf{c}$ are vectors in $\mathbf{R}^n$. Furthermore, a and d must be sums of scalars, bivectors and n-vectors so that they have equal bivector parts and equal or opposite n-vector parts.

One can say even more:

(1) $a,\ d \in \mathbf{R} + C\ell_n^2$

(2) $a + d^* = 0$

(3) $\mathbf{b},\ \mathbf{c} \in \mathbf{R}^n$.

The Lie algebra is spanned by the matrices

$$\mathrm{L}_{\mu\nu} = \begin{pmatrix} -\frac{1}{2}\mathbf{e}_{\mu\nu} & 0 \\ 0 & -\frac{1}{2}\mathbf{e}_{\mu\nu} \end{pmatrix} \qquad \mathrm{P}_{\nu} = \begin{pmatrix} 0 & \mathbf{e}_{\nu} \\ 0 & 0 \end{pmatrix}$$

$$\mathrm{D} = \begin{pmatrix} \frac{1}{2} & 0 \\ 0 & -\frac{1}{2} \end{pmatrix} \qquad \mathrm{K}_{\nu} = \begin{pmatrix} 0 & 0 \\ \mathbf{e}_{\nu} & 0 \end{pmatrix}.$$

These matrices represent rotations, translations, dilatations and transversions.

2.4. Möbius transformations of Cauchy-Riemann equations

Let $\mathbf{X}$ be a domain in $\mathbf{R}^n$. A mapping $u: \mathbf{X} \to C\ell_n$ satisfying the Cauchy-Riemann equations

$$\Big(\mathbf{e}_1 \frac{\partial}{\partial x_1} + \mathbf{e}_2 \frac{\partial}{\partial x_2} + \ldots + \mathbf{e}_n \frac{\partial}{\partial x_n}\Big)\, u(\mathbf{x}) = 0$$

is *monogenic* at $\mathbf{x} = x_1\mathbf{e}_1 + x_2\mathbf{e}_2 + \ldots + x_n\mathbf{e}_n \in \mathbf{X}$. For instance, monogenic functions are obtained by taking symmetrized products of the polynomials $s_\nu = x_\nu + y\mathbf{i}_\nu$, where $\mathbf{i}_\nu = \mathbf{e}_\nu\mathbf{e}_n$ for $\nu = 1, 2, \ldots, n\text{-}1$, and $y = x_n$. The following holds for a Vahlen matrix $\binom{a\ \ b}{c\ \ d}$ of $\mathbf{R}^n$: If u is monogenic at every $g\mathbf{x} = (a\mathbf{x} + b)(c\mathbf{x} + d)^{-1} \in \mathbf{G} = g\mathbf{X}$ then the function v defined by

$$v(\mathbf{x}) = \frac{(c\mathbf{x} + d)^*}{|c\mathbf{x} + d|^n}\, u(g\mathbf{x}) \qquad \text{is monogenic in } \mathbf{X}.$$

3. THE MINKOWSKI SPACE AND THE DIRAC ALGEBRA

The Minkowski space of special relativity has an indefinite quadratic form $\mathbf{x}^2 = x_1^2 + x_2^2 + x_3^2 - x_4^2$. The orthogonal unit vectors $\mathbf{e}_1, \mathbf{e}_2, \mathbf{e}_3, \mathbf{e}_4$ satisfy the relations $\mathbf{e}_1^2 = \mathbf{e}_2^2 = \mathbf{e}_3^2 = 1$, $\mathbf{e}_4^2 = -1$, and $\mathbf{e}_\mu\mathbf{e}_\nu = -\mathbf{e}_\nu\mathbf{e}_\mu$, $\mu \neq \nu$. The Minkowski space is denoted by $\mathbf{R}^{3,1}$ and its Clifford algebra by $C\ell_{3,1}$. It is well known that $C\ell_{3,1} \simeq \mathbf{R}(4)$.

The Lipschitz group $\mathbf{\Gamma}_{3,1}$ consists of products of invertible vectors in $\mathbf{R}^{3,1}$. We shall also need the set $\mathbf{\Pi}_{3,1}$ consisting of products of vectors in $\mathbf{R}^{3,1}$ whether invertible or not.

A matrix $g = \binom{a\ \ b}{c\ \ d}$ is called a Vahlen matrix of $\mathbf{R}^{3,1}$ if the following conditions are fulfilled:

(1) $a, b, c, d \in \mathbf{\Pi}_{3,1}$
(2) $ad^* - bc^* = \pm 1$
(3) $ab^*,\ b^*d,\ dc^*,\ c^*a \in \mathbf{R}^{3,1}$.

The matrix $g = \binom{a\ \ b}{c\ \ d}$ is made to act on vectors $\mathbf{x}$ in $\mathbf{R}^{3,1}$ according to the rule $g\mathbf{x} = (a\mathbf{x} + b)(c\mathbf{x} + d)^{-1}$. If $c\mathbf{x} + d$ is not invertible, then $\mathbf{y} = g\mathbf{x}$ means that $\mathbf{y}(c\mathbf{x} + d) = a\mathbf{x} + b$.

The Vahlen matrices of $\mathbf{R}^{3,1}$ form a four-component group. In contrast to the euclidean case, all these four components are needed to construct a four-fold covering of the conformal group of $\mathbf{R}^{3,1}$, or more precisely, the compactification $S^3 \times S^1/\{\pm 1\}$ of $\mathbf{R}^{3,1}$. This is due to the fact that all the four pre-images

$$\pm\begin{pmatrix} 1 & 0 \\ 0 & 1 \end{pmatrix}, \qquad \pm\begin{pmatrix} \mathbf{j} & 0 \\ 0 & -\mathbf{j}^{\iota} \end{pmatrix} \qquad (\mathbf{j} = \mathbf{e}_1\mathbf{e}_2\mathbf{e}_3\mathbf{e}_4)$$

of the identity are in the identity component of the Vahlen group of $\mathbf{R}^{3,1}$. This identity component is isomorphic to the matrix group $SU(2,2)$. It generates the *Dirac algebra*. The Dirac algebra consists of 2×2-matrices with diagonal entries in $C\ell_{3,1}^{(0)}$ and off-diagonal in $C\ell_{3,1}^{(1)}$. It is well known that the Dirac algebra is isomorphic to $\mathbf{C}(4)$.

The Dirac spinors are elements of a minimal left ideal of the Dirac algebra; they are of the form

$$\begin{pmatrix} u_0 & 0 \\ u_1 & 0 \end{pmatrix} e$$

where $u_i \in C\ell_{3,1}^{(i)}$ and $e = \frac{1}{2}(1 + \mathbf{e}_{34})$ is a primitive idempotent of $C\ell_{3,1}^{(0)}$.

In general, the conformal group of $\mathbf{R}^{p,q}$ has two components (either p or q even) or four components (both p and q odd). If both p and q are odd, then all the four components of the Vahlen group of $\mathbf{R}^{p,q}$ are needed to cover the conformal group of $\mathbf{R}^{p,q}$.

References

[1] Ahlfors, L.: 'Möbius transformations and Clifford numbers'. I. Chavel, H. M. Farkas (ed.): *Differential Geometry and Complex Analysis*. Dedicated to H. E. Rauch. Springer-Verlag, Berlin, 1985, pp. 65-73.

[2] Anglés, P.: 'Construction de revêtements du groupe conforme d'un espace vectoriel muni d'une "metrique" de type (p,q)'. *Ann. Inst. H. Poincaré. Sect.* A **33** (1980), 33-51.

[3] Brackx, F., R. Delanghe, F. Sommen: *Clifford analysis*. Pitman Books, London, 1982.

[4] Crumeyrolle, A.: *Algèbres de Clifford et spineurs*. Université Paul Sabatier, Toulouse, 1974.

[5] Fillmore, J. P.: 'The fifteen-parameter conformal group'. *Internat. J. Theoret. Phys.* **16** (1977), 937-963.

[6] Gilbert, R. P., J. L. Buchanan: *First order elliptic systems, a function theoretic approach*. Academic Press, New York, 1983.

[7] Greider, K. R.: 'A unifying Clifford algebra formalism for relativistic fields'. *Found. Phys.* **14** (1984), 467-506.

[8] Hestenes, D.: 'Vectors, spinors, and complex numbers in classical and quantum physics'. *Amer. J. Phys.* **39** (1971), 1013-1027.

[9] Hile, G. N.: 'Representations of solutions of a special class of first order systems'. *J. Differential Equations* **25** (1977), 410-424.

[10] Lounesto, P.: 'Conformal transformations and Clifford algebras'. *Proc. Amer. Math. Soc.* **79** (1980), 533-538.

[11] Maass, H.: 'Automorphe Funktionen von mehreren Veränderlichen und Dirichletsche Reihen'. *Hamburg Math. Abh.* **16** (1949), 72-100.

[12] Porteous, I.R.: *Topological geometry*. Van Nostrand-Reinhold, London, 1969. Cambridge University Press, Cambridge, 1981.

[13] Riesz, M.: *Clifford numbers and spinors*. University of Maryland, 1958.

[14] Ryan, J.: 'Conformal Clifford manifolds arising in Clifford analysis'. *Proc. Roy. Irish Acad. Sect.* A **85** (1985), 1-23.

[15] Vahlen, K. Th.: 'Über Bewegungen und komplexe Zahlen'. *Math. Ann.* **55** (1902), 585-593.

Part II

Complex Analytic Geometry

edited by

Pierre Dolbeault

Mathématiques, L.A. 213 du C.N.R.S.
Université Pierre et Marie Curie
Paris, France

Julian Ławrynowicz

Institute of Mathematics
Polish Academy of Sciences
Łódź, Poland

and

Edoardo Vesentini

Classe di Scienze
Scuola Normale Superiore
Pisa, Italy

DOUBLES OF ATOROIDAL MANIFOLDS, THEIR CONFORMAL UNIFORMIZATION AND DEFORMATIONS

Boris Nikolaevich Apanasov
Institute of Mathematics
Siberian Branch of the
USSR Academy of Sciences
SU-630090 Novosibirsk 90, USSR

ABSTRACT. We investigate doubles of compact irreducible n-manifolds (orbifolds), whose boundary consists of a finite sum of (n-1)-dimensional tori and their interior admits a conformal structure which is uniformizable by some Möbius b-group. On the obtained closed manifolds we introduce a uniformizable conformal structure and investigate the space of its deformations.

INTRODUCTION

As classical uniformization Klein-Poincaré theorem shows, the investigation of conformal structures on surfaces is equivalent to the study of their geometrical structure - hyperbolic (if genus $g > 1$), euclidean (if $g = 1$) and spherical (if $g = 0$), that is to the study of their Teichmüller spaces (see [12] and [18]). For the manifolds $n \geq 3$ this approach turns out to be impossible because of two reasons. Firstly, while the geometrical structures of $n \geq 3$ have a known rigidity (Mostow [23]), the conformal structures (even hyperbolic manifolds) admit non-trivial deformations (Apanasov-Tetenov [11], Apanasov [2] and [3]). Secondly, in many cases on the conformal manifold it is impossible to introduce a geometrical structure. Thus, we first have to cut a manifold into pieces (along non-trivial imbedded (n-1)-dimensional spheres and tori), on which, as uniformization Thurston theorem (see Thurston [26] and Morgan [22]) shows for all known 3-manifolds, we introduce (hypothetically) the geometry of one of canonical homogeneous spaces. In this way conformal structures on n-manifolds for $n \geq 3$ are more useful tools of uniformization than geometrical ones. The first results in this field enable us to formulate most likely the following conjecture (see for details: chapt. 8, English version of [6]): Let a compact manifold M satisfy the structure hypothesis of Thurston [26]. Let further assume that by canonical cutting its interior desintegrates into components which admit the uniformizable conformal structures. Then the interior of the manifold M admits a conformal structure which turns out to be uniformizable.

J. Ławrynowicz (ed.), Deformations of Mathematical Structures, 93–114.

There are two purposes of the present paper, firstly, introducing uniformizable conformal structures on a class of n-manifolds (especially on parabolic doubles of hyperbolic manifolds), which by canonical cutting disintegrate into two components, secondly, constructing geometrical deformations of the obtained conformal structure and obtaining estimations of the dimension of the space formed by them. We get the means for the first purpose from combination theorems by Maskit (see [21], also [6]).

The main idea of the method of geometrical deformations of a conformal structure lies in the investigation of hyperbolic geometry of (n+1)-dimensional manifold whose boundary components turn out to be conformally equivalent to the uniformizable manifold. Moreover, the estimation of the dimension of the deformation space of the conformal structure of the doubles depends on the number of some totally geodesic submanifolds of co-dimension 1 of the given (n+1)-dimensional hyperbolic manifold.

Let us observe that the obtained space of conformal structures on n-dimensional manifold has for $n \geq 3$ significant differences in comparison with its plane analogons. This fact emphasises once more the existence of singularities discovered recently by Johnson-Milson (see [15]).

2. CONFORMAL STRUCTURES, COMBINATION THEOREMS AND SOME INTRODUCTORY INFORMATION

2.1. Let us denote by $M_n = \text{Möb}_+(n)$ the group of all conformal automorphisms of conformal sphere $S^n = \mathbb{R}^n \cup \{\infty\}$, which preserve the orientation. Let $G \subset M_n$ be a Kleinian group, i.e. a discrete subgroup in M_n with the limit set $\Lambda(G) \neq S^n$ (= the set of density points of orbits $G(x)$, $x \in S^n$) and the discontinuity set $\Omega(G) = S^n \setminus \Lambda(G) \neq \emptyset$. A set $D \subset S^n$ is called strictly invariant with respect to a subgroup $H \subset G$ if $h(D) = D$ for any $h \in H$ and $g(D) \cap D = \emptyset$ for any $g \in G \setminus H$.

2.2. <u>THEOREM</u> (Maskit's first combination). *Let* $G_1, G_2 \subset M_n$ *be Kleinian groups with an amalgamated subgroup* $H = G_1 \cap G_2$, $S \subset \mathbb{R}^n$ *- Jordan hyperplane dividing* $\mathbb{R}^n$ *into two domains* D_1 *and* D_2, *whose closures are strongly invariant in* G_1 *and* G_2 *(respectively) with respect to the subgroup* H, *and let the following conditions be satisfied:*

(i) *If* Δ, F_1, F_2 *are fundamental domains of the groups* H, G_1, G_2, *then there exists a neighbourhood* V *of a surface* S *such that*

$$\Delta \cap V \subset F_i, \quad i = 1, 2;$$

and

(ii) $\Delta \cap \bar{D}_i = \bar{D}_i \cap F_i \neq F_i, \quad i = 1, 2.$

Then the following statements are satisfied:

(a) *the group* $G = \langle G_1, G_2 \rangle$ *is a Kleinian group and is isomorphic to the free product of the groups* G_1 *and* G_2 *with an amalgamated subgroup* H;

(b) $F_1 \cap F_2$ *is a fundamental domain of the group* G;

(c) *if for the Lebesgue measure* μ_n *we have* $\mu_n(\Lambda(G_i)) = 0$, $i = 1, 2$, *then* $\mu_n(\Lambda(G)) = 0$;

(d) *if* $g \in G$ *is an elliptic or parabolic element, then* g *is conjugate in* G *to an element of* $G_1 \cup G_2$.

2.3. THEOREM (Maskit's second combination). *Suppose* Γ, $\Gamma \subset M_n$ *to be a Kleinian group with subgroups* H_1 *and* H_2 *such that some closed domains* B_1 *and* B_2, *bounded by Jordan surfaces* S_1 *and* S_2, *are strictly invariant with respect to* H_1 *and* H_2. *Moreover, suppose for some* $A \in M_n$, $A(\operatorname{int} B_1) = \operatorname{ext} B_2$ *and the following conditions:*

(I) *If* Δ_1, Δ_2 *and* $F_o \subset \Delta_1 \cap \Delta_2$ *are fundamental domains of the groups* H_1, H_2 *and* Γ, *respectively, then there exist neighbourhoods* V_1 *and* V_2 *of* S_1 *and* S_2 *with the properties*

(i) $\Delta_i \cap V_i \subset F_o$, $i = 1, 2$;

(ii) $\Delta_i \cap B_i = F_o \cap B_i$, $i = 1, 2$;

(iii) $F = \operatorname{int}\{F_o \cap \operatorname{ext}(B_1 \cup B_2)\} \neq \emptyset$;

(iv) $A H_1 A^{-1} = H_2$;

(v) $\gamma(B_1) \cap B_2 = \emptyset$ *for any* $\gamma \in \Gamma \setminus \mathrm{id}$.

Then the following conclusions hold:

(a) $G = \langle \Gamma, A \rangle$ *is a Kleinian group and* F *its fundamental domain;*

(b) *the group* G *is isomorphic to* HNN-*extension of* Γ *with respect to* H_1, $H_2 \subset \Gamma$, *and the element* A;

(c) *if* $\mu_n(\Lambda(\Gamma)) = 0$, *then* $\mu_n(\Lambda(G)) = 0$;

(d) *if* $g \in G$ *is an elliptic or parabolic element, then* g *is conjugate in* G *with an element from* Γ.

The given combination theorems have been proved by Maskit [21] for Kleinian groups on the plane; the case $n \geq 3$ is analogous to the plane case if we replace Koebe theorem with its spatial anologon (see [6]). Moreover, though we have formulated theorems for the case of the groups acting on the sphere S^n it is easy to reformulate them for discreete subgroups of the hyperbolic space H^{n+1}.

2.4. Following W. Thurston ([25], ch. 13), we introduce the object which generalizes the notion of a manifold. Thus an orbifold means a space which is locally a ball with respect to the modulus of a finite group of homeomorphisms or, precisely, the orbifold M is the Hausdorff space X with an open covering $\{U_i\}$, closed with respect to finite intersections. With each U_i we associate the finite group Γ_i acting on open $\tilde{U}_i \subset \mathbb{R}^n$ and the homeomorphism $\phi_i : U_i \to \tilde{U}_i/\Gamma_i$. Moreover, if $U_i \subset U_j$, then there exist a monomorphism $f_{ij} : \Gamma_i \to \Gamma_j$ and an imbedding $\tilde{\phi}_{ij} : \tilde{U}_i \hookrightarrow \tilde{U}_j$, such that $\tilde{\phi}_{ij}(\gamma(x)) = f_{ij}(\gamma)\tilde{\phi}_{ij}(x)$ for

$\gamma \in \Gamma_i$ and the following diagram is commutative:

$$\begin{array}{ccccc} \tilde{U}_j & \longrightarrow & U_j/f_{ij}(\Gamma_i) \to U_j/\Gamma_j & \xleftarrow{\ \phi_j\ } & U_j \\ \uparrow \tilde{\phi}_{ij} & & \nwarrow \ \phi_{ij} = \tilde{\phi}_{ij}/\Gamma_i & & \cup \\ \tilde{U}_i & \longrightarrow & \tilde{U}_i/\Gamma_i & \xleftarrow{\ \phi_i\ } & U_i \end{array}$$

where $\tilde{\phi}_{ij}$ is defined modulo superpositions with Γ_j and f_{ij} modulo conjugacy by elements of Γ_i. Two coverings give one structure of the orbifold if their sum satisfies the conditions given in the above definition. The orbifold M has conformal, hyperbolic or another G-structure (see [25] and [6], ch. 2), if all mappings and groups of the definition in question are in the G-category, especially are conformal or hyperbolic isometries. We will deal with orbifolds of the form $M = \Omega/G$, where the simply connected domain $\Omega \subset \Omega(G) \subset S^n$ is invariant with respect to the Kleinian group $G \subset M_n$. In this case the deck transformation group of the universal orbifold $\tilde{M}$ ($\cong \Omega$) is called the fundamental group $\pi_1(M)$ of the orbifold M (should not be confused with the fundamental group $\pi_1(X)$ of the space X defining the orbifold M) and is always isomorphic with the group G.

2.5. We say that a hyperbolic orbifold M has a geometrically finite structure if a discreete subgroup $G \subset \mathrm{Isom}\, H^n$ which is an image of $\pi_1(M)$ by the holonomy map $d^*: \pi_1(M) \to \mathrm{Isom}\, H^n$ (induced by the developing mapping $d: \tilde{M} \to H^n$) is geometrically finite, i.e. it has finite-sided fundamental polyhedron $P(G) \subset H^n$. If the orbifold M can finitely be covered by a manifold or, equivalently, if the group G has a finite index subgroup without elements of finite order (due to A.I. Malcev [20] it is always fulfilled for any finitely generated group), then the existence of a geometrical finite structure on M is equivalent to the finiteness of the volume of ε-neighbourhood of minimal convex retract $M_G \subset M$. Then, in turn, the finiteness is equivalent to the compactness of the suborbifold $[M_G]_{[\varepsilon,\infty)} \subset M_G$, consisting of all points $x \in M_G$ through which non-trivial loops of the length less than ε do not pass. It follows from the results of the author in [4-6] for geometrically finite groups without torsion and from the fact that the finite extension of the group do not change the character of its limit points which are the points of conic approximation and parabolic vertices.

2.6. <u>Remark.</u> Directly from Theorems 2.2 and 2.3 it follows that the property of geometrical finiteness of the groups G_1 and G_2 implies geometrical finiteness of the group $G = \langle G_1, G_2 \rangle$ or $G = \langle G_1, \tilde{A} \rangle$.

2.7. By a parabolic vertex (a cusp) we mean (cf. [4]) a point z, fixed for a parabolic element of the group $G \subset M_n$ (more precisely, we have to take the orbit $G(z)$), if for this point one of the following conditions is fulfilled:

(i) An isotropy group $G_z = \{g \in G: g(z) = z\}$ has a free abelian subgroup of rank n;

(ii) There exists an open subset $U_z \subset \Omega(G)$ strictly invariant to the subgroup $G_z \subset G$ and such that for some $t > 0$ and some integer k, $1 \le k \le n-1$, which equals the rank of the maximal free abelian subgroup in the isotropy group G_z, there exists $h \subset M_n$, $h(z) = \infty$, for which

$$\{x \in \mathbb{R}^n: \sum_{i=k+1}^{n} x_i^2 > t\} \subset h(U_z).$$

According to the conditions (i) or (ii), we say that the point z has the rank n or k.

2.8. Let M be an n-dimensional orbifold with two conformal (or hyperbolic) atlases A_1 and A_2 (see 2.4). We say that they are equivalent to each other if $A_1 \cup A_2$ is again conformal (hyperbolic) atlas. A class of equivalent atlases, i.e. a maximal atlas, is called the conformal (hyperbolic) structure on M. A class of equivalent conformal (hyperbolic) structures on M is called the marked structure on M. Here we say that two structures s_1 and s_2 on M are equivalent if there is conformal (isometric, respectively) homeomorphism $(M, s_1) \to (M, s_2)$ which homotopic to the identity. We denote the set of these structures by $\mathcal{C}(M)$ and $\mathcal{T}(M)$ respectively.

2.9. In order to investigate $\mathcal{C}(M)$ and $\mathcal{T}(M)$, we can pass to the spaces of conjugate classes of representations of the group $G \cong \pi_1(M)$ in the group $\text{Möb}(n) \cong SO(n+1, 1)$ of conformal automorphisms of the sphere S^n with respect to the group $\text{Isom}\, H^n \cong SO(n, 1)$ of isometries of the hyperbolic space H^n. This possibility results from the following construction.

Let $\mathcal{S}(M)$ be a space of marked locally homogeneous structures on M with a modelling homogeneous space $Y = G/H$. When considering any chart (U_o, ϕ_o) on M and all extensions for ϕ_o on M, we get in a natural way any $s \in \mathcal{S}(M)$, the developing map $d: \tilde{M} \to Y$, and the holonomy homomorphism $d^*: \pi_1(M) \to G$. In this way, for $\Gamma \subset G$, $\Gamma \cong \pi_1(M)$, we can define the mapping

$$\text{hol}: \mathcal{S}(M) \to \text{Hom}(\Gamma, G)/G = \text{Def}\, \Gamma\ , \tag{2.10}$$

where $\text{hol}(s)$ is the orbit of the representation $d^*: \Gamma \to G$ by the conjugacy with elements G.

Now we endow the representation space $\text{Hom}(\Gamma, G)$ with the usual topology of algebraical convergence, and $\text{Hom}(\Gamma, G)/G$ with the induced quotient topology. We equip the space $\mathcal{S}(M)$ with the topology which, after lifting to the covering space $\mathcal{D}(M)$ of developments d, becomes the topology of convergence of developments on compact subsets. Then an important property of the mapping (2.10) is given by the following holonomy theorem [19; Th. 1.11]:

2.11. <u>THEOREM</u>. *The mapping* hol *in* (2.10) *is an open mapping which can be lifted to a local homeomorphism:* $\mathcal{D}(M) \to \text{Hom}(\Gamma, G)$ (cf. Remark 5.20 below).

2.12. Let, for two conformal orbifolds M_1 and M_2 with the structures C_1 and C_2, a local homeomorphic map $f: M_1 \to M_2$ be given, which preserves orbifold structures in the sense of 2.4. We say that f is conformal if f is an (S^n, M_n)-mapping, i.e. if the conformal structure on M_1, obtained by the lifting with the use of the mapping f of the conformal structure C_2, is equivalent to the conformal structure C_1.

We call a conformal structure C on M uniformizable if there exists a Kleinian group $G \subset M_n$ such that the structure C is conformally equivalent to the natural structure of some connected component of $\Omega(G)/G$, induced by the natural projection $\Omega(G) \to \Omega(G)/G$. Moreover, we will most often be in the situation when the given component of $\Omega(G)/G$ corresponds to G-invariant simply connected $\Omega_0 \subset \Omega(G)$, which in the case of irreducible M is contractible.

3. FUNCTIONAL GROUPS AND PARABOLIC COMBINATION

3.1. Although it is not essential for the results of this section, here we will consider geometrically finite functional groups $G \subset M_n$, i.e. Kleinian groups whose discontinuity set has G-invariant connected component $\Omega_0 \subset \Omega(G)$. Besides, we will assume that an invariant component Ω_0 of the group G is contractible. In this case the orbifold Ω_0/G is irreducible (see [25], ch. 13) and has only finite number of ends, each of which is homeomorphic to $T^{n-1} \times [0,\infty)$, where T^{n-1} is an $(n-1)$-dimensional torus (see [5, 6]). Parabolic vertices which correspond to these ends form a finite number of disjoint G-orbits and in each of these vertices z there exists a peak domain U_z (see 2.7). We always have $\operatorname{rank} z = n-1$ and $U_z = U_z^o \cup U_z^1$, where U_z^i is a tangent balls at the point z, $U_z^o \subset \Omega_0$, $U_z^1 \subset \Omega(G) \setminus \Omega_0$. This follows from Lemmas 3 and 4 of [24] and from the fact that for $\operatorname{rank} z \leq n-1$ the peak domain U_z is connected.

3.2. We define, for a parabolic vertex z of the group $G \subset M_n$, a radius of the ball $B = U_z^o \subset \Omega_0$ which is strongly invariant with respect to an isotropy subgroup $G_z \subset G$ (see [6], ch. 6). Thus, let us consider the maximal ball $B_o \subset \Omega_0$ (also $B_o = \Omega_0$) which is invariant with respect to G_z. Let us also consider a mapping $w \in M_n$:

$$w(B_o) = \mathbb{R}^n_+, \quad w(z) = \infty, \quad w(B_1) = \{x \in \mathbb{R}^n: \ x_n > 1\}, \ B_1 \subset B_o,$$

where B_1 is a maximal strictly invariant ball with respect to G_z:

$$\gamma(B_1) = B_1, \quad g(B_1) \cap B_1 = \emptyset \quad \text{for} \quad \gamma \in G_z, \quad g \in G \setminus G_z.$$

The mapping w is defined up to euclidean isometries which preserve $\mathbb{R}^n_+$. That is why we can define the function

$$\rho_z(x) = \exp[-(w(x), e_n)] \tag{3.3}$$

where $(*,*)$ is a scalar product in $\mathbb{R}^n$, and treat a number $r > 0$, with a radius of a horoball $B \subset \Omega_o$ at the parabolic point z, such that

$$B = B_z(r) \equiv \{x \in \Omega_o : \rho_z(x) < r\}. \tag{3.4}$$

$B_z(1/e)$ is the maximal strictly invariant horoball with respect to G_2. Moreover, for any $r_1, r_2 \leq 1/e$, strictly invariant horoballs are disjoint at different parabolic points z_1, z_2 of the group G (see [6], Theorem 6.15):

$$B_{z_1}(r_1) \cap B_{z_2}(r_2) = \emptyset; \quad r_1, r_2 \leq 1/e. \tag{3.5}$$

We observe that the property (3.5) for $n = 3$ is fulfilled with no assumption that the group G is geometrically finite. This fact is connected with the fact that parabolic subgroups of rank 2 in the group M_3 are conjugated with the groups of euclidean translations. When $n \geq 4$ this property does not hold and we can give examples of Kleinian groups infinitely generated, whose parabolic points do not fulfil the property (3.5) (see [7]).

3.6. Now we shall describe a process of parabolic combination. Without changing the notation we shall treat the group M_n and its subgroups, which act in $\mathbb{R}^{n+1}_+$, as the groups of isometry of the Poincaré model H^{n+1} of the hyperbolic space.

Let $G_o \subset M_n$ be a geometrically finite functional group with invariant contractible component $\Omega_o \subset \Omega(G_o)$ which has parabolic vertices z, $U_z \cap \Omega_o \neq \emptyset$, on its boundary. By assuming that one of them is ∞, we choose in H^{n+1} a Dirichlet polyhedron with the centre in some point $x \in H^{n+1}$, which has not equivalent parabolic points on its boundary. Let its parabolic vertices be $z_1 = \infty, z_2, \ldots, z_m$. Then the polyhedron P^n, fundamental for G_o on $S^n \cong \bar{\mathbb{R}}^n$, has the same property

$$P^n = P^n(G_o) = \operatorname{int}(\bar{P}^{n+1}(G_o) \cap \bar{\mathbb{R}}^n).$$

As it was observed in 3.1, at each parabolic point $z_1, \ldots z_m$ each domain of the cycles $U_{z_1}, \ldots, U_{z_m}$ has two components which are the balls, tangent at z_1, with the radius $r_1 \leq 1/e$ (see 3.5). Let us denote by $B_i(r_i)$ components of the peak domain U_{z_i}, $i = 1, \ldots, m$, which are contained in Ω_o, and by $S_i = S_i(r_i)$ n-dimensional planes in $H^{n+1} \cong \mathbb{R}^{n+1}_+$ such that

$$\partial S_i(r_i) = \partial B_i(r_i). \tag{3.7}$$

Let $J_i = J_i(r_i)$ be reflections in the hyperbolic space H^{n+1} with respect to n-dimensional planes $S_i(r_i)$, $i = 1, \ldots, m$.

Let us consider the Kleinian group $J_1 G_o J_1 \subset M_n$. Then this group and G_o have an amalgamated subgroup G_∞, which together with the sphere ∂B_1 and fundamental domains P^n and $J_1(P^n)$ fulfil the hypotheses of Theorem 2.2. As the result of applying this theorem we get the Kleinian group

$$G_1 = G_1(r_1) = G_0 \underset{G_\infty}{*} J_1 G_0 J_1, \tag{3.8}$$

which has an invariant component $\Omega_0(G_1) \subset \Omega_0 \cap J_1(\Omega_0)$. The fundamental polyhedron

$$P^{n+1}(G_1) = P^{n+1}(G_0) \cap J_1(P^{n+1}(G_0)) \tag{3.9}$$

has on its boundary (nonequivalent) parabolic vertices $z_{i1} = z_i$ and $z_{i2} = J_1(z_i)$, $i = 2, \ldots, m$, which are also vertices of the polyhedron $P^n(G_1) \cap \Omega_0(G_1)$. Moreover, $B_{i1}(r_i) = B_i$ and $B_{i2}(r_i) = J_1(B_i)$ are the connected components of the peak domains of these vertices and are contained in a G_1-invariant component of $\Omega_0(G_1) \subset \Omega(G_1)$.

Let us observe that the balls B_{i1} are mapped by $J_1 J_i \in M_n$ into exterior of the closed balls B_{i2} for each $i = 2, \ldots, m$; what more an isotropy group of the point z_{i1} in the group G_1 is mapped by conjugacy with an element $J_1 J_i$ in isotropy subgroup of the point z_{i2}. It is also easy to check that for each $i = 2, \ldots, m$ the tuples

$$G_{z_{i1}}, G_{z_{i2}} \subset G_1, \quad f_i = J_1 J_i, \quad S_1 = \partial B_{i1}, \quad S_2 = \partial B_{i2},$$

which depend on parameter $r = (r_1, r_2, \ldots, r_m) \subset (0, 1/e)^m$, fulfil the conditions of Theorem 2.3. When applying this theorem $(m-1)$-times, we obtain the Kleinian group $G \subset M_n$:

$$G = G(r) = \langle G_0, J_1 G_0 J_1, J_1 J_2, \ldots, J_1 J_m \rangle, \tag{3.10}$$

which has an invariant component $\Omega_0(G) \subset \Omega_0(G_1) \subset \Omega_0 \cap J_1(\Omega_0)$.

3.11. Remark. On the one hand, the process of parabolic combination, giving the Kleinian group (3.10), brings the closed orbifold $\Omega_0(G)/G$ and, on the other hand, gives a family of quasihomogeneous domains $\Omega_0(r)$, r $(0, 1/e)^n$, in the sense of [16], which turn out to be contractible domains $\Omega_0(G) \subset S^n$, as it will be shown further on. Namely, for such a domain there exists a compact set $F_0 \subset \Omega_0(G)$ whose images by $g \in G$ fill the whole domain $\Omega_0(G)$. It is obvious that any fundamental polyhedron $F_0 \subset \Omega_0(G)$ may be taken as this compact set for the action of the group G in $\Omega_0(G)$.

3.12. Remark. We will use the process of parabolic combination to an arbitrary functional group $G_0 \subset M_n$, having fixed parabolic points z at which G_z invariant balls are tangent externally to each other. Such points are called parabolic Ω-peaks of the group (see [6], pp. 154-155). The relation (3.5) holds not only for parabolic vertices but for Ω-peaks as well (see [6], Theorem 6.15).

3.13. An important subclass of geometrically finite functional groups $G \subset M_n$ with contractible invariant components, described above, is formed by groups whose all the remaining components of the discontinuity set are contractible domains. Such groups will be called b--groups according to suitable Kleinian groups on the plane (see [12] and [18]). From the results of A.V. Tetenov [24] it follows that each b-group $G \subset M_n$, $n \geq 3$ either has exactly one invariant component $\Omega_0 \subset \Omega(G)$ or its discontinuity set is composed of two invariant contractible components (the case of a quasi-Fuchsian group). The following theorems show that the class of b-groups is closed with respect to the process of parabolic combination in question.

3.14. THEOREM. *Let b-groups* $G_1, G_2 \subset M_n$, $n \geq 2$, *with invariant components* Ω_1 *and* Ω_2, *respectively, have the common maximal parabolic subgroup* $H = H_z = G_1 \cap G_2$. *Let further an* $(n-1)$*dimensional sphere* $S \subset \bar{\mathbb{R}}^n$, $S \setminus \{z\} \subset \Omega_1 \cap \Omega_2$, *disjoin* $\bar{\mathbb{R}}^n$ *into balls* D_1 *and* D_2 *which fulfil the conditions of Theorem* 2.2. *Then the group* $G = G_1 \underset{H}{*} G_2$ *is a b-group.*

3.15. THEOREM. *Let a b-group* $G_1 \subset M_n$, $n \geq 2$, *have maximal parabolic subgroups* H_1, H_2, *according to which the balls* B_1 *and* B_2 *from the invariant component* Ω_0, *are strictly invariant. Further, let it be* $A \in M_n$ *such that* $A(B_1) = \text{ext}\,\bar{B}_2$ *and let a group* $G = \langle G_1, A \rangle$ *be the HNN-extension of the group* G_1 *which, in turn, is obtained according to the second Maskit combination* (*Theorem* 2.3). *Then* G *is a b-group*

3.16. Remark. As it is shown below, Theorems 3.14 and 3.15 will remain true if we change S and B_1, B_2 into a topological sphere and topological balls, respectively, and b-groups into groups with an invariant contractible component.

3.17. Proof of Theorem 3.14. From the hypotheses of Theorem 2.2 it follows that the intersection of some neighbourhood of the sphere S with the fundamental domain of the subgroup H is contained in a component $F_o \subset \Omega_1 \cap \Omega_2$ of the fundamental domain $F(G)$ of the group G. Hence and from the invariance of Ω_1 and Ω_2 due to G_1 and G_2, respectively, it follows directly that for any fixed point $x_0 \in S \cap F_o$ and for any generator g of G, points x_0 and $g(x_0)$ can be joined by a path in $\Omega(G)$. Therefore, (see [6], Lemma 3.7) the set

$$\Omega_0 = \text{int} \bigcup_{g \in G} g(\bar{F}_o) \qquad (3.18)$$

is a G-invariant connected component of $\Omega(G)$. We shall prove that Ω_0 is a contractible domain.

We express any $g \in G \setminus H$ in the normal form:

$$g = g_n \cdot \ldots \cdot g_1, \qquad (3.19)$$

where $n > 0$ and either $g_{2i+1} \in G_1 \setminus H$ and $g_{2i} \in G_2 \setminus H$ or $g_{2i+1} \in G_2 \setminus H$ and $g_{2i} \in G_1 \setminus H$. From the group theory it is known that the length $n > 0$ is defined uniquely by an element $g \in G$; let us denote it by $|g|$ and let us set $|g| = 0$ for $g \in H$. We will call g positive, denoting it by $g > 0$, (negative - $g < 0$), if in (3.19) $g_1 \in G_1 \setminus H$ ($g_1 \in G_2 \setminus H$, respectively).

For any fixed decomposition of G

$$G = H \cup (\bigcup_{n,m} a_{nm} H) \cup (\bigcup_{n,m} b_{nm} H), \tag{3.20}$$

where $|a_{nm}| = |b_{nm}| = n$, $a_{nm} > 0$, $b_{nm} < 0$, and, for fixed $n > 0$, we shall consider sets

$$T_n = \bigcup_m (a_{nm}(D_1) \cup b_{nm}(D_2)) \tag{3.21}$$

and without loss of generality we shall assume that $\infty \in \Omega(G)$ and $\infty \notin G(S)$. Then, from the proof of Theorem 2.2 (see [21], pp. 302-308 or [6], pp. 72-76), we have the following properties:

1. $T_n \subset T_{n-1}$ for $n > 1$;
2. Connected components T_{nm} of a set T_n are exactly disjoint balls of the form $a_{nm}(D_1)$ or $b_{nm}(D_2)$;
3. $\sum_{i,j} d^n(T_{ij}) = \sum_{i=1}^{\infty} \sum_j d^n(T_{ij}) < \infty$.

The last property of euclidean diameters $d(T_{ij})$ of balls T_{ij} implies the equality

$$\lim_{i \to \infty} \sum_j d^n(T_{ij}) = 0. \tag{3.22}$$

We denote $\Omega_i' = \Omega(G_i) \setminus \Omega_i$, $i = 1, 2$. From the hypotheses of the theorem it follows that $\Omega_1' \cap \Omega_2' = \emptyset$, $\Omega_1' \subset D_2$, $\Omega_2' \subset D_1$, $\Omega_0 \cap (\Omega_1' \cup \Omega_2') = \emptyset$. Besides, we can say that Ω_1' is tangent to Ω_2' at the point z which is fixed for the parabolic subgroup H in the sense that there exist balls $B_1 \subset \Omega_1'$, $B_2 \subset \Omega_2'$ which are strictly invariant with respect to H in G_1 and G_2, respectively, and tangent to each other at the point z (see 3.1).

We consider together with sets T_n from (3.21) the following sets:

$$O_n = \bigcup_m (a_{nm}(\Omega_2') \cup b_{nm}(\Omega_1')). \tag{3.23}$$

From the properties of Ω_i' and T_n properties of sets O_n follow directly:

1) $O_n \subset T_n \subset T_{n-1} \setminus O_{n-1}$;

2) Connected components of the set O_n are contractible domains, which generate disjoint families O_{nm} of the form $a_{nm}(\Omega_2') \subset a_{nm}(D_1)$ or $b_{nm}(\Omega_1') \subset b_{nm}(D_2)$. Moreover, the unique component $a_{nm}(\Omega_1')$ (i.e. the one containing the ball $a_{nm}(B_1)$) is tangent to the unique component $b_{n-1,m}(\Omega_1')$ (i.e. the one containing the ball $b_{n-1,m}(B_1)$) at the unique point which is the image of z.

From Theorem 2.2 it follows that the fundamental domain $F(G)$ of the group G splits into three parts: $F = F_o \cup F_1 \cup F_2$, $F_o \subset \Omega_0$, $F_1 \subset \Omega_1'$, $F_2 \subset \Omega_2'$. What more, because we have $\Omega_i' = \operatorname{int} G_i(\overline{F}_i)$, $i = 1, 2$, then from (3.18) and the properties of O_i, $i \geq 1$, and $O_o = \Omega_1' \cup \Omega_2'$ given above it follows that

$$\Omega_0 = \operatorname{int}(\overline{\mathbb{R}}^n \setminus \bigcup_{i=0}^{\infty} O_i). \tag{3.24}$$

From the conditions of the theorem we have $S\setminus\{z\}\subset\Omega_1\cap\Omega_2=\Omega_*=$ $=\operatorname{int}\overline{\mathbb{R}}^n\setminus O_o$. Both this property and contractibility of invariant components Ω_1 and Ω_2 imply the contractibility of the domain Ω_*. Hence we get that in balls D_1 and D_2 there exist H-invariant balls $D_i'\subset D_i$, tangent to each other at the point z such that $\partial D_i'\setminus\{z\}\subset\Omega_*$. When assuming $z=\infty$ we can choose balls D_1' and D_2' (conjugating the group G, if necessary) so that their boundary spheres are in distance 1 from S and their complement is H-invariant neighbourhood V of the sphere $S\setminus\{z\}$ which is mentioned in the hypotheses of Theorem 2.2.

We denote by T_n' subsets in T_n obtained by changing in (3.21) balls D_i with balls D_i'. Moreover, the inclusions, analogous to the properties T_n,

$$T_{n+1}\subset T_n'\subset T_n\subset T_{n-1}',\quad O_n\subset T_n'\subset T_n \tag{3.25}$$

are evident. Let us consider standard closed balls in $\overline{\mathbb{R}}^n$, tangent to each other:

$$\begin{aligned}&\Delta_1=\{x\in\overline{\mathbb{R}}^n:\ x_n\le 0\}\subset\Delta=\{x\in\overline{\mathbb{R}}^n:\ x_n\le 1\},\\&\Delta_2=\{x\in\overline{\mathbb{R}}^n:\ x_n\ge 2\}.\end{aligned} \tag{3.26}$$

For these balls we define standard diffeomorphism (which commutes with the group of euclidean isometries in $\mathbb{R}^n$, which in turn, preserves $\mathbb{R}^n_+$):

$$\phi:\ \operatorname{ext}(\Delta_1\cup\Delta_2)\to\operatorname{ext}\Delta_1 \tag{3.27}$$

such that $\phi|_\Delta=\mathrm{id}$ and for every $y=\phi(x)$, $y'=\phi(x')$ we have the following equalities:

$$y_i=y_i'=x_i\quad\text{for}\quad i\ne n\quad\text{and}\quad y_n=y_n'=\phi_o(x_n),$$

where $\phi_o:(0,2)\to\mathbb{R}_+$ is a diffeomorphism, such that $\phi_o(x)=x$ for $x\in(0,1]$, and $\lim\limits_{x\to 2}\phi_o(x)=\infty$.

Now, we can define a sequence of diffeomorphisms

$$\phi_k:\ \operatorname{int}\overline{\mathbb{R}}^n\setminus\bigcup_{i=0}^{k}O_i\to\operatorname{int}\overline{\mathbb{R}}^n\setminus\bigcup_{i=0}^{k-1}O_i. \tag{3.28}$$

Each of ϕ_k is the identity on $\overline{\mathbb{R}}^n\setminus T_k$ and on each of disjoint balls T_{km} (which is a component of T_k, tangent to some component of $O_{k-1,\ell}$ and containing a family O_{km}) we construct it from a standard diffeomorphism (3.27) and applying conjugacy by a Möbius transformation h_{km} such that

$$\begin{aligned}&h_{km}(T_{km})=\overline{\mathbb{R}}^n\setminus\Delta_1,\qquad h_{km}(T_{km}')=\operatorname{int}\Delta_2,\\&h_{km}(\overline{T}_{km}\cap\overline{T}_{km}')=\infty.\end{aligned} \tag{3.29}$$

By diffeomorphisms ϕ_k we define the limit diffeomorphism

$$f = \lim_{k\to\infty} \phi_1 \circ \phi_2 \circ \ldots \circ \phi_k. \tag{3.30}$$

From (3.24) we can easily see that the diffeomorphism f maps the invariant component Ω_o of the group G on the contractible domain $\Omega_* = \Omega_1 \cap \Omega_2$. This shows that the group G is a b-group.

For the proof it is sufficient to show that the sequence $f_k = \phi_1 \circ \phi_2 \circ \ldots \circ \phi_k$ is convergent to the diffeomorphism f.

Therefore, let us consider, for any $\varepsilon > 0$, such a number k that for each $i \geq k$ and for each j, and for Lebesgue measure μ_m we have

$$\mu_n(T_i) < \varepsilon, \quad d(T_{ij}) < \varepsilon. \tag{3.31}$$

The existence of such k follows from (3.22). The property (3.31) implies that for every two numbers $m > \ell \geq k$ the mapping $f_\ell^{-1} \circ f_m = \phi_{\ell+1} \circ \ldots \circ \phi_m$ differs from the identity only on the set $T_{\ell+1}$, the measure of which is less than ε. Besides, since this mapping sends each component $T_{\ell+1,j}$ to itself, then from (3.31) we get

$$|f_\ell^{-1} f_m(x) - x| < \varepsilon \qquad \text{for } x \in T_{\ell+1}. \tag{3.32}$$

From (3.31) and (3.32) it follows that for each $\varepsilon > 0$ there exists such a number k, that for all $m > \ell \geq k$ diffeomorphisms f_m and f_ℓ differ from each other only on the set $T_{\ell+1}$, $\mu_n(T_{\ell+1}) < \varepsilon$, where they differ from each other not more than ε and $T_{i+1} \subset T_i$ for each i. This proves the uniform convergence of f_i to the homeomorphism (3.30). The smoothness of f follows from the fact that in sufficiently small neighbourhood of any point x the mapping f coincides with some diffeomorphism f_i for sufficiently large subscript i. This completes the proof of 3.14.

3.33. Proof of Theorem 3.15. It is based on the concepts of the proof of the previous theorem and Maskit's technique [21]. That is, each element $g \in G \setminus \{id\}$ is decomposed (not uniquely) to its normal form

$$g = A^{\alpha_n} \circ g_n \circ \ldots \circ A^{\alpha_1} \circ g_1, \tag{3.34}$$

where $g_i \in G_1$, $g_2, \ldots, g_n \neq id$, $\alpha_1, \ldots, \alpha_{n-1} \neq 0$; if $\alpha_i < 0$ and $g_{i+1} \in H_1$, then $\alpha_{i+1} \leq 0$; if $\alpha_i > 0$ and $g_{i+1} \in H_2$, then $\alpha_{i+1} \geq 0$. Moreover, for each such decomposition, the following number is invariant:

$$|g| = \sum_{i=1}^{n} |\alpha_i|, \tag{3.35}$$

which we shall call the length of the element g. It is obvious that the length $|g|$ is invariant with respect to superposition of g with any element $g_1 \in G_1$.

We shall call (3.34) a positive normal form $(g > 0)$ if $g_1 \in H_1$ and $\alpha_1 > 0$. In the opposite case we shall treat the form (3.34) as nonpositive one $(g \leq 0)$. Analogously, $g < 0$, if in (3.34) $g_1 \in H_2$ and

$\alpha_1 < 0$; in the opposite case g is nonnegative ($g \geq 0$). These properties of $g \in G \setminus \{id\}$, the same as its length, do not depend on the choice of decomposition (3.34) (see [21]).

We shall consider the decompisition of G modulo the subgroup H_1 (see 3.20). Then all the representatives of the fixed class gH_1 have the same length and simultaneously they are positive or nonpositive. We shall select from each nonpositive class of the length n its representative a_{nm}. Analogously, for the decomposition G modulo subgroup H_2 from each nonnegative class gH_2 of the length n, we choose a representative b_{nm}. For each fixed $n = 0, 1, 2, \ldots$, we shall consider the following sets, similar to (3.21):

$$T_n = \bigcup_m a_{nm}(B_1) \cup b_{nm}(B_2). \tag{3.36}$$

It is known [21] that $T_n \subset T_{n-1}$ for $n > 0$ and all the connected components T_{nm} of the set T_n, $n > 0$, are disjoint balls of the form $a_{nm}(B_1)$ or $b_{nm}(B_2)$. Let us assume, without the loss of generality, that $\infty \in \Omega(G)$ and $\infty \notin G(\partial B_1 \cup \partial B_2)$; hence we have also the following property of euclidean diameters $d(T_{ij})$ of components $T_{ij} \subset T_i$ (see (3.22)):

$$\lim_{i \to \infty} \sum_j d^n(T_{ij}) = 0. \tag{3.37}$$

Together with the sets (3.36) we shall consider for $n = 0$ the set $0_o = \Omega' = \Omega(G_1) \setminus \Omega_0$ and for $n \geq 1$ the sets

$$0_n = \bigcup_m (a_{nm}(\Omega') \cup b_{nm}(\Omega')). \tag{3.38}$$

Besides, we shall consider smaller balls $B_i' \subset B_i$, $i = 1, 2$, which are strictly invariant with respect to subgroups $H_i \subset G_1$ and such that their images $A(B_1')$ and $A^{-1}(B_2')$ are strictly invariant with respect to subgroups H_2 and H_1, respectively. We define subsets $T_n' \subset T_n$, $n \geq 0$, by changing in definition (3.36) the balls B_i with the balls B_i', $i = 1, 2$. From the definition of T_n, T_n' and 0_n, inclusions for $n \geq 0$ follow:

$$0_{n+1} \subset T_n' \subset T_n, \quad T_{n+1} \subset T_n' \setminus 0_{n+1}. \tag{3.39}$$

Besides, let us observe that the family 0_n of contractible domains generates subfamilies 0_{nm} of the form $a_{nm}(\Omega')$ or $b_{nm}(\Omega')$. Moreover, if components of Ω' to which the balls B_1 and B_2 are tangent (in the sense that there exist balls $B_i^* \subset \Omega'$, tangent to balls B_i at parabolic vertices z_i, invariant with respect to subgroups H_i, $i = 1, 2$; see (3.1)) are denoted by Ω_1 and Ω_2, then 0_{nm} have the following property: For every $n \geq 0$ in $a_{n+1,m}(\Omega')$ there exists the unique component ($= a_{n+1,m}(\Omega_2)$), tangent to the family $a_{nm}(\Omega')$, namely tangent to its unique component $a_{nm}(\Omega_1)$. Analogously, $b_{n+1,m}(\Omega')$ "is tangent" to the family $b_{nm}(\Omega')$ at the unique point and this tangency defines components $b_{n+1,m}(\Omega_1)$ and $b_{nm}(\Omega_2)$. Moreover, the tangency points are parabolic fixed points of subgroups which are conjugate in G to the subgroups H_1 and H_2.

Now, in the formulation of Theorem 2.2, let F_* be a component of the fundamental set $F(G) = \operatorname{int}\{F_o \cap \operatorname{ext}(\bar{B}_1 \cup \bar{B}_2)\}$, adjacent to the balls B_1 and B_2. Then by [6, Lemma 3.7], analogously as (3.18), it follows that the set

$$\Omega_* = \operatorname{int} \bigcup_{g \in G} g(\bar{F}_*) \tag{3.40}$$

is a G-invariant connected component of the discontinuity set $\Omega(G)$. We will show that Ω_* is diffeomorphic with the domain Ω_0. It will prove that G is a b-group.

As in the proof of Theorem 3.14 (see 3.24) we have from the properties of the sets O_k (see 3.38 and 3.39) the equality

$$\Omega_* = \operatorname{int} \bar{\mathbb{R}}^n \setminus \bigcup_{i=0}^{\infty} O_i . \tag{3.41}$$

Analogously as in (3.28) and (3.29), we will define a sequence of diffeomorphisms ϕ_k, $k \geq 1$:

$$\phi_k : \operatorname{int} \bar{\mathbb{R}}^n \setminus \bigcup_{i=0}^{k} O_i \to \operatorname{int} \bar{\mathbb{R}}^n \setminus \bigcup_{i=0}^{k-1} O_i , \tag{3.42}$$

out of which each is an identity on the set $\mathbb{R}^n \setminus T_{k-1}$ and on each disjoint ball $T_{k-1,m}$ (a component of the set T_{k-1}) we define each ϕ_k from the standard diffeomorphism (3.27) by conjugacy with the help of some Möbius mapping $h_{km} \in M_n$ such that

$$h_{km}(T_{k-1,m}) = \bar{\mathbb{R}}^n \setminus \Delta_1 , \quad h_{km}(T'_{k-1,m}) = \operatorname{int} \Delta_2 ,$$

$$h_{km}(\bar{T}_{k-1,m} \cap \bar{T}'_{k-1,m}) = \infty .$$

In this way, the defined diffeomorphisms ϕ_k give the limit mapping

$$f = \lim_{i \to \infty} \phi_1 \circ \phi_2 \circ \ldots \circ \phi_i : \Omega_* \to \Omega_0 . \tag{3.42}$$

Moreover, f as a limit of diffeomorphism $f_k = \phi_1 \circ \ldots \circ \phi_k$, is a diffeomorphism. This fact can be proved analogously to (3.30) (when changing the properties (3.22) to the property (3.37) of the set T_n from (3.36)). This completes the proof of Theorem 3.15.

4. DOUBLES OF ATOROIDAL MANIFOLDS AND UNIFORMIZATION BY B-GROUPS

4.1. Let M be an irreducible homotopic atoroidal compact n-dimensional ($n \geq 3$) manifold (orbifold) whose boundary consists of the sum of the finite number of (n-1)-dimensional tori T^{n-1} and let $\hat{M}$ be a closed manifold (orbifold) which is a double of M. In connection with the problem of conformal uniformization, the following question arises. How to define a conformal structure on the double $\hat{M}$, which, what more, could be uniformizable with the use of a uniformizable conformal structure on $\operatorname{int} M$?

Let us observe that for a three-dimensional manifold M with the prescribed properties, the uniformizable conformal structure on $\operatorname{int} M$ can be defined by a hyperbolic metric transforming $\operatorname{int} M$ into the complete hyperbolic manifold with a finite volume. It results from the uniformization Thurston theorem (see [26] and [22]). However, the double $\hat{M}$ is not (for all $n \geq 3$) a hyperbolic manifold, because $\pi_1(\hat{M})$ contains free abelian subgroup of rank $n-1$, and $\hat{M}$ is a closed manifold.

If a conformal structure on $\operatorname{int} M$ is quasiconformally equivalent to unique conformal structure defined by a hyperbolic metric, then there exists a quasi-Fuchsian geometrically finite group $G \subset M_n$ with the components Ω_0 and Ω_1 of the discontinuity set $\Omega(G) = \Omega_0 \cup \Omega_1$ such that $\operatorname{int} M$ is conformally equivalent to a manifold (orbifold) Ω_0/G with a natural conformal structure. The generalization of such a situation are conformal structures on $\operatorname{int} M$, uniformizable by b-groups, i.e. the structures conformally equivalent to the structure Ω_0/G, where $\Omega_0 \subset \Omega(G)$ is an invariant component of the geometrically finite Kleinian group, whose discontinuity set consists of contractible components.

For such conformal structures on M the problem of conformal uniformization of the double $\hat{M}$ is solved by the following

THEOREM 4.2. *Let on the interior of an irreducible homotopically atoroidal compact* n*-dimensional* $(n \geq 3)$ *orbifold* M*, having as the boundary* ∂M *- the sum of* m $(m \geq 1)$ *of* $(n-1)$*-dimensional tori, a conformal structure* C_0*, uniformizable by a* b*-group* $G_0 \subset M_n$*, be given. Then on the double* $\hat{M}$ *of the orbifold* M *there exists the conformal structure* C*, uniformizable by a* b*-group* $G \subset M_n$*. Moreover, in* $(\hat{M}, C)$ *there exists a family of conformally imbedded mutually disjoint* $(n-1)$*-dimensional euclidean tori* $T_1, \ldots, T_m$ *such that* $(\hat{M} \setminus \bigcup_{i=1}^{m} T_i, C)$ *is the sum of two conformal orbifolds equivalent conformally to the orbifold* $(\operatorname{int} M, C_0)$.

4.3. Proof. It suffices to use the results obtained in Section 3, concerning the parabolic combination of groups with respect to which the class of b-groups is closed. Indeed, by the hypotheses of the theorem, we have that the orbifold $(\operatorname{int} M, C_0)$ is conformally equivalent to the orbifold Ω_0/G_0, where $G_0 \subset M_n$ is a b-group with the invariant component $\Omega_0 \subset \Omega(G_0)$ which has m classes of conjugacy of maximal parabolic subgroups of rank $n-1$.

We shall use to the group G_0 the procedure of parabolic combination, described in 3.6. Then the direct conclusion of Theorems 3.14 and 3.15 is the fact that the Kleinian group $G \subset M_n$, resulted from this procedure and given in (3.10), is a b-group with the invariant component $\Omega_* \subset \Omega(G)$.

From the construction (3.10) of the group G it is easily seen that the orbifold Ω_*/G is homeomorphic with the double $\hat{M}$ in the orbifold category. The canonical projection $\pi: \Omega_* \to \Omega_*/G$ induces on the orbifold a conformal structure uniformizably by the b-group G.

We can observe that, by the projection π, conformally imbedded (n-1)-dimensional tori $T_1, \dots T_m$ (non-contractible in Ω_*/G, i.e. inducing monomorphisms $\pi_1(T_1) \to G$), whose images are mutually disjoint in Ω_*/G, correspond to G-orbits of punctured strictly G_{z_i}-invariant spheres $\partial B_i(r_i) \setminus \{z_i\}$, having the property (3.7).

By getting rid of G-orbits of all punctured spheres from Ω_* and considering the group of isotropy Γ of any obtained connected component Ω_Γ, we get orbifolds Ω_Γ/Γ, each of which turns out to be one of two connected components of $(\Omega_*/G) \setminus \bigcup_{i=1}^m T_i$.

Moreover, we should note that the conformal structure of an orbifold Ω_Γ/Γ is conformally equivalent to the conformal structure C_o of the orbifold $M \cong \Omega_0/G_o$. It follows from the contruction of diffeomorphisms $f\colon \Omega_\Gamma \to \Omega_0$ in the proof of Theorems 3.14 and 3.15 (see (3.30), (3.42)) which agree with an action of the group Γ and G_o (by properties of the standard diffeomorphism (3.37)). In other words, the group Γ acts in the conformal structure of Ω_Γ, lifted from Ω_0 by the diffeomorphism f, as a group of conformal automorphisms. This proves that the components of the conformal orbifold $(\hat{M} \setminus \bigcup_{i=1}^m T_i, C)$ with the structure C, introduced by the group G, are conformally equivalent to the conformal orbifold (M, C_o).

4.4. Remark. As it is seen from the proofs of Theorems 3.14 and 3.15, the class of Kleinian groups G, having invariant contractible components $\Omega_0 \subset \Omega(G)$, is closed under the process of parabolic combination. This shows that Theorem 4.2 is also true for conformal structures uniformizable by such Kleinian groups.

5. GEOMETRICAL DEFORMATIONS OF CONFORMAL STRUCTURES

5.1. Let M be a closed irreducible conformal n-orbifold, $n \geq 3$, which is a parabolic double of an orbifold $\Omega_0(G_o)/G_o = M_o$, where $G_o \subset \mathcal{M}_n$ is a b-group (see Theorem 4.2). Moreover, the conformal structure on M is canonically induced by the projection

$$\pi\colon \Omega_0 \to M = \Omega_0/G,$$

where $\Omega_0 \subset \overline{\mathbb{R}}^n$ is an invariant component of the b-group $G = G(r^o)$ (see 3.10), $r^o = (r_1^o, r_2^o, \dots, r_m^o) \in I^m = (0, 1/e)^m$, and m is the number of parabolic ends in $\Omega_0(G_o)/G_o = M_o$. When changing the parameter $r \in I^m$, we obtain various conformal structures $C(r) \in \mathcal{C}(M)$ on the orbifold M. It turns out that these structures are quasiconformally equivalent and depend smoothly on the parameter r.

Let us assume that groups $G = G(r^o)$ and $G(r)$ from (3.10), where $r \in I^m \setminus \{r^o\}$, act in the hyperbolic space $H^{n+1} \cong \mathbb{R}_+^{n+1}$. Let us consider their fundamental polyhedrons $P^{n+1} = P^{n+1}(G)$ and $P_r^{n+1} = P^{n+1}(G(r))$, symmetric with respect to n-planes $S_1 = S_1(r^o)$ and $S_1(r)$, and having as their faces other planes $S_i, J_1 J_i(S_i)$ and $S_i(r), J_1(r)J_i(r)(S_i(r))$, $i = 2, \dots, m$, respectively, In other words, these polyhedrons are obtained by combination from the only one polyhedron $P^{n+1}(G_o)$ (see (3.9)). It is obvious that the mapping

$$\phi_r: P^{n+1}(G) \to P^{n+1}(G(r)),$$

in the neighbourhoods of the above faces turns out to be the uniform stretching along horocycles of the space H^{n+1} which are at the same time orthogonal to $S_1(r^o)$ and $S_1(r)$, $S_i(r^o)$ and $S_i(r)$, $J_1 J_i(S_i(r^o))$ and $J_1 J_i(S_i(r))$, as well as the identity inside those neighbourhoods. The mapping ϕ_r is a $q(r)$-quasiconformal mapping and compactible with the action of the group G:

$$\phi_r(g(x)) = \chi(g)(\phi_r(x)), \tag{5.2}$$

where $g \in G$ idetifies some faces of P^{n+1}; $x, \phi_r(x) \in \bar{P}^{n+1}$, and $\chi: G \to G(r)$ is the canonical isomorphism.

Hence the mapping ϕ_r can be extended by a symmetry (using (5.2)) on the whole space H^{n+1} to a $q(r)$-quasiconformal mapping $\tilde{f}_r$: $H^{n+1} \to H^{n+1}$ which is compatible with the action of G. Moreover, the coefficient of quasiconformality $q(r)$ is a continuous function of r and $\lim_{r \to r^o} q(r) = 1$. The mapping $\tilde{f}_r$ can be extended on the absolute $\partial H^{n+1} = \bar{\mathbb{R}}^n$ to $q(r)$-quasiconformal automorphism $f_r: \bar{\mathbb{R}}^n \to \bar{\mathbb{R}}^n$, which gives us deformation of the group G:

$$f_r^*: G \to G(r) = f_r\, G\, f_r^{-1}. \tag{5.3}$$

5.4. <u>LEMMA</u>. *The mapping* $\Phi_o: I^m \to \mathrm{Def}\, G$, $\Phi_o(r) = f_r^*$, *is real-analytic.*

5.5. Proof. We will write down f_r^* in a more explicit form. Let (as in 3.6) $J_i = J_i(r_i^o)$, $J_i(r_i)$ be reflections with respect to n-planes $S_i = S_i(r_i^o)$ and $S_i(r_i)$ with radii r_i^o and r_i from (3.4), tangent to parabolic points z_i of the group G_o, where $r = (r_1, \ldots, r_n) \in I^m = (0, 1/e)^m$. Then the dependence on r_i is defined by the deformation (3.8):

$$f_{r_1}^*: G_1 = G_1(r_1^o) = \langle G_o, J_1 G_o J_1 \rangle \to G_1(r_1),$$

which can also be defined as follows:

$$f_{r_1}^*(g_o) = g_o, \quad f_{r_1}^*(g) = W(r_1) \circ g \circ W(r_1)^{-1}, \tag{5.6}$$

where $g_o \in G_o$, $g \in J_1 G_o J_1$, and the parabolic mapping $W(r_1) = J_1 \circ J_1(r_1) \in M_n$ with an fixed point z_1 is conjugate in M_n to the mapping

$$x \to x + e_n \cdot \ln(r_1^o/r_1)^2. \tag{5.7}$$

The dependence on $r_2, \ldots, r_m$ is given by the deformations of HNN-extensions of the group G_1 by the mappings $J_1 J_i$, $i = 2, \ldots, m$, to the group (3.10) and hence these deformations can be defined in the form:

$$f^*_{r_i}(g_1) = g_1, \quad f^*_{r_i}(J_1 J_i) = J_1 \circ W(r_i) \circ J_i \circ W(r_i)^{-1}, \tag{5.8}$$

where $g_1 \in G_1$, $i = 2, \ldots, m$, and the parabolic mappings $W(r_i) \in M_n$ with fixed points z_i map S_i into $S_i(r_i)$ and are conjugate in M_n to

$$x \to x + e_n \cdot \ln(r_i^o / r_i). \tag{5.9}$$

Now the statement of Lemma results from (5.6) and (5.9) for the deformation (5.3) and real-analytic dependence of (5.7) and (5.9) in the cube I^m with respect to each variable r_i.

5.10. THEOREM. *Let a conformal orbifold* M *be the parabolic double of the orbifold* $M_o = \Omega_0 / G_o$ *(according to its* m *parabolic ends* $m \geq 1$*), where* $G_o \subset M_n$, $n \geq 3$, *is a* b*-group with an invariant component* Ω_0. *Then for* b*-group* $G \subset M_n$ *from* (3.10), *uniformizing* M, *there exists real-analytic imbedding* $\Phi_o : I^m \to \mathrm{Def}\, G = \mathrm{Hom}(G, M_n)/M_n$, *inducing the imbedding of* m*-dimensional ball in the space* C(M) *of marked conformal structures on* M.

5.11. P r o o f. By Lemma 5.4 for the proof it suffices to explain the triviality conditions of the deformation f^*_r. From the construction of the group of the type (3.10) by combination and from the proofs of Theorems 3.14 and 3.15 we can see that the limit sets of the groups $G = G(r^o)$ and $G(r) = f^*_r(G)$ are symmetric with respect to the sphere ∂S_i and $\partial S_i(r)$, $i = 1, \ldots, m$, respectively. Hence and from [14, Proposition 12], there results the existence of a loxodromic element $g_i \in G$, whose axis ℓ_i intersects transversely S_i. For the element g_i and for $z \in \ell_i \subset H^{n+1}$ we have the following inequality:

$$d(z, g_i(z)) \gtrless d(z, f^*_{r_i}(g_i)(z)),$$

where the inequality direction depends on the sign of $(r_i - r_i^o)$. This shows that the length of hyperbolic translation g_i decreases or increases when $r_i \neq r_i^o$, i.e. $f^*_{r_i}$ is a non-trivial deformation. The last statement of Theorem concerning the imbedding into C(M) results directly from the holonomy theorem of Goldman-Lock (Theorem 2.11) about the local homeomorphism of the space C(M) and Def G.

5.12. In our case the statement of Theorem 5.10 about the imbedding $I^m \to C(M)$ can be more clarified. It means that a conformal structure $C \in C(M)$, which is conformally equivalent to the structure $\Omega_0(G)/G$, corresponds by this imbedding to each $r \in I^m \setminus \{r^o\}$, i.e. to each non-trivial deformation $f^*_r \in \mathrm{Def}\, G$. This deformation can be defined in the following way:

Let $M^{n+1} = [H^{n+1} \cup \Omega(G)]/G$ be an orbifold with a boundary, whose component is our double M. From 4.3 it follows that in M there exists a family of conformally imbedded, non-contractible, mutually disjoint euclidean $(n-1)$-dimensional tori $T_1, \ldots, T_m \cong \mathbb{R}^{n-1}/\mathbb{Z}_{n-1}$ into which the spheres $\Omega(G) \cap \partial S_i = \partial S_i \setminus \{z_i\}$ are mapped by the projection $\pi: \Omega(G) \to \Omega(G)/G$. These spheres are strictly invariant with respect to parabolic subgroups $G_{z_i} \subset G$. This shows that in M^{n+1} we have imbedded m mutually disjoint non-contractible n-dimensional totally geodesic (in $\mathrm{int}(M^{n+1})$) cylinders:

$$N_i = [S_i \cup (\partial S_i \setminus \{z_i\})]/G_{z_i}, \qquad \partial N_i = T_i. \tag{5.13}$$

The deformation f_r^* has been given by the quasiconformal mapping $\tilde{f}_r: \bar{H}^{n+1} \to \bar{H}^{n+1}$ compatible with the group G. This shows that to f_r^* there corresponds the quasiconformal mapping

$$F_r: M^{n+1} \to M^{n+1}(r) = [H^{n+1} \cup \Omega(G(r))]/G(r),$$

which in small neighbourhoods $U(N_i)$ is a uniform stretching along horocycles in M^{n+1}, orthogonal to the geodesic cylinders $N_i \subset M^{n+1}$ from (5.13), as well as the identity in the complement of $U(N_i)$. The restriction $f_r|\partial M^{n+1}$ does give us the needed quasiconformal deformation of the structure $C \in \mathcal{C}(M)$.

5.14. Remark. The mapping F_r of the orbifold M^{n+1} gives us also the quasiconformal deformation of the hyperbolic structure of $\mathrm{int}\, M^{n+1} = H^{n+1}/G$. In this sense we can say that the imbeddings Φ_o: $I^m \to \mathrm{Def}\, G$ and $I^m \to \mathcal{C}(M)$ are induced by deformations of the hyperbolic orbifold H^{n+1}/G along totally geodesic, mutually disjoint cylinders $N_i \cap \mathrm{int}\, M^{n+1} = N_i \setminus \partial N_i$.

5.15. Let us assume that a conformal structure of the orbifold $M_o = \Omega_o/G_o$ ($G_o \subset \mathcal{M}_n$ is a b-group with the invariant component Ω_o) from Theorem 5.10 is quasiconformally equivalent to a conformal structure of hyperbolic orbifold H^n/G_* for some discrete subgroup $G_* \subset \mathrm{Isom}\, H^n$, $G_* \cong G_o$. Such conformal structures on the orbifold M_o will be called quasiconformally hyperbolized. For them, the estimation of the dimension of the ball imbedded in the space $\mathcal{C}(M)$ of conformal structures on the double M of the orbifold M_o, given by 5.10, can be sharpened in a siquificant way:

5.16. THEOREM. *Let the conformal orbifold* M *be a parabolic double of the orbifold* M_o *with quasiconformally hyperbolized conformal structure (with respect to its* m *parabolic ends,* $m \geq 1$). *Then for* b-*group* $G \subset \mathcal{M}_n$ *from* (3.10), *which uniformizes* M, *there exists a real-analytic imbedding* $\Phi: I^{m+k} \to \mathrm{Def}\, G = \mathrm{Hom}(G, \mathcal{M}_n)/\mathcal{M}_n$ *inducing the imbedding* I^{m+k} *in* $\mathcal{C}(M)$. *Moreover, the number* k *is equal to the number of mutually disjoint, non-contractible, totally geodesic* $(n-1)$-*dimensional suborbifolds in* M_o *with respect to the hyperbolic metric in question.*

5.17. The proof of the Theorem can be obtained by the construction of an imbedding $\Phi = (\Phi_o, \Phi_1)\colon I^m \times I^k \to \mathrm{Def}\, G$, where $\Phi_o\colon I^m \to \mathrm{Def}\, G$ is the imbedding from Theorem 5.10, and really analytical imbedding $\Phi_1\colon I^k \to \mathrm{Def}\, G_o \subset \mathrm{Def}\, G$ is defined by the imbedding $\Phi_1^*\colon I^k \to \mathrm{Def}\, G^*$, $G^* \subset \mathrm{Isom}\, H^n$. Moreover, the group G^* is considered as the subgroup of M_n (acting in the upper space $\mathbb{R}_+^n$), quasiconformally equivalent to $G_o \subset M_n$, where $M_o = \Omega_o/G_o$.

If we want to construct Φ_1^*, we have the choice of a number of independent approaches. For the first time such deformations of the conformal structure of an hyperbolic orbifold of the dimension $n \geq 3$ have been constructed by the author in [11, 2, 3]. The development of this geometrical approach (which makes use of the geometry of accute fundamental polyhedron of the group G_* - see [1]) is the construction of the Φ_1^*-imbedding (see [8, 10])(where, however, the number k is given in another sense, although close to the above one).

Another approach to the construction of a Φ_1^*-imbedding, which makes use not of the geometry of fundamental polyhedron of the group G_* but of the geometry of the manifold H^n/G_* as well as of the limiting process on a set of quasiconformal mappings of hyperbolic space, has been given by Kourouniotis (see [17]).

Finally, the existence of such a Φ_1^*-imbedding in the case of closed M_o has been announced (without proof) in the paper by Johnson and Millson (see [15]).

5.18. Let us observe that in the case of 2-dimensional surfaces M_o, M, the dimension $m+k$ of the ball imbedded (from Theorem 5.16) in $C(M)$ has the following geometrical sense: the number $m \geq 1$ is the number of points (or closed discs) whose cutting out from the surface S_g gives us the surface $\mathrm{int}\, M_o$. Moreover, the number k equals $3g-3$, where g is the genus of the surface S_g, i.e. the number of mutually disjoint, closed geodesics on S_g.

5.19. When examining the geometry of convex hull $H_G \subset H^{n+1}$, of the limit set of the group G_o M_n from Theorem 5.16 ($M_o = \Omega_o/G_o$), we can show that in the case $n \geq 3$ there exist other deformations of the conformal structure of M_o (i.e. of the structure of double M, as well), which are lying in the image of $\Phi_1^*(I^k)$ and of $\Phi(I^{m+k})$. In the case $n=2$ it is evident from 5.18. It will to appear elsewhere.

5.20. <u>Remark</u> to Theorem 2.11 (added in proof). In spite of the fact that, in general (for representations ρ having an isotropy subgroup $Z(\rho) \subset G$, different from the centre of G), the mapping hol is not a local homeomorphism, it is a local homeomorphism for conformal structures with developments d which map $\tilde{M}$ onto a domain $d(\tilde{M}) \subset \mathbb{R}^n$, whose complement does not lie on the sphere S^k, $1 \leq k \leq n-2$ (cf. [15]).

References

[1] ANDREEV, E.M.: 'The intersection of the planes of the faces of polyhedra with sharp angles' [in Russian], *Mat. Zametki* 8 (1970), no. 4, 521-527.

[2] APANASOV, B.N.: 'Kleinian groups, Teichmüller space and Mostow's rigidity theorem' [in Russian], *Sibirsk. Mat. Zh.* 21 (1980), no. 4, 3-15; English transl. in *Siberian Math. J.* 21 (1980).

[3] ——: 'Nontriviality of Teichmüller space for Kleinian group in space', in: *Riemann surfaces and related topics: Proceedings of the 1978 Stony Book Conference* (Ann. of Math. Studies 97), Princeton Univ. Press, Princeton 1981, pp. 21-31.

[4] ——: 'Geometrically finite transformation groups of the space' [in Russian], *Sibirsk. Mat. Zh.* 23 (1982), no. 6, 16-27; English transl. in *Siberian Math. J.* 23 (1982).

[5] ——: 'Geometrically finite hyperbolic structures on manifolds', *Annals of Global Analysis and Geometry* 1 (1983), no. 3, 1-22.

[6] ——: *Discrete transformation groups and structures of manifolds* [in Russian], Izd. 'Nauka', Sibirsk. Otdel., Novosibirsk 1983; extended English transl. by Reidel Publ. Co., to appear.

[7] ——: 'Cusp ends of hyperbolic manifolds', *Annals of Global Analysis and Geometry* 3 (1985), no. 1, 1-12.

[8] ——: 'Thurston's bends and geometric deformations of conformal structures', in *Complex analysis and applications'85*, ed. by L. Iliev and I. Ramadanov, Publishing House of Bulgarian Acad. Sci., Sofia 1987.

[9] ——: 'The effect of dimension four in Aleksandrov's problem of filling a space by polyhedra', *Annals of Global Analysis and Geometry* 4 (1986), no. 2, 1-22.

[10] ——: 'Deformations of conformal structures on manifolds connected with totally geodesic submanifolds' [in Russian], *Dokl. Akad. Nauk SSSR*, 293 (1987), no. 1, 14-17.

[11] —— and TETENOV, A.V.: 'The existence of nontrivial quasiconformal deformations of Kleinian groups in space' [in Russian], *ibid.* 239 (1978), no. 1, 14-17; English transl. in *Soviet Math. Dokl.* 19 (1978), 242-245.

[12] BERS, L.: 'On Hilbert's 22nd problem', in: *Mathematical developments arising from Hilbert problems* (Proc. Symposia of Pure Math. 28, part 2), Amer. Math. Soc., Providence 1976, pp. 559-609.

[13] GOLDMAN, W.M.: *Affine manifolds and projective geometry on surfaces*, Senior Thesis, Princeton Univ. Press, Princeton 1977.

[14] GREENBERG, L.: 'Discrete subgroups of the Lorentz group', *Math. Scand.* 10 (1962), no. 1, 85-107.

[15] JOHNSON, D. and MILLSON, J.J.: 'Deformation spaces associated to compact hyperbolic manifolds', in: *Discrete groups in geometry and analysis*, Proc. of the Conf. at Yale Univ. in honour of G.D. Mostow (Progress in Math.), Birkhäuser Verlag, Basel, to appear. (+ *Bull. Amer. Math. Soc.* 14 (1986), no. 1, 99-102)

[16] KIMEL'FEL'D, B.N.: 'Quasihomogeneous domains on a conformal sphere'

[in Russian], *Uspekhi Mat. Nauk* 33 (1978), no. 2 (200), 193-194.

[17] KOUROUNIOTIS, C.: 'Deformations of hyperbolic structures', *Math. Proc. Cambridge Philos. Soc.* 98 (1985), 247-261.

[18] KRUSHKAL', S.L., APANASOV, B.N., and GUSEVSKIĬ, N.A.: *Kleinian groups and uniformization in examples and problems* [in Russian], Izdat. 'Nauka', Sibirsk. Otdel., Novosibirsk 1981; English transl: Transl. Math. Monographs 62, Amer. Math. Soc., Providence 1986.

[19] LOCK, W.A.: *Deformations of locally homogeneous flat spaces and Kleinian groups*, Dr. Thesis, Columbia Univ., New York 1984.

[20] MAL'CEV, A.I.: 'On isomorphic matrix representations of infinite groups' [in Russian], *Mat. Sbornik* 8 (50) (1940), 405-422.

[21] MASKIT, B.: 'On Klein's conbination theorem III', in: *Advances in the theory of Riemann surfaces*, Princeton Univ. Press, Princeton 1971, pp.297-316.

[22] MORGAN, J.W.: 'On Thurston's uniformization theorem for three--dimensional manifolds', in: *The Smith conjecture*, Academic Press, 1984, pp. 37-125.

[23] MOSTOW, G.D.: *Strong rigidity of locally symmetric spaces*, Princeton Univ. Press, Princeton 1974.

[24] TETENOV, A.V.: *On the number of invariant components of Kleinian groups in space* [in Russian], Inst. of Math. Siberian Branch of the Acad. Sci. USSR preprint, Novosibirsk 1982.

[25] THURSTON, W.: *The geometry and topology of 3-manifolds*, ch. 1-9, 13, Princeton Univ. Lecture Notes, Princeton 1978-1980.

[26] ——: 'Three-dimensional manifolds, Kleinian groups and hyperbolic geometry', *Bull. Amer. Math. Soc.* 6 (1982), no. 3, 357-381.

HYPERBOLIC RIEMANN SURFACES WITH THE TRIVIAL GROUP OF AUTOMORPHISMS

Aleksandr D. Mednykh
Institute of Mathematics
Omsk State University
SU - 644077 Omsk 77, USSR

ABSTRACT. In [2] for the first time there have been given a complete proof of the following statement of A. Hurwitz: For any integer $g > 2$ there exists a compact Riemann surface of genus g, whose group of all conformal automorphisms is trivial. Then Greenberg [3] has shown that for $g > 2$ almost all points in the Teichmüller space T_g, except perhaps for an analytic subset, correspond to Riemann surfaces with the trivial group of conformal automorphisms. Nevertheless, only a few constructive examples of such Riemann surfaces are known. One of them is given by Accola [1]. However, the method of Accola does not let us describe analytically the fundamental set of a Fuchsian group which uniformizes that surface. In the present paper, announced in [5], we construct in an explicit way the fundamental set of a Fuchsian group which uniformizes a compact Riemann surface with the trivial group of conformal automorphisms.

1. Consider the unit disc $D = \{z \in \mathbb{C}: |z| < 1\}$ as a model of the Lobachevskiĭ plane with metric $d\sigma^2 = 4(1 - |z|^2)^{-2}|dz|^2$. Given a prime number $p > 3$ we define three substitutions:

$$\zeta = (p-1, p-2, \ldots, 4, 3, 1, p, 2), \quad \eta = (1, 4, 6, p-1, 3, 5, \ldots, p, 2), \quad \xi = (1, 2, \ldots, p-1, p) \tag{1}$$

satisfying the condition $\zeta\eta = \xi$.

Next, let $\theta_1, \theta_2, \ldots, \theta_r$ be cycles of length p with the property

$$\theta_1 \theta_2 \ldots \theta_r = \xi, \tag{2}$$

where r is an integer $\geq 2p$. The substitutions θ_j, $j = 1, 2, \ldots, r$, will be specified below.

Within D we construct a noneuclidean triangle OAB with vertex O at $z = 0$ and angles at O, A, and B equal π/pr, π/p, and π/r, respectively. Obviously

J. Ławrynowicz (ed.), Deformations of Mathematical Structures, 115–125.

$$\pi/p\,r + \pi/p + \pi/r < \pi, \tag{3}$$

so such a triangle always exists. It will be denoted by F, while its sides AB and OB by I and J, respectively. Their reflections in the side OA will be denoted by F^-, I^-, and J^-, respectively.

Consider the elliptic transformation $y = \exp(2\pi i/p\,r)z$ of D of order $p\,r$ and the noneuclidean polygon

$$F_o = \bigcup_{\ell=0}^{p\,r-1} y^{\ell}(F \cup F^-) \tag{4}$$

whose $2\,p\,r$ sides are determined by the sides of F and F^- according to the formulae

$$I^+_{k,j} = y^{jr+k-1}(I), \quad I^-_{k,j} = y^{jr+k-1}(I), \quad \begin{array}{l} k = 1, 2, \ldots, r, \\ j = 1, 2, \ldots, p. \end{array} \tag{5}$$

Denote by x_k the elliptic transformation of order p that maps the side $I^+_{k,p}$ onto $I^-_{k,p}$ for $k = 1, 2, \ldots, r$. Then, set $x_{r+1} = y^r$ and consider the polygon

$$F_1 = \bigcup_{\ell=0}^{r-1} y^{\ell}(F \cup F^-), \tag{6}$$

shown in Fig. 1.

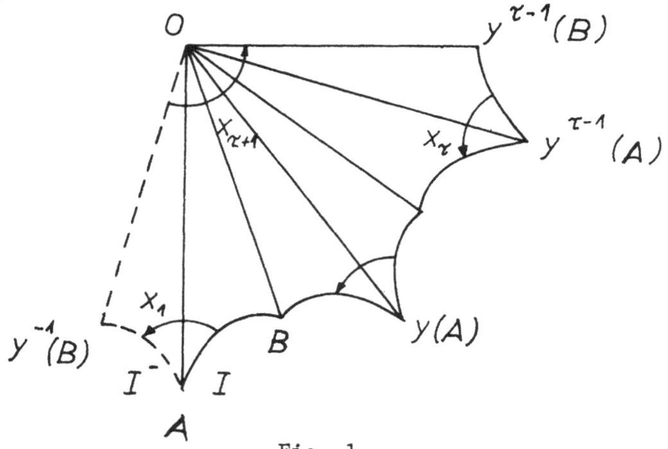

Fig. 1

One verifies directly that

$$F_o = \bigcup_{j=0}^{p-1} x^j_{r+1}(F_1) \tag{7}$$

(cf. Fig. 2, where $I^+_{1,p} = I$, $I^-_{1,p} = I^-$) and the ratio of noneuclidean

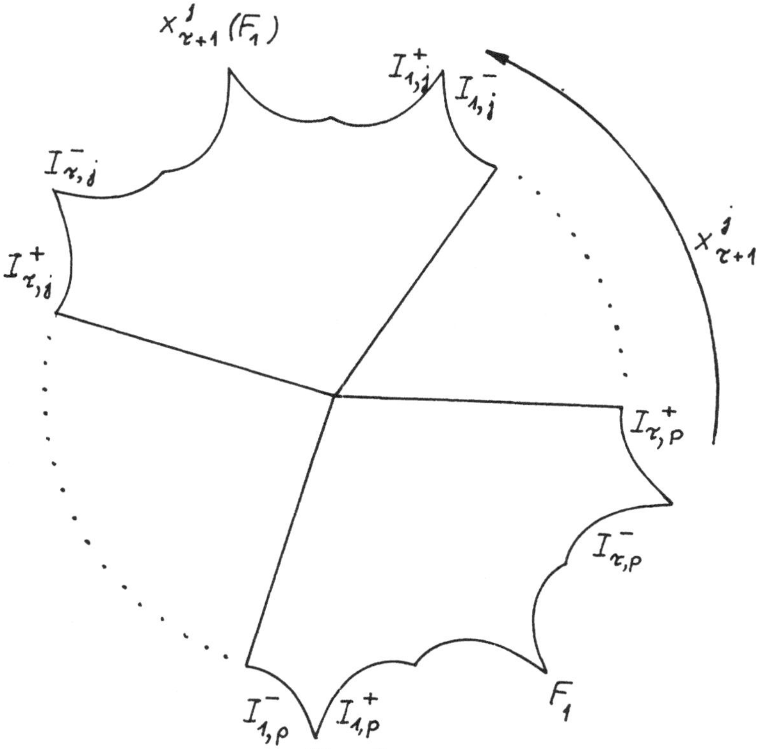

Fig. 2

areas $\mu(F_i)$, $i = 0, 1$, of the polygons F_i in the metric $d\sigma^2$ equals

$$\mu(F_o): \ \mu(F_1) = p. \tag{8}$$

The elliptic mapping x_{r+1} is of order p and maps the side I^- of F_1 onto the side $x_{r+1}(I^-)$.

2. By the theorem of Poincaré ([4], p. 227), the transformations x_1, $x_2, \ldots, x_{r+1}$ generate a Fuchsian group Δ with F_1 as its fundamental polygon. With help of the method given in [5], p. 234, we are going to derive for Δ the totality of determining relations.

The boundary of F_1 contains one excidental cycle given by its vertices

$$y^{-1}(B), \ y^{r-1}(B), \ y^{r-2}(B), \ldots, B \tag{9}$$

and $r+1$ one-point elliptic cycles

$$A, \ y(A), \ldots, y^{r-1}(A), \ 0. \tag{10}$$

Indeed, we have

$$x_{r+1}(y^{-1}(B)) = y^{r-1}(B),$$
$$x_r(y^{r-1}(B)) = y^{r-2}(B), \quad \ldots, \quad x_1(y(B)) = y^{-1}(B). \tag{11}$$

Hence

$$x_1 x_2 \ldots x_r x_{r+1}(y^{-1}(B)) = y^{-1}(B) \tag{12}$$

and the relationship yielded by the cycle (9) has the form

$$x_1 x_2 \ldots x_r x_{r+1} = 1. \tag{13}$$

The points given in (10) appear to be invariant for the p-th order elliptic elements $x_1, x_2, \ldots, x_{p+1}$, respectively, and determine the relations

$$x_1^p = x_2^p = \ldots = x_{r+1}^p = 1. \tag{14}$$

According to the theorem of [4], p. 234, (13) and (14) express a basis of generating relations in the group Δ, so it has the representation

$$\Delta = \{x_1, x_2, \ldots, x_{r+1}: \ x_1^p = x_2^p = \ldots = x_{r+1}^p = x_1 x_2 \ldots x_r x_{r+1} = 1\}. \tag{15}$$

Now we are going to construct the group Γ_o whose fundamental set is the polygon F_o. Thus, with the help of substitutions in (2) we define mappings which pairwise identify the sides of F_o:

$$T_{k,j} = x_{r+1}^{\theta_k(j)} \, x_k \, x_{r+1}^{-j}, \quad k = 1, \ldots, r, \quad j = 1, \ldots, p. \tag{16}$$

We notice that $T_{k,j}$ maps the side $I^+_{k,j}$ onto $I^-_{k,\theta_k(j)}$. By the theorem of Poincaré the group

$$\Gamma_o = \{T_{k,j}: k = 1, \ldots, r, \ j = 1, \ldots, p\}, \tag{17}$$

generated by the transformations $T_{k,j}$, has F_o as its fundamental polygon. Moreover, Γ_o does not contain elliptic elements and determines a compact Riemann surface $T = D/\Gamma_o$.

With help of the method of Reidemeister and Schreier we can derive the complete system of determining relations for the group Γ_o, namely

$$T_{k,\theta_k^{p-1}(p)} \ldots T_{k,\theta_k^2(p)} \, T_{k,\theta_k(p)} \, T_{k,p} = 1, \quad k = 1, 2, \ldots, r; \tag{18}$$

$$T_{1,\theta_2\theta_3\ldots\theta_r(j)} \, T_{2,\theta_3\ldots\theta_r(j)} \ldots T_{r-1,\theta_r(j)} \, T_{r,j} = 1, \quad j = 1, 2, \ldots, p. \tag{19}$$

Let G be the transitive group of substitutions, generated by the cycles $\theta_1, \theta_2, \ldots, \theta_r$ of (2). We define an epimorphism θ between the groups Δ and G by the rule

$$\theta(x_k) = \theta_k, \quad k = 1, 2, \ldots, r, \qquad \theta(x_{r+1}) = \xi^{-1}. \tag{20}$$

By (2) and (20) we have

$$\theta(x_1 x_2 \ldots x_r x_{r+1}) = \theta_1 \theta_2 \ldots \theta_r \xi^{-1} = 1, \tag{21}$$

so the epimorphism is well-defined.

3. Suppose G_o is a subgroup of G which leaves the symbol p unchanged. By [6] the group $\Gamma = \theta^{-1}[G_o]$ has signature $(\gamma, -)$, where γ is given by the Riemann-Hurwitz formula

$$2\gamma - 2 = p[2 \cdot 0 - 2 + (r-1)(1 - 1/p)], \tag{22}$$

and hence

$$\gamma = \tfrac{1}{2}(p-1)(r-1). \tag{23}$$

LEMMA 1. $\Gamma = \Gamma_o$.

Proof. By construction of Γ, its index in the group Δ is equal to $|\Delta : \Gamma| = p$. On the other hand, by relation (8) for the groups Δ and Γ_o, we have

$$|\Delta : \Gamma_o| = \mu(F_o) : \mu(F_1) = p. \tag{24}$$

Therefore, in order to prove the lemma it is sufficient to show that $\Gamma_o \leq \Gamma$ or, equivalently that $\theta(\Gamma_o) \leq G_o$.

We are going to check the latter inequality. To this, let

$$t_{k,j} = \theta(T_{k,j}), \quad k = 1, 2, \ldots, r, \quad j = 1, 2, \ldots, p. \tag{25}$$

For the sake of simplicity we suppose that the substitutions within the group G are well-defined with values on the set $N_p = \{1, 2, \ldots, p\}$ consisting of the resudues classes modulo p. With this assumption, for ξ given by (1) and any integers s and α, we have

$$\xi^{\alpha}(s) = s + \alpha \pmod p. \tag{26}$$

By (16), (20), (25), and (26) we get

$$\begin{aligned} t_{k,j}(s) = \xi^{-\theta_k(j)} \theta_k \xi^j(s) \equiv \xi^{-\theta_k(j)} \theta_k(s+j) \equiv \\ \theta_k(s+j) - \theta_k(j) \pmod p. \end{aligned} \tag{27}$$

In particular, for $s = p$, (27) becomes

$$t_{k,j}(p) \equiv \theta_k(p+j) - \theta_k(j) \equiv p \pmod p, \tag{28}$$

which means that $t_{k,j}(p) = p$, and hence $t_{k,j} \in G_o$ for $k = 1, \dots, r$ and $j = 1, \dots, p$. To end the proof it suffices to notice that the elements $t_{k,j}$ generate the group $\theta(\Gamma_o)$.

4. Suppose now that $T = D/\Gamma$ and $S = D/\Delta$ are compact Riemann surfaces obtained from the polygons F_o and F_1 by identifying their corresponding sides, as it has been described above. The surface T is of genus γ given by (23). On applying (15) we get that S has genus 0 with $r+1$ distinguished points being themselves projections of invariant points of the elliptic elements $x_1, x_2, \dots, x_{r+1}$. The covering $\phi: T \to S$, induced by the group inclusion $\Gamma \leq \Delta$, is branched over $r+1$ points indicated above and under each of them it has the branching order p.

Following Accola [1], a p-sheeted covering $\phi: T \to S$ is said to be *strongly branched* if

$$r_b > 2\,p(p-1), \tag{29}$$

where $r_b = \Sigma_{t \in T}[\beta_\phi(t) - 1]$ is the total ramification, i.e. the sum of indices for all the branch points of the covering ϕ. In our case

$$r_b = (r+1)(p-1), \tag{30}$$

so the condition for strong branching may be rewritten as $r+1 > 2p$. Since r and p are integers, the condition is equivalent to

$$r \geq 2p,$$

as supposed at the beginning of this paragraph.

5. Denote by $\mathrm{Aut}(T)$ the group of all conformal automorphisms of a Riemann surface T and by $N(\Gamma)$ the normalizer of Γ within the group $\mathrm{Aut}(D)$. By general results of the theory of coverings we evidently have

LEMMA 2. $\mathrm{Aut}(T) \cong N(\Gamma)/\Gamma$.

We are going to show that

LEMMA 3. $N(\Gamma) \leq N(\Delta)$.

Proof. Suppose E is a subset of S consisting of $r+1$ points over which ϕ has branch points of order p. By Lemma 2 any mapping $g \in N(\Gamma)$ is lowered to a conformal automorphism g_T of the surface T. In turn, since p is prime, we see that the covering $\phi: T \to S$ appears to be the maximal strongly branched covering in the sense of [7], so, any element $g_T \in \mathrm{Aut}(T)$ is lowered to an automorphism g_S of the surface S for which the set E stays invariant.

On the other side, by Lemma 2, g_S is lifted to some $\tilde{g}$ of the group $N(\Delta)$. The mappings $\tilde{g}$ and g are obtained by lifting the same conformal automorphism g_S, so $\tilde{g} = g\gamma$, where $\gamma \in \Delta$. Since

$\tilde{g} \in N(\Delta)$, we conclude that $g \in N(\Delta)$ as well.

Now we are going to calculate the normalizer $N(\Delta)$. From the geometrical meaning of the mappings x_k, $k = 1, \ldots, r+1$ (cf. Fig. 1) one verifies directly that

$$\begin{array}{ll} y\, x_1\, y^{-1} = x_2, & \\ y\, x_2\, y^{-1} = x_3, & y\, x_r\, y^{-1} = x_{r+1}\, x_1\, x_{r+1}^{-1}, \\ \ldots\ldots\ldots\ldots & y\, x_{r+1}\, y^{-1} = x_{r+1}. \\ y\, x_{r-1}\, y^{-1} = x_r, & \end{array} \tag{32}$$

Let $\Delta_1 = \langle \Delta, y \rangle$ be the group determined by the elements of Δ and y. By (32) we have

$$\Delta_1 \leq N(\Delta). \tag{33}$$

We proceed to show that in (33) the equality sign is attained.

To this end we have to describe the minimal system of generators and determining relations within the group Δ_1. We eliminate, subsequently, the terms $x_1, x_2, \ldots, x_r$ from (32) and take into account that $x_{r+1} = y^r$. Hence, by (13) it follows that

$$x_1\, y\, x_1\, y^{-1}\, y^2\, x_1\, y^{-2} \ldots y^{r-1}\, x_1\, y^{1-r}\, y^r = 1, \tag{34}$$

which, after simplification, becomes

$$(x_1\, y)^r = 1. \tag{35}$$

Therefore, by $x_1^p = y^{p\,r} = 1$ we conclude that

$$\Delta_1 = \{x_1, y:\ x_1^p = y^{p\,r} = (x_1\, y)^r = 1\}. \tag{36}$$

We will show that the triangular group Δ_1 is the maximal Fuchsian group, i.e. there does not exist a Fuchsian group Δ' such that $\Delta_1 \leq \Delta'$ and $\Delta_1 \neq \Delta'$. Now, by virtue of Greenberg's results [3], we only have to check the finite maximality of Δ_1, that is to show that there does not exist a Fuchsian group Δ such that the index of $|\Delta' : \Delta_1|$ is finite and $\Delta_1 < \Delta'$.

From Theorem 3 B in [3] it follows that the signatures of all triangular groups, which are not finite maximal, are of three possible kinds:

M 1. (m, m, n);

M 2. $(2, n, 2n)$;

M 3. $(3, n, 3n)$.

By the condition for a number p given at the very beginning and by (31) it follows that

$$p > 3 \tag{37}$$

and

$$r \geq 2p > 6. \tag{38}$$

The signature of Δ_1 has the form (p, r, pr), so by (37) and (38) it comes out that it does not satisfy the conditions M1 - M3. Hence Δ_1 is a Fuchsian group which is maximal and by (38) we get

$$N(\Delta) = \Delta_1. \tag{39}$$

Moreover, by (36) we have the decomposition

$$N(\Delta) = \Delta + y\Delta + \ldots + y^{r-1}\Delta \tag{40}$$

for the disjoint classes of contiguity.

6. We have concluded that Γ contains no elliptic elements and $|\Delta : \Gamma| = p$. Hence

$$\Delta = \Gamma + x_{r+1}\Gamma + \ldots + x_{r+1}^{p-1}\Gamma. \tag{41}$$

LEMMA 4. $N(\Gamma) \subseteq \sum_{\ell=0}^{pr-1} y^{\ell}\Gamma$.

Proof. Substituting (41) for Δ in (40) and making use of $x_{r+1} = y^r$, we get

$$N(\Delta) = \sum_{n=0}^{r-1} y^n \sum_{m=0}^{p-1} x_{r+1}^m \Gamma = \sum_{n=o}^{r-1} \sum_{m=0}^{p-1} y^{mr+n}\Gamma. \tag{42}$$

Then, by Lemma 3 and (42) it follows that

$$N(\Gamma) \subseteq \sum_{n=0}^{r-1} \sum_{m=0}^{p-1} y^{mr+n} = \sum_{\ell=0}^{pr-1} y^{\ell}\Gamma, \tag{43}$$

as desired.

To obtain the normalizer $N(\Gamma)$ we observe that, by Lemma 4, it is sufficient to determine the integers ℓ, $0 \leq \ell < pr$, for which $y^{\ell} \in N(\Gamma)$. In this direction we have

LEMMA 5. *An element* y^{mr+n} *with* $m = 0, 1, \ldots, p-1$ *and* $n = 0, 1, \ldots, r-1$ *belongs to the class* $N(\Gamma)$ *if and only if*

$$\xi^m \theta_k \xi^{-m} = \theta_{k+n}, \quad k+n \leq r,$$
$$\xi^{m+1} \theta_k \xi^{-m-1} = \theta_{\overline{k+n}}, \quad k+n > r, \tag{44}$$

where $k = 1, 2, \ldots, r$ *and* $\overline{k+n} = k+n-r$.

Proof. By (33) a transformation $y \in N(\Delta)$ and thus also its conjugate with respect to the element $\nu = y^{mr+n}$ give rise to an automorphism of the group Δ. By this the condition $\nu \in N(\Gamma)$ is equivalent to

$$\nu T_{k,j} \nu^{-1} \in \Gamma, \quad k = 1, \ldots, r, \quad j = 1, \ldots, p, \tag{45}$$

where $T_{k,j}$ are generators of the group Γ, determined by (16). We are going to show that the conditions (45) imply (44).

To this let us note that each elliptic element of order p in the group Δ is uniquely determined by its own invariant point inside D. Hence we have

$$\begin{aligned} y^n x_k y^{-n} &= x_{k+n}, \quad k+n \leq r, \\ y^n x_k y^{-n} &= x_{r+1} x_{\overline{k+n}} x_{r+1}^{-1}, \quad k+n > r \end{aligned} \tag{46}$$

whenever $k = 1, 2, \ldots, r$ and $n = 0, 1, \ldots, r-1$. Taking into account that $\nu = x_{r+1}^m y^n$ from (46) we get

$$\begin{aligned} \nu x_k \nu^{-1} &= x_{r+1}^m x_{k+n} x_{r+1}^{-m}, \quad k+n \leq r, \\ \nu x_k \nu^{-1} &= x_{r+1}^{m+1} x_{\overline{k+n}} x_{r+1}^{m-1}, \quad k+n > r. \end{aligned} \tag{47}$$

Now, by (46) and (47), since $\nu x_{r+1} \nu^{-1} = x_{r+1}$ for $k+n \leq r$, we have

$$\nu T_{k,j} \nu^{-1} = x_{r+1}^{\theta_k(j)} x_{r+1}^m x_{k+n} x_{r+1}^{-m-1} = x_{k+n,\, j+m}^{\theta_k(j)+m-\theta_{k+n}(m+j)}, \quad j = 1, 2, \ldots, p. \tag{48}$$

Here, if $j+m > p$, then $j+m$ has to be replaced by the residue modulo p. In analogy, if $k+n > r$, then by (47) we have

$$\nu T_{k,j} \nu^{-1} = x^{\theta_k(j)+m-1-\theta_{\overline{k+n}}(m+j+1)} T_{\overline{k+n},\, j+m+1}, \quad j = 1, 2, \ldots, p. \tag{49}$$

Yet the group Γ contains no elliptic elements, so - by (45), (48), (49) and the equality $x_{r+1}^p = 1$ - we obtain

$$\begin{aligned} &\theta_k(j) + m - \theta_{k+n}(m+j) = 0 \pmod p, \quad k+n \leq r, \\ &\theta_k(j) + m + 1 - \theta_{\overline{k+n}}(m+j+1) = 0 \pmod p, \quad k+n > r, \quad j = 1, 2, \ldots, p. \end{aligned} \tag{50}$$

With the use of (26), the above can be rewritten as

$$\begin{aligned} &\xi^m \theta_k(j) \equiv \theta_{k+m} \xi^m(j) \pmod p, \quad k+n \leq r \\ &\xi^{m+1} \theta_k(j) \equiv \theta_{\overline{k+n}} \xi^{m+1}(j) \pmod p, \quad k+n > r, \quad j = 1, 2, \ldots, p. \end{aligned} \tag{51}$$

By the assumptions introduced after Lemma 1, we see that the relations (51) and (44) are equivalent to each other.

Conversely, by (44) in the form (50) as well as by (48) and (49), we have

$$\begin{aligned} \nu T_{k,j}\nu^{-1} &= T_{k+n,\, j+m}, \quad k+n=r, \quad k=1,2,\ldots,r, \\ \nu T_{k,j}\nu^{-1} &= T_{\overline{k+n},\, j+m+1}, \quad k+n>r, \quad j=1,2,\ldots,p. \end{aligned} \tag{52}$$

In particular, we see that $\nu \in N(\Gamma)$.

7. Now we are going to show the explicit form of the substitutions in (2) which are determining a Riemann surface $T = D/\Gamma$ with trivial group of conformal automorphisms. By Lemma 1 this group is generated by the transformations (16) and has F_o as its fundamental set. Let

$$\theta_k = \xi^{(-1)^k}, \quad k=1,2,\ldots,r-2, \quad \theta_{r-1} = \zeta, \quad \theta_r = \eta \tag{53}$$

for r even and

$$\begin{aligned} \theta_k = \xi^{(-1)^k}, \quad k=1,2,\ldots,r-4, \quad \theta_{r-3} &= \xi^2, \quad \theta_{r-2} = \xi^{-1}, \\ \theta_{r-1} &= \zeta, \quad \theta_r = \eta \end{aligned} \tag{53´}$$

for r odd. By the explicit form of sibstitutions (1) we see that $\zeta\eta = \xi$ and (2) is satisfied for every r.

THEOREM. *Let* p *be a prime number,* $p > 3$, *and let* $r \geq 2p$. *If* T *is a Riemann surface described by* (53) *or* (53´), *then* $\mathrm{Aut}(T) \cong 1$.

Proof. By virtue of the isomorphism $\mathrm{Aut}(T) \cong N(\Gamma)/\Gamma$, established in Lemma 2, it is sufficient to show that $N(\Gamma) = \Gamma$. Now, by Lemma 4 we have

$$N(\Gamma) \subseteq \sum_{n=0}^{r-1} \sum_{m=0}^{p-1} y^{mr+n}\Gamma . \tag{54}$$

Let $y^{mr+n} \in N(\Gamma)$. Then, by Lemma 5 we have equalities (44). At first we shall show that $m = 0$.

By contradiction, let $m > 0$. From (44) with $k = r$ we get

$$\xi^{m+1}\theta_r\xi^{-m-1} = \theta_n, \quad r+n > r. \tag{55}$$

In the both cases (53) and (53´) we have $\theta_r = \eta$, by which

$$\xi^{m+1}\eta\xi^{-m-1} = \theta_n, \quad n=1,2,\ldots,r-1. \tag{55´}$$

Since $\theta_n \neq \eta$ for $n < r$ and $\xi^{m+1}\eta\xi^{-m-1} \neq \xi, \xi^{-1}, \xi^2$, then for the equality (55) there exists only one possibility $\xi^{m+1}\eta\xi^{-m-1} = \zeta$.

Taking into account that $\zeta = \theta_{r-1}$ and $n = r-1$, from (44) with $k = 1$ we obtain

$$\xi^m \theta_1 \xi^{-m} = \theta_r .$$

Substituting for θ_1 and θ_r the values given by (53) and (53') we have $\xi^m \xi^{-1} \xi^{-m} = \eta$, which contradicts $\xi^{-1} \neq \eta$. Hence $n = 0$ and by (44) with $k = 1$ we have

$$\xi^m \theta_k \xi^{-m} = \theta_k, \qquad k = 1, 2, \ldots, r. \tag{56}$$

In particular, for $k = r$,

$$\xi^m \eta \xi^{-m} = \eta. \tag{57}$$

Yet, this is impossible for $m \neq 0$, so η cannot be a degree of the substitution ξ. Consequently, $m = 0$ and $y^{mr+n} = 1$.

By Lemma 5 we have $N(\Gamma) = \Gamma$, and hence

$$\mathrm{Aut}(T) \cong N(\Gamma)/\Gamma = 1.$$

8. At the end we remark that the genus of a compact Riemann surface T is determined by (23): $\gamma = \frac{1}{2}(p-1)(r-1)$. In our Theorem the minimum value of γ is 18; it is obtained when $p = 5$ but $r = 10$. With this we get T from the noneuclidean 100-cornered polygon F_0 by pairwise identifications of its sides with help of the fractional linear transformations (16), determined by the substitutions (53). Making use of the formulae (1) one can get an explicit form of such mappings.

References

[1] ACCOLA, R.D.M.: 'Strongly branched coverings of closed Riemann surfaces', *Proc. Amer. Math. Soc.* 26, no. 2 (1970), 315-322.

[2] BAILY, W.: 'On the automorphism group of a generic curve of genus > 2', *J. Math. Kyoto Univ.* 1 (1961/1962), 101-108; Correction, p.325.

[3] GREENBERG, L.: 'Maximal Fuchsian groups', *Bull. Amer. Math. Soc.* 69 (1963), 569-573.

[4] LEHNER, J.: 'Discontinuous groups and automorphic functions', *Amer. Math. Soc.* , 1964.

[5] MEDNYH, A.D.: 'On an example of a compact Riemann surface with the trivial group of automorphisms', *Dokl. Akad. Nauk SSSR* 237, no. 1 (1977), 32-34.

[6] SINGERMAN, D.: 'Subgroups of Fuchsian groups and finite permutation groups', *Bull. London Math. Soc.* 2 (1970), 319-329.

ON THE HILBERT SCHEME OF CURVES IN A SMOOTH QUADRIC

Edoardo Ballico *
Classe di Scienze
Scuola Normale Superiore
I-56100 Pisa, Italy

ABSTRACT and INTRODUCTION. Let X be a projective homogeneous manifold over $\mathbb{C}$, say $X = G/P$, with G being a connected linear algebraic group and P a parabolic subgroup of it. As in the case $X = \mathbb{P}^n$, it seems interesting to study the Hilbert scheme $\text{Hilb}(X)$ of X and, in particular, the part of $\text{Hilb}(X)$ related to curves in X.

Here we consider the case $X = Q$, Q being a smooth quadric. We use projective techniques: deformation theory and degeneration of a smooth curve to a reducible one [6]. Fix a smooth quadric Q, $\dim(Q) = n-1$. Let $Z^*(d, g, n, Q)$ be the subset of $\text{Hilb}(Q)$ formed by the smooth nondegenerate connected curves of degree d and genus g. Let $Z(d, g, n, Q)$ be the closure of $Z^*(d, g, n, Q)$ in $\text{Hilb}(Q)$. The main result of this paper is the following theorem (over $\mathbb{C}$):

THEOREM 1. *If* $n \geq 7$ *and* $g \leq (n/2) - 1$, *then* $Z^*(d, g, n, Q)$ *is smooth and irreducible.*

The proof of Theorem 1 given in Sect. 1 shows how to use the techniques of [6] to prove the existence of many reducible elements in $Z(d, g, n, Q)$ (see Remark 1.1). In Sect. 2 we use this result (following [5], [7]) to show that the postulation of a general element of $Z^*(d, 0, n, Q)$ is as good as possible.

1. PROOF OF THEOREM 1.

Fix a smooth quadric Q, $\dim(Q) = n-1 \geq 6$, and a curve C in $Z^*(d, g, n, Q)$. First assume that n is odd, $n = 2m-1$. By Bertini's theorem [8] there is a linear space $V \subset Q$, $\dim(V) = m-1$, such that $V \cap \text{Sec}(C) = \emptyset$. We may choose coordinates $x_1, \ldots, x_m, y_1, \ldots, y_m$ such that V is determined by the equations $y_1 = \ldots = y_m = 0$ and Q by the equation $x_1y_1 + \ldots + x_my_m = 0$. Let q be the rational projection of $\mathbb{P}^n$ from V into the linear space W with equations $x_1 = \ldots = x_m = 0$. Define $q_t \in \text{Aut}(Q)$ by $q_t((x_1, \ldots, x_m, y_1, \ldots, y_m)) = (tx_1, \ldots, tx_m, y_1, \ldots, y_m)$.

*) Partially supported by M.P.I.

J. Ławrynowicz (ed.), Deformations of Mathematical Structures, 127–132.

Since $V \cap \mathrm{Sec}(C) = \emptyset$, q induces an embedding p of C into W. Since Hilb(Q) is proper, the family $\{q_t(C)\}$ has a limit in Hilb(Q) for $t=0$; it is necessarilly p(C). By semicontinuity, to prove the smoothness of $Z^*(d,g,n,Q)$ at C, it suffices to check that p(C) is a smooth point of Hilb(Q). Thus it suffices to check that $0 = H^1(p(C), N_{p(C),Q})$, where $N_{A,B}$ is the normal bundle (or normal sheaf) of the subvariety A in B. Note that $N_{W,Q} \cong \Omega_W(2)$ (see for instance [1], Prop. 2). By [4], Remark (2) p. 498, $h^1(p(C), \Omega_W(2)|p(C)) = 0$.

Since the hyperplane section of p(C) is not special, $H^1(p(C), N_{p(C),W}) = 0$. Hence $H^1(p(C), N_{p(C),Q}) = 0$. Note that the set $Z^*(d,g,W)$ of smooth connected (nondegenerate) curves of degree d and genus g in W is smooth and irreducible if $d \geq 2g-1$. Furthermore, we may also assume $W \cap \mathrm{Sec}(C) = \emptyset$ and consider, instead of p, the projection p´ of C from W into V. Hence, although the (m-1)-planes in Q from 2 irreducible families, $Z^*(d,g,n,Q)$ is irreducible.

Now assume that n is even, say: $n = 2m$. Let S be the set of linear spaces U in $\mathbb{P}^n$, $\dim(U) = m$, such that $U \cap Q$ is a double linear space. Set $x = \dim(S)$. Fix a point P in $\mathbb{P}^n \setminus Q$ and set $S(P) = \{U \in S: P \in U\}$. Note that $\dim(S(P)) = x - m$ (use the action of the orthogonal group). Hence, by Bertini's theorem and a dimensional count, we may find U in S with $U \cap \mathrm{Sec}(C) = \emptyset$. Choose homogeneous coordinates $x_1, \ldots, x_m, y_1, \ldots, y_m, z$, such that U has equations $y_1 = \ldots = y_m = 0$ and Q is determined by the equation $x_1y_1 + \ldots + x_my_m + z^2 = 0$. Let W be the linear space with equations $x_1 = \ldots = x_m = z$. Let q be the projection from U into W´. Define q_t in Aut(Q) by $q_t((x_1, \ldots, x_m, y_1, \ldots, y_m, z)) = (t^2x_1, \ldots, t^2x_m, y_1, \ldots, y_m, tz)$, Then copy the previous proof. The proof of Theorem 1 is completed.

By a *tree* we understand any curve Y in Q, which is reduced, connected, with only ordinary double points, with lines as irreducible components, and with arithmetic 0.

<u>Remark 1.1</u>. From the proof of Theorem 1 and [6], we obtain that $Z(d,0,n,Q)$ contains all the trees of degree d in Q, even the degenerate ones.

In this paper we have used the assumption that C is a base field, only when quotating Remark (2) of [4], p. 498. In particular, for $g=0$ everything works over any algebraically closed field $\mathbb{K}$ if $\mathrm{ch}(\mathbb{K}) = 2$ (see [3], p. 34).

2. POSTULATION OF RATIONAL CURVES IN Q

In this section we show how to use the results of Sect. 1 to obtain the existence of many rational smooth curves Y contained in a smooth quadric Q and with "good" postulation. The method was introduced in [5], [7] and used several times (for instance in [2]).

Let Y be a curve contained in a smooth quadric Q. Let $r_{Y,Q}$: $H^0(Q, \mathcal{O}_Q(k)) \to H^0(Y, \mathcal{O}_Y(k))$ be the restriction map. If $\dim(Q) = n-1$, we will often write $r_{Y,n}(k)$ instead of $r_{Y,Q}(k)$ if there is no danger of misunderstanding. This section is devoted to the proof of the following result.

PROPOSITION 2.1. *For every* $n \geq 7$, $d > n$, *a general element of* $Z(d, 0, n, Q)$ *has the maximal rank.*

Let Q be a smooth quadric, T a curve in Q, and H a linear section of Q. We say that k lines $L_1, \ldots, L_k$ in H are *good secants* to T if each L_i intersects T quasi-transversally exactly at 2 points, the lines L_i are disjoint and the union of T, $L_1, \ldots, L_k$ has k connected components less than T. Define integers $r(k, n)$, $q(k, n)$, $k \geq 1$, $n \geq 3$, by the following relations:

$$k\,r(k, n) + 1 + q(k, n) = \binom{n+k}{n} - \binom{n+k-2}{n}, \quad 0 \leq q(k,n) \leq k-1. \qquad (1)$$

The essential step for the proof of 2.1 will be the inductive proof in 2.3 of the following assertions $H(k, n)$, $A(k, n)$: $H(k, n)$, $k \geq 1$, $n \geq 5$. There is a smooth quadric Q, $\dim(Q) = n-1$, a smooth hyperplane section H of Q and a reduced curve Y in Q such that:

1. $\deg(Y) = r(k, n)$; Y has $q(k, n) + 1$ trees as connected components;
2. $\dim(Y \cap H) = 0$; H contains $q(k, n)$ good secants to Y;
3. $r_{Y,Q}(k)$ is bijective.

$A(k, n)$, $k \geq 1$, $n \geq 4$. There is a curve Y in a smooth quadric Q, $\dim(Q) = n-1$, $\deg(Y) = r(k, n) - 1 + 1$, Y - union of k disjoint trees such that $r_{Y,Q}(k)$ is surjective. Note that a curve Z in $Z^*(r(k, n), Q, n, Q)$ or a tree W of degree $r(k, n)$ have maximal rank if and only if $r_{Z,n}(k-1)$ is injective and $h^0(Q, J_{Y,Q}(k)) = q(k, n)$ (and the same conditions for W) by Castelnouvo-Mumford's lemma [9], p. 99.

We need the following numerical lemma:

LEMMA 2.2. *For all integers* $k \geq 2$, $n \geq 5$, *we have*

$$r(k, n) \geq r(k-1, n) + k - 1, \qquad r(k, n) \geq 2,$$

$$r(k, n) - r(k-1, n) \leq r(k, n-1) - k + 1,$$

and

$$r(k, 4) \geq r(k-1, 4) + 2,$$

$$r(k, 4) \leq r(k-1, 4) + k + 1.$$

Proof. By the definitions (1) of $r(k, n)$, $r(k-1, n)$, we obtain:

$$k(r(k, n) - r(k-1, n)) + r(k-1, n) + q(k, n) - q(k-1, n) = \binom{k+n-1}{n-1} - \binom{k+n-3}{n-1}. \qquad (2)$$

We prove, for instance, the third inequality of 2.2, the remaining ones being easier. Assume $r(k,n)-r(k-1,n)\geq r(k,n-1)-k+2$. By (2) we obtain

$$k(k-2)+\binom{k+n-1}{n-1}-\binom{k+n-3}{n-1}-(k-1)+r(k-1,n)-(k-2)\leq$$
$$\leq\binom{k+n-1}{n-1}-\binom{k+n-3}{n-1}.$$

This inequality is false by the definition of $r(k-1,n)$ if $k\geq 2$ and $n\geq 5$.

LEMMA 2.3. $H(k,n)$ *holds true for every* $k\geq 1$, $n\geq 5$; $A(k,n)$ *holds true for every* $k\geq 1$, $n\geq 4$.

Proof. $A(1,n)$ $(=H(1,n))$ is true for every $n\geq 4$. First we prove $A(k,4)$ by induction on k. Fix $k\geq 2$ and assume $A(k-1,4)$. Let $T\subset Q$, $\dim(Q)=3$, be a curve satisfying $A(k-1,4)$. Take a general hyperplane section H of Q. Consider the union Z of $x:=r(k,4)-r(k-1,4)$ disjoint lines in H; exactly $x-1$ of these lines intersect T. By 2.2, $r_{Z,H}(k)$ is surjective. Deforming T, if necessary, we may assume that $T\cap H$ is formed by general points of H. Hence, by (2), we obtain the surjectivity of $r_{Z\cup(T\cap H),H}(k)$. In a standard way (or see toward the end of the proof of this lemma) we obtain the surjectivity of $r_{Z\cup T,Q}(k)$; hence $A(k,4)$. Fix $k\geq 2$, $n\geq 5$, and assume $A(k,n-1)$, $A(k-1,n)$, $H(k-1,n)$. We will prove $H(k,n)$. The same argument can give $A(k,n)$; hence the inductive proof of the lemma will be over.

Take a smooth quadric Q, $\dim(Q)=n-1$, and T in Q, $\deg(T)=r(k-1,n)$, T satisfying $H(k-1,n)$ with respect to the smooth linear section H of Q. We may deform T to W, W being isomorphic to T, so that H contains exactly $\max(0,q(k-1,n)-q(k,n))$ good secants to W. By 2.2 and $A(k,n-1)$, we may find $A\subset H$, A being the union of k disjoint trees, $\deg(A)=r(k,n)-r(k-1,n)$, with $r_{A,H}(k)$ surjective. Furthermore, moving A, we may assume that $W\cup A$ is the union of $q(k,n)+1$ disjoint trees. By moving W we may assume that the points in $W\cap(H\setminus A)$ are general: given a point P and a line D in Q, there is a line R in Q containing P and intersecting D. By (2) we may assume the bijectivity of $r_{A\cup(H\cap W),H}(k)$. We claim that $r_{W\cup A,Q}(k)$ is injective (i.e. by (1) bijective). Take $f\in H^0(Q,\mathcal{J}_{W\cup A,Q}(k))$. Then $f|H$ vanishes on $A\cup(W\cap H)$. Hence f is divided by the equation z of H. Since f/z vanishes on W, we have $f=0$.

Now we show the existence of an isotrivial deformation Y of $W\cup A$ satisfying $H(k,n)$ with respect to H, i.e. with $\dim(H\cap Y)=0$, H containing $q(k,n)$ good secants to Y. Fix a point P in $W\cap(H\setminus A)$ and a line L in a connected component C of A with $C\cap W=\emptyset$. There is a line D in H with $P\in D$, $L\cap D\neq\emptyset$. Set $y=L\cap D$. We may move L to a line L' in Q with $y=L'\cap H$. We deform at the same time isotrivially C to C', C' containing L'. By semicontinuity we find A' isomorphic to A such that $W \cup A'$ satisfies $H(k,n)$ with respect to H.

Proof of Proposition 2.1. Fix integers $n \geq 7$ and $d \geq n+1$. Let k be the integer such that $r(k-1,n) < d \leq r(k,n)$. By [9], p. 99, semicontinuity and the irreducibility of $Z(d,0,n,Q)$, it is sufficient to find X, Y in $Z(d,g,n,Q)$ such that $r_{X,Q}(k-1)$ is injective and $r_{X,Q}(k)$ is surjective. First assume $d \geq r(k-1,n)+$ $+q(k-1,n)$. Take W satisfying $H(k-1,n)$. Consider a curve X in $Z(d,0,n,Q)$, X union of W and $d-r(k-1,n)$ suitable lines; obviously $r_{X,Q}(k-1)$ is injective. Now assume $d < r(k-1,n)+$ $+q(k-1,n)$, and in particular $k \geq 3$. The proof of 2.3 shows how to construct, starting with a curve E satisfying $H(k-2,n)$, a curve W in $Z(r(k-1,n)+1,0,n,Q)$, with $r_{W,Q}(k-1)$ injective. We take as X the union of W and $d-r(k-1,n)-1$ suitable lines.

Now we consider the surjectivity part. First assume $d \geq r(k-1,n)+$ $+q(k-1,n)$. Take W satisfying $H(k-1,n)$; use again 1.1 and the proof of 2.3 to construct Y in $Z(d,0,n,Q)$ with $r_{Y,Q}(k)$ surjective. Now assume $d < r(k-1,n)+q(k-1,n)$, hence $k \geq 3$. Using E satisfying $H(k-2,n)$, 1.1 and the proof of 2.3, we prove the existence of a curve I in $Z(r(k-1,n),0,n,Q)$ with $r_{I,Q}(k)$ surjective. By the proof of 2.3 (once more) we may construct Y in $Z(d,0,n,Q)$ with $r_{Y,Q}(k)$ surjective.

References

[1] BALLICO, E.: 'Normal bundles to curves in quadrics', *Bull. Soc. Math. France* 109 (1981), 69-80.

[2] —— and ELLIA, Ph.: 'The maximal rank conjecture for non-special curves in $\mathbb{P}^3$', *Invent. Math.* 79 (1985), 541-555.

[3] DIEUDONNÉ, J.: 'La géométrie des groupes classiques', *Ergebnisse der Math.* 5 (1955), Springer-Verlag, Berlin.

[4] GRUSON, L. LAZARSFELD, R. and PESKINE, C.: 'On a theorem of Castelnuovo, and the equations describing space curves', *Invent. Math.* 72 (1983), 591-506.

[5] HARTSHORNE, R. and HIRSCHOWITZ, A.: 'Droites en position général dans l'espace projectif', in: *Algebraic Geometry, Proceedings La Rabida*, p. 169-189, Lecture Notes in Math. 961, Springer-Verlag, 1982.

[6] —— and ——: 'Smoothing algebraic space curves', in: *Algebraic Geometry - Sitgers 1983*, p. 98-131, Lecture Notes in Math. 1124, Springer-Verlag, 1984.

[7] HIRSCHOWITZ, A.: 'Sur la postulation générique des courbes rationelles', *Acta Math.* 146 (1981), 209-230.

[8] KLEIMAN, S.: 'The transversality of a general translate', *Compo-*

sitio Math. 38 (1974), 287-297.

[9] MUMFORD, D.: 'Lectures on curves over an algebraic surface', *Annals of Math. Studies*, vol. 59, Princeton Univ. Press, Princeton N. J. 1966.

A CONTRIBUTION TO KELLER'S JACOBIAN CONJECTURE II

Zygmunt Charzyński, Jacek Chądzyński and Przemysław Skibiński
Institute of Mathematics, Łódź University
ul. S. Banacha 22
90-238 Łódź
Poland

SUMMARY. In this paper we give two necessary and sufficient conditions for a complex polynomial Q in two variables x,y to exist, such that $P_xQ_y - P_yQ_x = 1$ holds for the given analogous polynomial P.

INTRODUCTION.

In the circle of the famous Keller's Jacobian conjecture (cf. [5] and [2]), polynomial mappings with the constant non-zero Jacobian are considered.

In this paper we investigate polynomial mappings $H: \mathbb{C}^2 \ni (x,y) \to (P(x,y),Q(x,y)) \in \mathbb{C}^2$ fulfilling the condition

$$P_xQ_y - P_yQ_x = 1 \tag{1}$$

in the aspect of some properties of the coordinates of the mapping H. The constancy of the Jacobian implies that the coordinates P,Q cannot be arbitrary - there are various necessary conditions for them (cf. e.g. [1], p.138, [3], p.50, [6], p.262). In connection with this, there arises a natural problem of a full characterization of the coordinates in the form of some necessary and sufficient conditions.

Here such a characterization is given. We formulate some necessary and sufficient conditions which must be satisfied by the given polynomial P in order that another polynomial Q exist such that their Jacobian is equal to 1 (Theorems 1 and 2). The announced characterization is preceded by an inductive definition of a sequence of some pairs of polynomials appearing in these theorems (Observation 3).

1. TERMS AND AUXILIARY FACTS

(a) In the sequel, by $\mathbb{C}[x,y]$ we shall denote the ring of polynomials in two variables x,y over the field $\mathbb{C}$ of complex numbers. The analogous notation is introduced for one and three variables. If $A \in \mathbb{C}[x,y]$, then by deg A we shall denote the degree of the polynomial A and by

J. Ławrynowicz (ed.), Deformations of Mathematical Structures, 133–140.

$\deg_x A$ - the degree of the polynomial A with respect to x, with the agreement that the polynomial identically equal to zero has the degree $-\infty$.

(b) A polynomial mapping $H = (P,Q) : \mathbb{C}^2 \to \mathbb{C}^2$ fulfilling identically condition (1) is called Keller's mapping and P,Q - Keller's coordinates of the mapping H. In the light of the above, for a polynomial P to be a Keller's coordinate, it is necessary and sufficient that there exist some polynomials Q such that (1) holds.

Moreover, if I_o is such a polynomial, then any other polynomial Q fulfilling (1) has the form

$$Q = I_o + \Phi(P) \tag{1'}$$

where $\Phi \in \mathbb{C}[x]$.

Indeed, in this situation we have

$$P_x(Q - I_o)_y = P_y(Q - I_o)_x = 0,$$

i.e. the polynomials $Q - I_o$ and P are algebraically dependent, and since, in view of (1), P_x and P_y are relatively prime in $\mathbb{C}[x,y]$, therefore this is the explicit dependence $Q - I_o = \Phi(P)$ (cf. [4]).

Thus, the family of polynomials Q fulfilling (1) depends, by (1'), on arbitrary parameters involved in Φ.

Using, if necessary, the composition with a suitable linear automorphism of $\mathbb{C}^2$, we can confine ourselves to Keller's mappings such that $\deg_x P_x = \deg P - 1$ and $\deg_x P_y = \deg P - 1$.

(c) Let $M \in \mathbb{C}[x,y]$, $N \in \mathbb{C}[x,y]$. By the division with remainder we mean the representation $M = DN + E$ where $D,E \in \mathbb{C}[x,y]$ and $\deg_x E < \deg_x N$. At the same time, D and E are called the quotient and the remainder of this division, respectively.

Below, we shall give some auxiliary facts.

Let $P \in \mathbb{C}[x,y]$, $m = \deg P$. Throughout this section, we shall assume that $m > 1$ and that the following conditions hold:

(i) $\deg_x P_x = m - 1$, $\deg_x P_y = m - 1$,

(ii) the partial derivatives P_x, P_y have no zeros in common in $\mathbb{C}^2$.

OBSERVATION 1. Let $U,V,W \in \mathbb{C}[x,y]$, $\deg_x W \leq 2m - 3$ and

$$P_x V - P_y U = W. \tag{2}$$

Then the quotients of the divisions of U by P_x and of V by P_y are equal.

To justify the above observation, it suffices to compare the degrees with respect to x on both sides of (2) and take account of the assumption and condition (i).

OBSERVATION 2. For the given polynomial P and for an arbitrary polynomial $W \in \mathbb{C}[x,y]$ such that $\deg_x W \leq 2m - 3$, there exists exactly one pair of polynomials $A, B \in \mathbb{C}[x,y]$ such that the following linear representation holds:

$$P_x B - P_y A = W, \quad \deg_x A \leq m - 2, \quad \deg_x B \leq m - 2. \tag{3}$$

Indeed, in the case $W = 1$, according to theorems 9.2 and 9.6 from [7] (pp.23 and 25), there exists exactly one pair of polynomials $\tilde{A}, \tilde{B} \in \mathbb{C}[x,y]$ such that

$$P_x \tilde{B} - P_y \tilde{A} = 1, \quad \deg_x \tilde{A} \leq m - 2, \quad \deg_x \tilde{B} \leq m - 2. \tag{4}$$

If W is an arbitrary polynomial, then from (4) we get

$$P_x(\tilde{B}W) - P_y(\tilde{A}W) = W. \tag{5}$$

Next, we use the division of the expressions in brackets by P_y and P_x, respectively:

$$\tilde{A}W = IP_x + A, \quad \tilde{B}W = JP_y + B. \tag{6}$$

Hence, by Observation 1, we have $I = J$. Putting (6) into (5), we obtain (3).

This representation is unique. In the opposite case, we easily obtain the contradiction with unique representation (4).

Directly from this observation we obtain

OBSERVATION 3. For the given polynomial P, there exists exactly one sequence of pairs of polynomials

$$(A_k, B_k), \quad \deg_x A_k \leq m - 2, \quad \deg_x B_k \leq m - 2, \quad k = 1,2,\ldots, \tag{7}$$

defined inductively by the conditions

$$P_x B_1 - P_y A_1 = 1, \tag{8_1}$$

$$P_x B_{k+1} - P_y A_{k+1} = B_{kx} - A_{ky}, \qquad k = 1,2,\ldots\ . \tag{8_{k+1}}$$

We say that sequence (7) breaks off if there exists an index s such that

$$B_{sx} - A_{sy} = 0. \tag{9}$$

This easily implies

OBSERVATION 4. If sequence (7) breaks off, then there exists a minimal index r such that $B_{rx} - A_{ry} = 0$, $(A_r, B_r) \neq (0,0)$, $(A_k, B_k) =$

$= (0,0)$ for $k > r$.

At last, we easily check the following

OBSERVATION 5. The system of equations $I_x = F$, $I_y = G$, where $F, G \in \mathbb{C}[x,y]$ and $G_x - F_y = 0$, has a solution $I \in \mathbb{C}[x,y]$ (unique up to a constant).

2. CHARACTERIZATION OF THE COORDINATES

In this section, $P \in \mathbb{C}[x,y]$ will denote an arbitrary polynomial of degree $m > 1$, satisfying the condition

(i) $\deg_x P_x = m - 1$, $\deg_x P_y = m - 1$ (cf. sec. 1 (b)).

THEOREM 1. *The polynomial P is a Keller coordinate if and only if the following conditions are satisfied:*

(ii) the partial derivatives P_x, P_y have no zeros in common in $\mathbb{C}^2$ (this gives the existence of sequence (7) described in Observation 3),

(iii) sequence (7) breaks off.

Proof. First, let us assume that P is a Keller coordinate. Then there exists a polynomial $Q \in \mathbb{C}[x,y]$ such that

$$P_x Q_y - P_y Q_x = 1. \tag{10}$$

Hence we immediately see that condition (ii) holds.

Now, we define by induction two sequences of divisions of polynomials in the sense of section 1 (c):

$$Q_x = I_1 P_x + A_1, \qquad Q_y = J_1 P_y + B_1, \tag{11$_1$}$$

$$I_{kx} = I_{k+1} P_x + A_{k+1}, \qquad J_{ky} = J_{k+1} P_y + B_{k+1}, \qquad k = 1,2,\ldots . \tag{11$_{k+1}$}$$

Let us first notice that $I_k = J_k$ for $k = 1,2,\ldots$. Indeed, the equality $I_1 = J_1$ follows from (10), (11_1) and Observation 1. If $I_l = J_l$ for $l = 1,\ldots,k$, then

$$I_{k-1\,x} = I_k P_x + A_k, \qquad \deg_x A_k \le m - 2,$$

$$I_{k-1\,y} = I_k P_y + B_k, \qquad \deg_x B_k \le m - 2.$$

Hence, differentiating and taking account of the equality of mixed partial derivatives, we obtain

$$P_x I_{ky} - P_y I_{kx} = B_{kx} - A_{ky}, \tag{12}$$

whence, by (11_{k+1}) and Observation 1, we get $I_{k+1} = J_{k+1}$.

Let us next notice that the pairs

$$(A_k, B_k), \quad k = 1,2,\ldots, \tag{13}$$

from (11_1), (11_{k+1}) fulfil the conditions of Observation 3. Indeed, putting (11_1) into (10) and using the equality $I_1 = J_1$, we obtain

$$P_x B_1 - P_y A_1 = 1, \quad \deg_x A_1 \leqq m-2, \quad \deg_x B_1 \leqq m-2,$$

which, according to Observation 2, means that (A_1, B_1) is the first pair from Observation 3. Assume that the pair (A_k, B_k) from (11_{k+1}) is identical with the corresponding pair from sequence (7). By (11_{k+1}) and the equality $I_k = J_k$, we have

$$I_{kx} = I_{k+1}P_x + A_{k+1}, \quad I_{ky} = I_{k+1}P_y + B_{k+1}, \tag{14}$$

$$\deg_x A_{k+1} \leqq m-2, \quad \deg_x B_{k+1} \leqq m-2.$$

Putting (14) into (12), we obtain

$$P_x B_{k+1} - P_y A_{k+1} = B_{kx} - A_{ky},$$

which, according to Observation 2, gives that (A_{k+1}, B_{k+1}) is the $(k+1)$-st term of sequence (7) from Observation 3.

In the light of this, it suffices to show that sequence (13) breaks off. Indeed, since $m - 1 > 0$ and the sequence $(\deg_x I_k)$, $k = 1,2,\ldots,$ is decreasing up to an index s such that $\deg_x I_s = -\infty$, therefore

$$I_{s-1\,x} = 0 \cdot P_x + A_s, \quad I_{s-1\,y} = 0 \cdot P_y + B_s. \tag{15}$$

From (15), by the equality of mixed partial derivatives, we obtain

$$B_{sx} - A_{sy} = 0,$$

and so, sequence (13) breaks off. Thus conditions (ii) and (iii) are satisfied, as described.

Now, let us assume that these conditions are satisfied.

We begin, using the decreasing induction, with showing that there exist some polynomials $I_o,\ldots,I_{s-1}$ belonging to $\mathbb{C}[x,y]$ such that

$$P_x I_{oy} - P_y I_{ox} = 1, \qquad (16_o)$$

$$P_x I_{ly} - P_y I_{lx} = B_{lx} - A_{ly}, \qquad l = 1,\ldots,s-1, \quad \text{if} \quad s > 1. \qquad (16_l)$$

According to the assumption, equality (9) holds. Thus, in virtue of Observation 5, there exists a polynomial $I_{s-1} \in \mathbb{C}[x,y]$ such that

$$I_{s-1\,x} = A_s, \qquad I_{s-1\,y} = B_s. \qquad (17_{s-1})$$

Putting (17_{s-1}) into (8_s), we obtain (16_{s-1}). Let now $I_k \in \mathbb{C}[x,y]$ fulfil (16_k). Then $(I_kP_x + A_k)_y = (I_kP_y + B_k)_x$. Consequently, in virtue of Observation 5, there exists a polynomial $I_{k-1} \in \mathbb{C}[x,y]$ such that

$$I_{k-1\,x} = I_kP_x + A_k, \qquad I_{k-1\,y} = I_kP_y + B_k. \qquad (17_{k-1})$$

Putting (17_{k-1}) into (8_k), we obtain (16_{k-1}).

By this procedure, we have obtained the existence of a polynomial I_o satisfying (16_o). So P is a Keller coordinate, as desired (cf. sec. 1 (b)).

This ends the proof of the theorem.

Let us notice that the first essential fact in the characterization considered is condition (ii) which permits the construction of sequence (7).

Let us notice that the second essential fact in the given characterization is condition (iii). It means that in sequence (7) there must appear a term (A_s,B_s) such that A_s and B_s are coefficients of a total differential.

Let us notice that the coefficients of the polynomials A_1,B_1 from formula (8_1) in Observation 3 can be expressed effectively polynomially by the coefficients of the polynomial P (cf. [7], p.25). Consequently, the same occurs to the further polynomials $(A_2,B_2),\ldots,$ in particular, to the last pair (A_s,B_s). In this situation, condition (ii), that is, (8_1), and condition (iii) of the breaking off of sequence (7), i.e. (9), can be written as a system of the finite number of effective polynomial equations connecting the coefficients of the polynomial P.

By these equations, in view of (16_o), (16_l), the coefficients of the corresponding polynomials $I_{s-1},I_{s-2},\ldots,$ in particular, the coefficients of the last one - I_o, can be expressed effectively polynomially by the coefficients of the polynomial P. In other words, the condition of the constancy of the Jacobian, connecting the coefficients of both the coordinates of the mapping, can be replaced by a system of algebraic equations connecting the coefficients of one coordinate only; the coefficients of the other coordinate can already be expressed polynomially by the coefficients of the first one.

Summing up, the coefficients of one coordinate of the Keller mapping under consideration run over effectively determined algebraic sets, whereas the coefficients of the other coordinate run over the images of the above-mentioned algebraic sets under an effective polynomial mapping depending also on arbitrary parameters (cf. sec. 1 (b)).

THEOREM 2. The polynomial P is a Keller coordinate if and only if the following condition is satisfied:

(iv) there exist polynomials $S,T \in \mathbb{C}[x,y,\lambda]$ such that

$$P_xT - P_yS = 1 + \lambda(T_x - S_y), \quad \deg_x S \leqq m - 2, \quad \deg_x T \leqq m - 2. \tag{18}$$

Proof. At first, we shall show the necessity of the condition. Let P be a Keller coordinate. According to Theorem 1, sequence (7) from Observation 3 breaks off. Let r be the number indicated in Observation 4. Let us define the desired polynomials in variables x,y,λ as follows

$$S = \sum_{k=1}^{r} A_k\lambda^k, \qquad T = \sum_{k=1}^{r} B_k\lambda^k. \tag{19}$$

Hence, for $r > 1$, by the definition of (A_1,B_1), we have

$$P_xT - P_yS = 1 + \sum_{k=1}^{r-1} (P_xB_{k+1} - P_yA_{k+1})\lambda^{k+1}. \tag{20}$$

On the other hand, in view of Observation 4, we obtain

$$\lambda(T_x - S_y) = \sum_{k=1}^{r} (B_{kx} - A_{ky})\lambda^{k+1} = \sum_{k=1}^{r-1} (B_{kx} - A_{ky})\lambda^{k+1}. \tag{21}$$

From (20), (21) and (8_{k+1}) we obtain (18). For $r = 1$, the left-hand side of (21) vanishes, therefore (19) and (8_1) immediately give (18). Thus, condition (iv) is fulfilled.

Now, we shall show the sufficiency of the condition. Let (iv) be fulfilled. Let us represent the polynomials S,T in the form

$$S = \sum_{k=1}^{s} A_k\lambda^k, \qquad T = \sum_{k=1}^{s} B_k\lambda^k \tag{22}$$

where $s = \max(\deg_\lambda S, \deg_\lambda T)$ and $A_k,B_k \in \mathbb{C}[x,y]$, $k = 1,2,\dots,s$, denote some polynomials which, for the sake of simplicity and because of what follows, are denoted by the same symbols as in (7). Putting (22) into (18) and comparing the coeeficients at the corresponding powers of the variable λ, we state that the following conditions are satisfied:

$$P_x B_1 - P_y A_1 = 1, \quad \deg_x A_1 \leqq m-2, \quad \deg_x B_1 \leqq m-2, \qquad (23_1)$$

$$P_x B_{k+1} - P_y A_{k+1} = B_{kx} - A_{ky}, \qquad (23_{k+1})$$

$$\deg_x A_k \leqq m-2, \quad \deg_x B_k \leqq m-2, \quad k = 1,\dots,s-1,$$

and

$$B_{sx} - A_{sy} = 0. \qquad (24)$$

From (23_1), (23_{k+1}) and Observation 3 it follows that the polynomials A_k, B_k, $k = 1,\dots,s$, from (23_1) and (23_{k+1}) are identical with the corresponding polynomials from sequence (7). Furthermore, sequence (7), in view of (24), breaks off. So, according to Theorem 1, P is a Keller coordinate.

This ends the proof of the theorem.

REFERENCES

[1] S.S. Abhyankar, *Expansion technique in algebraic geometry*, Tata Institute of Fundamental Research, Bombay, 1977.

[2] H. Bass, E.H. Connel and D. Wright, 'The Jacobian conjecture: reduction of degree and formal expansion of the inverse', *Bull. of Amer. Math. Soc.* 7, No. 2, (1982), pp.287-330.

[3] Z. Charzyński, J. Chądzyński and P. Skibiński, 'A contribution to Keller's Jacobian conjecture', Seminar on Deformations, Proceedings, Łódź-Warsaw 1982/84, *Lecture Notes in Math.* 1165, pp.36-51, Springer-Verlag, Berlin - Heidelberg, 1985.

[4] Z. Charzyński and P. Skibiński, 'A criterion for explicit dependence of polynomials', *Bull. Soc. Sci. Lettres Łódź* (to appear).

[5] O.H. Keller, 'Ganze Cremona-Transformationen', *Monats. Math. Physik* 47 (1939), pp.299-306.

[6] A. Magnus, 'Volume-preserving transformations in several complex variables', *Proc. Amer. Math. Soc.* 5 (1954), pp.256-266.

[7] R.J. Walker, *Algebraic curves*, Springer-Verlag, New York-Heidelberg-Berlin, 1978.

LOCAL PROPERTIES OF INTERSECTION

Tadeusz Winiarski
Institute of Mathematics
Jagiellonian University
PL-30-059 Kraków, Poland

ABSTRACT. Rouche's and Hurwitz's type theorems can be used in a non-compact case to prove theorems on existence of zero points or points of intersection. One of them is the local version of Bézout's theorem for intersection of analytic subsets of an open bounded subset of $\mathbb{C}^n$ presented in [14]. The other one is presented in this paper.

1. INTRODUCTION

The majority of theorems concerning the theory of intersection of analytic subsets of a non-compact manifold are stated for the points of their intersection. However, the answer to the question of the existence of their common points does not appear in the literature. The similar situation concerns theorems on the existence of a zero point of a holomorphic mapping. The case of the intersection of algebraic sets is clearer and "almost all" items of information connected with intersection theory in algebraic geometry can be found in Fulton's book [4].

In order to follow the programme given in Abstract, denote by $X_1,\dots,X_\nu$ pure-dimensional analytic subsets of an open subset U of $\mathbb{C}^n$ with $\sum_{i=1}^{\nu} \dim X_i = (\nu-1)n$. Assume P is an isolated point of $X_1 \cap \dots \cap X_\nu$ and

$$\dim((C_p(V_1)) \cap \dots \cap (C_p(V_\nu))) \geq 1,$$

where $C_p(V_j)$ is the tangent cone to V_j at the point P. We prove (Theorem 3) that for every $r>0$ there exist neighbourhoods $\mathcal{U}_1,\dots,\mathcal{U}_\nu$ (in the topology of local uniform convergence) of $X_1,\dots,X_\nu$, respectively, such that

$$(B(0,r)\setminus\{0\}) \cap (Y_1 \cap \dots \cap Y_\nu) \neq \emptyset,$$

whenever $Y_1 \in \mathcal{U}_1,\dots,Y_\nu \in \mathcal{U}_\nu$ are pure-dimensional analytic subsets of U with $\dim Y_1 = \dim X_1,\dots,\dim Y_\nu = \dim X_\nu$,

$$\deg_P Y_j \leq \deg_P X_j \quad \text{for} \quad j=1,\dots,\nu$$

J. Ławrynowicz (ed.), Deformations of Mathematical Structures, 141–150.

and

$$(C_P Y_1) \cap \ldots \cap (C_P Y_\nu) = \{0\}.$$

To prove this theorem we present some preparatory material supported by Whitney's and Draper's ideas as well as the ideas presented in [11] and [14].

As a simple corollary to the presented definition of the degree of an analytic set at P we obtain the following theorem: if $X_1,\ldots,X_\nu$ are pure-dimensional analytic subsets of U, $\Sigma_{i=1}^{\nu} \dim X_i = (\nu-1)n$ and P is an isolated point of $X_1 \cap \ldots \cap X_\nu$, then

$$i(X_1 \cdot \ldots \cdot X_\nu;\ P) \geq \deg_P X_1 \cdot \ldots \cdot \deg_P X_\nu$$

with equality if and only if $(C_P X_1) \cap \ldots \cap (C_P X_\nu) = \{0\}$. Without difficulties a similar theorem can be formulated for the cycle intersection.

A special case of this theorem (except [4], Corollary 12.4.) appears frequently in the literature (for more detailed information see [4], p. 234).

We complete this paper with the case of the isolated zero point of a holomorphic mapping $f = (f_1,\ldots,f_n)$ of an open subset of $\mathbb{C}^n$ into $\mathbb{C}^n$ and indicate how this technique can be used to construct a simple proof of Cih Južakov's theorem (see [5], [2], and [7]) which establishes the relation between the multiplicity of f at its zero point and the degrees of the leading forms of the expansions of $f_1,\ldots,f_n$ as the series of homogeneous polynomials.

2. CONTINUITY OF INTERSECTION

In this section we review some of the known facts on the topology of local uniform convergence, especially some results concerning theorems on continuity of intersection.

Let X be a metric space. Let $\mathcal{F}_X$ be the family of all closed subsets of X. We endow $\mathcal{F}_X$ with the topology $\mathcal{F}_X$ generated by the sets

$$\mathcal{U}(S, K) = \{F \in \mathcal{F}_X : F \cap K = \emptyset \text{ and } F \cap U \neq \emptyset \text{ for } U \in S\}$$

corresponding to all compact subsets $K \subset X$ and all finite families S of open subsets of X. We call this topology the topology of local uniform convergence.

Let Ω be an open subset of $\mathbb{C}^n$. By $A_p(\Omega)$ we will denote the subset of $\mathcal{F}_\Omega$ consisting of all purely p-dimensional analytic subsets of Ω. We will suppose that $\emptyset \in A_p(\Omega)$ for $p = 0,1,\ldots,n$.

<u>THEOREM 1</u> ([11], Th. 3). *Let* $V_o \in A_p(\Omega)$, $W_o \in A_q(\Omega)$, *and* $p+q \geq n$. *If* $V_o \cap W_o \in A_{p+q-n}(\Omega)$ *then the mapping*

$$\cap : A_p(\Omega) \times A_q(\Omega) \ni (V,W) \to V \cap W \in \mathcal{F}_\Omega$$

is continuous at the point (V_o, W_o).

Let M be a complex vector space of dimension $n+1$ and let $\mathbb{P}(M)$ denote the projective space of M. The mapping

$$\mathbb{P} : M \setminus \{0\} \ni x \to \mathbb{C}x \in \mathbb{P}(M)$$

is holomorphic and denoted by the same letter $\mathbb{P}$ for all vector spaces. The points of Grassmann manifold $G_p(M)$ are regarded as $(p+1)$-dimensional linear subspaces of M or p-planes in $\mathbb{P}(M)$.

For all $p \in \{1,\dots,n\}$ the set $F_p(M)$, defined by

$$F_p(M) = \{(a,\zeta) \in \mathbb{P}(M) \times G_p(M) : a \in \zeta\},$$

is a compact connected submanifold of $\mathbb{P}(M) \times G_p(M)$ of dimension $(p+1)(n-p)+p$. We shall use two natural projections p_1, p_2 defined by

$$p_1 : F_p(M) \ni (a,\zeta) \to a \in \mathbb{P}(M)$$

and

$$p_2 : F_p(M) \ni (a,\zeta) \to \zeta \in G_p(M).$$

It is easily seen that p_1, p_2 are holomorphic proper surjective submersions.

If X is a non-empty purely k-dimensional locally analytic subset of the projective space $\mathbb{P}(M)$, we define

$$D = D(X) = \{\zeta \in G_{n-k}(M) : \zeta \cap \partial X = \emptyset\},$$

where $\partial X = \bar{X} \setminus X$. We now turn to one theorem presented in [14].

THEOREM 2. *If* $D(X)$ *is a non-empty subset of* $G_{n-k}(M)$, *then*
1. $D(X)$ *is open in* $G_{n-k}(M)$,
2. if D' *is a connected component of* $D(X)$ *then*
a) the set A' *defined by*

$$A' = \{\zeta \in D' : \# \zeta \cap X = \infty\}$$

is a proper analytic subset of D' *and*

b) the function

$$D' \setminus A' \ni \zeta \to \deg(\zeta \cdot X) = \Sigma_{a \in \zeta \cap X}\, i(\zeta \cdot X; a) \in \mathbb{Z}$$

is constant, where $i(\zeta \cdot X; a)$ *is the intersection multiplicity of* ζ *and* X *at* a *(in the sense given in [3]).*

Part 1.b) of the above theorem can be deduced from Draper's theorem ([3], Th. 5.4, p. 192) or independently as it was presented in [14].

3. PLANES HAVING COMMON POINTS WITH THE BORDER

We may consider $\mathbb{C}^n$ an open subset of $\mathbb{P}(\mathbb{C}^{n+1})$ by means of the mapping

$$\phi : \mathbb{C}^n \ni x \to \mathbb{P}(x, 1) \in \mathbb{P}(\mathbb{C}^{n+1}).$$

If $\zeta \in G_{n-k-1}(\mathbb{C}^n)$ then $\phi(\zeta) \in G_{n-k}(\mathbb{C}^{n+1})$ and we can see that ϕ sets a one-to-one correspondence between the (n-k)-dimensional subspaces of $\mathbb{C}^n$ and (n-k+1)-dimensional subspaces of $\mathbb{C}^{n+1}$ passing through the line $\{0\} \times \mathbb{C}$, where $0 = (0,\dots,0) \in \mathbb{C}^n$. Then we may consider $G_{n-k-1}(\mathbb{C}^n)$ an algebraic subset of $G_{n-k}(\mathbb{C}^{n+1})$.

Let C be an analytic cone in $\mathbb{C}^n$ of pure dimension k. Recall that by Chow's theorem C is an algebraic subset of $\mathbb{C}^n \subset \mathbb{P}(\mathbb{C}^{n+1})$. Since p_1 is a submersion and p_2 is proper, we deduce that $p_2(p_1^{-1}(C))$ is an algebraic subset of $G_{n-k-1}(\mathbb{C}^n)$ of codimension ≥ 1, whenever $k < n$. Therefore, we have just proved a well-known

PROPOSITION 1. *If* C *is an analytic cone in* $\mathbb{C}^n$ *of pure dimension* $k < n$, *then*

(a) the set

$$\tilde{D}(C) = \{\zeta \in G_{n-k-1}(\mathbb{C}^n) : \zeta \cap C = \{0\}\}$$

is an open subvariety of $G_{n-k-1}(\mathbb{C}^n)$ *and*

(b) for every $\zeta \in D(C)$ *(following Th.2)*

$$i(C \cdot \zeta;\ 0) = \deg(C \cdot \zeta) = \deg C,$$

where $\deg C$ *is the degree of the Zariski closure of* C.

A little bit more delicate methods have to be used to analyse the structure of (n-k)-dimensional subspaces of $\mathbb{C}^n$ having common points with the border of a locally analytic subset of $\mathbb{C}^n$ of pure dimension k. Yet if we study the local properties of analytic sets we only need to cinsider the locally analytic subsets of $\mathbb{C}^n$ with real analytic borders.

Let

$$\Omega_k = \{(x,\zeta) \in (\mathbb{C}^n \setminus \{0\}) \times G_{n-k-1}(\mathbb{C}^n) : x \in \zeta\}.$$

The set Ω_k is a submanifold of $(\mathbb{C}^n \setminus \{0\}) \times G_{n-k-1}(\mathbb{C}^n)$ of dimension $(n-k) + \dim G_{n-k-1}(\mathbb{C}^n)$. Thus Ω_k is a real manifold of dimension twice as much as its complex dimension. It is easy to verify that the natural projections

$$\pi_1 : \Omega_k \to \mathbb{C}^n \setminus \{0\} \quad \text{and} \quad \pi_2 : \Omega_k \to G_{n-k-1}(\mathbb{C}^n)$$

are submersions and π_1 is a proper mapping.

LEMMA 1. *Let* $V \subset \mathbb{C}^n \setminus \{0\}$ *be a (real) subanalytic subset of* $\mathbb{C}^n \setminus \{0\}$ *treated as* $\mathbb{R}^{2n} \setminus \{0\}$ *of codimension* $1 \geq 2n - 2k + 1$. *Then we set* W *defined by*

$$W = \pi_2(\pi_1^{-1}(W))$$

is a subanalytic subset of $G_{n-k-1}(\mathbb{C}^n)$ *of codimension* ≥ 1.

P r o o f. Since π_1 is an analytic submersion, the set $\pi_1^{-1}(V)$ is a subanalytic subset of Ω_k and $\operatorname{codim} V = \operatorname{codim} \pi_1^{-1}(V)$. Since π_2 is proper, the set W is a subanalytic subset of $G_{n-k-1}(\mathbb{C}^n)$ and

$$\dim_{\mathbb{R}}(W) \leq \dim_{\mathbb{R}} \pi_1^{-1}(V).$$

Hence

$$\operatorname{codim}_{\mathbb{R}}(W) \geq 2\dim G_{n-k-1}(\mathbb{C}^n) - (2\dim G_{n-k-1}(\mathbb{C}^n) + 2(n-k) + 2n - 2k - 1) = 1,$$

which completes the proof.

4. DEGREE AT POINT

Let X be a pure k-dimensional locally analytic subset of $\mathbb{C}^n$. Without loss of generality we can assume that $0 \in X$. For $t \in C$ we define

$$X_t = \begin{cases} \{x \in \mathbb{C}^n : tx \in X\} & \text{if } t \neq 0, \\ C = C_0(X) & \text{if } t = 0, \end{cases}$$

where $C_0(X)$ is the tangent cone to X at the point 0 (in the sense of Whitney [13]).

For every $r > 0$ the family $(X_t \cap B(0,r))$, parametrized by t, is a continuous family of pure k-dimensional analytic subsets of the ball $B(0,r)$ for $|t|$ small enough. Hence, for every open cone U in $\mathbb{C}^n$ containing $C_0(X)$ there exists $r_0 > 0$ such that (comp. [13], Lemma 8.15.)

(i) X is analytic in an open set containing $\overline{B(0, 2r_0)}$ and

(ii) $(X \cap B(0,r)) \subset U$, whenever $r \leq r_0$.

Since $X \cap \partial B(0,r)$ is a compact real analytic subset of $\mathbb{C}^n$ of real codimension $1 \geq 2n - 2k + 1$, by using Lemma 1 we can see that the set $D_r(X)$ defined by

$$D_r(X) = \{\zeta \in G_{n-k-1}(\mathbb{C}^n) : \zeta \cap X \cap \partial B(0,r) = \emptyset\}$$

is an open dense subset of $G_{n-k-1}(\mathbb{C}^n)$ for $r \leq r_0$ and, following (ii), if $\zeta_1, \zeta_2 \in \tilde{D}(C_0(X))$ then there exists $r_1 \leq r_0$ such that ζ_1 and ζ_2 lie at the same connected component of $D_r(X)$ for $r \leq r_1$. Therefore (by Th. 1):

$$i(\zeta_1 \cdot X;\, 0) = i(\zeta_2 \cdot X;\, 0)$$

and if $\zeta \in \tilde{D}(C_0(X))$, $\zeta' \notin \tilde{D}(C_0(X))$ and $\zeta' \in D_r(X)$ then

$$i(\zeta \cdot X;\, 0) < i(\zeta' \cdot X;\, 0).$$

Moreover, there exists $0 < r < r_0$ such that $D_r(X)$ has only one component if and only if $X \cap B(0,r) \subset C_0(X)$. The degree of X at 0 is defined as

$$\deg_o X = i(\zeta \cdot X; 0) \quad \text{for} \quad \zeta \in \tilde{D}(C_o(X)).$$

Since the intersection multiplicity is invariant with respect to biholomorphic transformations we define

$$\deg_P X = \deg_o (X - p) \quad \text{for} \quad p \in X,$$

whenever the germ of X at P has pure dimension. More general definition of $\deg_P X$ and its comparison with the Lelong number and the multiplicity of the local rings may be found in [3] and [10].

Now, we give three properties of the degree of X at P.

Property 1. If $X = X_1 \cup \ldots \cup X_s$ is the decomposition of the germ of X at P on irreducible components, then

$$\deg_P X = \deg_P X_1 + \ldots + \deg_P X_s.$$

Property 2. If X_j has pure dimension at $P_j \in X_j$, $j = 1,\ldots,s$, $X = X_1 \times \ldots \times X_s$ and $P = (P_1,\ldots,P_s)$, then

$$\deg_P X = \prod_{j=1}^{s} \deg_P X_j.$$

Property 3. If $X_1,\ldots,X_\nu$ are pure dimensional analytic subsets of an open neighbourhood $U \subset \mathbb{C}^n$ of the point 0 such that

$$\Sigma_{i=1}^{\nu} \dim X_i = (\nu - 1)n,$$

then

$$i(X_1 \cdot \ldots \cdot X_\nu; 0) \geq \ _{i=1}^{\nu} \deg_o X_i$$

with equality if and only if $\mathbb{P}(C_o(X_1) \cap \ldots \cap C_p(X_\nu)) = \emptyset$.

Proof. Properties 1, 2, and 3 are immediate consequences of the definition of the degree of X at P and the following two simple facts:

1. $i(X_1,\ldots,X_\nu; P) = i((X_1 \times \ldots \times) \cdot \Delta; \delta(P))$,

where Δ is the diagonal in $(\mathbb{C}^n)^\nu$ and $\delta : \mathbb{C}^n \ni x \to (x,\ldots,x) \in \Delta$ is the natural imbedding of $\mathbb{C}^n$ onto Δ;

2. $C_{\delta(p)}(X_1 \times \ldots \times X_\nu) = C_p(X_1) \times \ldots \times C_p(X_\nu)$.

Note that by using Draper's definition of the cycle intersection it is easy to prove a theorem for the cycle intersection, analogous to Property 3.

5. EXISTENCE OF POINTS OF INTERSECTION

Let $f, f_\nu : U \to \mathbb{C}^n$, $\nu = 1,2,\ldots$, be holomorphic mappings defined in an open subset $U \subset \mathbb{C}^n$. Suppose that $P \in U$, f has an isolated zero point at P with its multiplicity $m(P,f) = k \geq 1$ and that $m(f_\nu,P) < k$ for $\nu = 1,2,\ldots$. Then, by Hurwitz's theorem, for every $r > 0$ there exists an integer ν_o such that f_ν has a zero point in $B(P,r)$ different from P, for $\nu \geq \nu_0$.

In this part we present an extension of this fact to the case of the intersection of analytic sets.

PROPOSITION 1. *Let* U *be an open neighbourhood of a point* $P \in \mathbb{C}^n$, $X \in A_k(U)$ *and let* ζ *be an* $(n-k)$*-dimensional affine subspace of* $\mathbb{C}^n$ *such that* P *is an isolated point of* $\zeta \cap X$. *Then for every* $r > 0$ *there exists a neighbourhood* $\mathcal{U}$ *of* X *(in the topology of local uniform convergence) such that for every* $Y \in \mathcal{U} \cap A_k(U)$ *if*

$$i(\zeta \cdot Y;\, P) < i(\zeta \cdot X;\, P).$$

Then

$$Y \cap \zeta \cap B(P,r) \setminus \{P\} \neq \emptyset.$$

Proof. There is no loss of generality in assuming that $P = 0$. Let η be a k-dimensional vector subspace of $\mathbb{C}^n$ such that $\mathbb{C}^n = \eta + \zeta$. Obviously, there exist two open connected neighbourhoods Ω_η, Ω_ζ of 0 in η, ζ, respectively, such that

1. $\overline{\Omega}_\eta$, $\overline{\Omega}_\zeta$ are compact subsets of $\mathbb{C}^n$,
2. $\overline{\Omega_\eta + \Omega_\zeta} \subset U$

and

3. $(\overline{\Omega}_\eta + \partial\Omega_\zeta) \cap X = \emptyset$ and $(\zeta \cap \overline{\Omega}_\zeta) \cap X = \{0\}$.

Let p be the natural projection of $\Omega_\eta + \Omega_\zeta$ onto Ω_η. Then the restriction $p|X$ is $s = i(\zeta \cdot X; 0)$-sheeted branched covering. Let us fix a point $x \in \Omega_\eta$ which is a regular point of the branched covering $p|X$. Thus

$$(x + \Omega_\zeta) \cap X = \{x + y_1, \ldots, x + y_s\},$$

where $y_j \in \Omega_\zeta \cap X$ for $j = 1, \ldots, s$ and $y_i \neq y_j$, whenever $i \neq j$. Therefore there exist open neighbourhoods $U_x \subset \eta$ of x and $V_j \subset \zeta$ of y_j, $j = 1, \ldots, s$ such that

$$\overline{U_x + V_j} \subset \Omega_\eta + \Omega_\zeta,$$

$\overline{V}_j$ is a compact subset of Ω_ζ for $j = 1, \ldots, s$ and

$$\overline{V}_i \cap \overline{V}_j = \emptyset \quad \text{for } i \neq j.$$

We now construct the required neighbourhood $\mathcal{U} = \mathcal{U}(S,K)$ by putting

$$S = \{U_x + V_1, \ldots, U_x + V_s\}$$

and

$$K = (\overline{\Omega}_\eta + \partial\Omega_\zeta) \cup (x + (\Omega_\zeta \setminus \textstyle\bigcup_{j=1}^{s} V_j)).$$

Now, if $Y \in A_k(U) \cap \mathcal{U}(S,K)$ then $p|Y$ is at least s-sheeted branched covering and $Y \cap \zeta \cap \Omega_\zeta \setminus \{0\} = \emptyset$ implies that $i(\zeta \circ Y; 0) \geq i(\zeta \cdot X; 0)$ and it contradicts our assumption.

THEOREM 3. *Let* $U \subset \mathbb{C}^n$ *be an open subset of* $\mathbb{C}^n$, $P \in U$ *and let* $X_1, \ldots, X_\nu$ *be analytic subsets of* U *of pure dimensions* $d_1, \ldots, d_\nu$, *respectively. If* $d_1 + \dots + d_\nu = (\nu - 1)n$, P *is an isolated point of* $X_1 \cap \dots \cap X_\nu$ *and* $\mathbb{P}(C_p(X_1) \cap \dots \cap C_p(X_\nu)) \neq \emptyset$ *then for every* $r > 0$ *there exist neighbourhoods* $\mathcal{U}_1, \dots, \mathcal{U}_\nu$ *(in the topology* T_U*) of* $X_1, \dots, X_\nu$, *respectively, such that if*

$$Y_1 \in A_{d_1} \cap \mathcal{U}_s, \dots, Y_\nu \in A_{d_\nu} \cap \mathcal{U}_\nu,$$

$$\mathbb{P}(C_p(Y_1) \cap \dots \cap C_p(Y_\nu)) = \emptyset \quad \textit{and} \quad \deg_p Y_j \leq \deg_p X_j, \quad j = 1, \dots, \nu,$$

then

$$(B(P,r) \setminus \{P\}) \cap (Y_1 \cap \dots \cap Y_\nu) \neq \emptyset.$$

Proof. The set $X = X_1 \times \dots \times X_\nu \subset \mathbb{C}^{n\nu}$ is an analytic subset of the set

$$\tilde{U} = U \times \dots \times U \subset \mathbb{C}^{n\nu}$$

of pure dimension $(\nu - 1)n$ and

$$\deg_{\tilde{P}} X = \prod_{j=1}^{\nu} \deg_p X_j,$$

where $\tilde{P} = (P, \dots, P)$. The diagonal

$$\Delta = \{(x, \dots, x) \in (\mathbb{C}^n)^\nu : x \in \mathbb{C}^n\}$$

has dimension n and $\tilde{P}$ is an isolated point of $\Delta \cap X$. Now, to end the proof of Theorem 3 it suffices to apply Proposition 1, with $\zeta = \Delta$, and the fact, easy to verify, that every neighbourhood of X in F_U contains a Cartesian product of neighbourhoods $\mathcal{U}_1, \dots, \mathcal{U}_\nu$ of $X_1, \dots, X_\nu$, respectively.

Now, we apply the above theorem to the case of holomorphic mappings. In order to present it we begin with some preparatory information.

Let $\mathbb{L} = \mathbb{L}(\mathbb{C}^n, \mathbb{C}^n)$ be the set of all linear mappings of $\mathbb{C}^n$ into $\mathbb{C}^n$, let $\mathbb{L}^n$ denote the k-fold Cartesian product and let $I = (I_1, \dots, I_n)$ be the identity mapping of $(\mathbb{C}^n)^n$.

Now, let $f = (f_1, \dots, f_n)$ be a holomorphic mapping defined in an open neighbourhood of the point 0 in $\mathbb{C}^n$ such that $f(0) = 0$. We can expand f_j into the series of homogeneous polynomials

$$f_j = f_j^* + f_{j1} + f_{j2} + \dots = f_j^* + h_j, \quad j = 1, \dots, n,$$

where $f_j^* \neq 0$ is a homogeneous polynomial and h_j consists only of homogeneous polynomials of degree greater than the degree of f_j^* for $j = 1, \dots, n$.

For the simplicity of notation we define f_L and f_L^* by

$$f_L = (f_1 \circ L_1, \dots, f_n \circ L_n)$$

and

$$f_L^* = (f_1^* \circ L_1, \dots, f_n^* \circ L_n)$$

for $L = (L_1, \dots, L_n) \in \mathbb{L}^n$.

We can easily prove that

a) the set A defined by

$$A = \{L \in \mathbb{L}^n : f_L^*(x) \neq 0 \text{ for an } x \neq 0\}$$

is an algebraic subset of $\mathbb{L}^n$ of codimension ≥ 1;

b) if $L_1,\ldots,L_n$ are linear automorphisms of $\mathbb{C}^n$, then

$$\deg_o(f_j \circ L_j)^{-1}(0) = \deg_o(f_j^{-1}(0)), \quad j = 1,\ldots,n.$$

An immediate application of Theorem 3 gives

COROLLARY. *If* 0 *is an isolated point of* $f^{-1}(0)$ *and* $\mathbb{P}((f^*)^{-1}(0)) \neq \emptyset$, *then for every* $r > 0$ *there exists an open neighbourhood* V *of* I *in* $\mathbb{L}^n$ *such that* f_L *has at least one zero point in* $B(0,r)\setminus\{0\}$ *for all* $L \in V \setminus A$.

If we combine the above corollary with Rouché's and Bezout's theorems we obtain

THEOREM 4 (see [5], [2] and also [7]). *If* 0 *is an isolated point of the zero set of* f, *then*

$$m(f,0) \geq \prod_{i=1}^{n} \deg f_i^*$$

with equality if and only if $\mathbb{P}((f^*)^{-1}(0)) = \emptyset$, *where* $m(f,0)$ *denotes the multiplicity of* f *at* 0.

References

[1] ANDREOTTI, A. and STOLL, W.: 'Analytic and Algebraic Dependence of of Meromorphic Functions', *Lecture Notes in Math.* 234 (1971).

[2] ČIRKA, E.M.: *Kompleksnyje Analitičeskije Množestva*, "Nauka", Moskva 1985.

[3] DRAPER, R.: 'Intersection Theory in Analytic Geometry', *Math. Ann.* 180 (1969), 175-204.

[4] FULTON, W.: *Intersection Theory*, Springer-Verlag, 1984.

[5] JUŽAKOV, A.P. and CIH, A.K.: 'O kratnosti nula sistemy golomorfnych funkcij', *Sib. Mat. Ž.* 19,3 (1978), 693-697.

[6] MILNOR, J.: *Singular points of complex hypersurfaces*, Princeton Univ. Press, 1968.

[7] MIODEK, A.: 'Generalized parametric multiplicity', *Bull. Soc. Sci. Lettres Łódź* (to appear).

[8] PŁOSKI, A.: 'Sur l'exponant d'une application analytique II', *Bull. Polish Acad. Sci. Math.* 32,3-4 (1985), 123-127.

[9] STOLL, W.: 'The Multiplicity of Holomorphic Map', *Inv. Math.* 2 (1976), 15-58.

[10] THIE, P.R.: 'The Lelong number of a point', *Math. Ann.* 172 (1967), 269-312.

[11] TWORZEWSKI, P. and WINIARSKI, T.: Continuity of intersection of analytic sets', *Ann. Polon. Math.* 42 (1983), 387-393.

[12] WHITNEY, H.: *Complex analytic varieties*, Addison-Wesley Publ. Comp., 1972.

[13] ———: 'Tangents to an analytic variety', *Ann. Math.* 81 (1965), 496-549.

[14] WINIARSKI, T.: 'Continuity of total number of intersection', *Ann. Polon. Math.* 47 (to appear).

GENERALIZED PADÉ APPROXIMANTS OF KAKEHASHI'S TYPE AND MEROMORPHIC CONTINUATION OF FUNCTIONS

Ralitza Krumova Kovacheva
Institute of Mathematics
Bulgarian Academy of Sciences
BG-1090 Sofia, Bulgaria

ABSTRACT. Converse theorems related to interpolating sequences of rational functions with a fixed number of free poles are proved. These theorems give sufficient conditions for the meromorphic continuability of functions.

1. Denote by $\bar{D}$ the closed unit disk. Let $N = 0, 1, 2, \ldots$. Let $\omega = \{\omega_n\}_{n \in N}$ be a sequence of polynomials, $\omega_n(z) = z^n + \ldots$, $\omega_n(z) \neq 0$ for $z \in \bar{D}$, such that

$$\lim \omega_n(z) \cdot z^{-n} = \lambda(z) \quad \text{as} \quad n \in N, \tag{1}$$

uniformly on compact sets in $D^c = C - \bar{D}$; the function $\lambda(z)$ is continuous in D^c, $\lambda(z) \neq 0$, $z \in D^c$. In our further considerations we shall use the notation $\omega \in K_\lambda(\bar{D})$.

We consider a function f holomorphic on $\bar{D}$ $(f \in H(\bar{D}))$. For each pair (n,m), $n, m \in N$, we denote by $R_{n,m} = R^{\omega}_{n,m}(f)$ the generalized Padé approximant to f of order (n,m) with respect to ω. It is known (see, for example, [1]) that $R_{n,m} = p/q$, where p and q are polynomials, the degrees of which do not exceed n and m, respectively, $q \not\equiv 0$, and are determined so that

$$(f.q - p) . \omega^{-1}_{n+m+1} \in H(\bar{D}). \tag{2}$$

We shall call the rational functions $R_{n,m}$, $n, m \in N$, *generalized Padé approximants of Kakehashi's type.*

We set

$$R_{n,m} = P_{n,m}/Q_{n,m},$$

where $P_{n,m}$ and $Q_{n,m}$ have no common divisor and $Q_{n,m}$ is monic. In the case when $\deg Q_{n,m} = m$, the rational function $R_{n,m}$ has exactly m finite (free) poles (as usual, we shall count the poles with their multiplicities) and (2) is valid with $p = P_{n,m}$ and $q = Q_{n,m}$. It is known that in spite of the fact that p and q are not unique,

J. Ławrynowicz (ed.), Deformations of Mathematical Structures, 151–159.

$R_{n,m}$, $P_{n,m}$ and $Q_{n,m}$ are unique.

In the present work we are going to prove three theorems.

THEOREM 1. *Let* $f \in H(\bar{D})$ *and* $\omega \in K_\lambda(\bar{D})$. *Suppose there exists a polynomial* Q *such that* $\deg Q = m$, $Q(z) \neq 0$ *for* $z \in \bar{D}$, *and*

$$\limsup_{n \in N} \|Q_{n,m} - Q\|^{1/n} = \delta < 1$$

(*the norm is understood as taken in the metric of the coefficients*).

Then the function $f.Q$ *admits a holomorphic continuation in the disk* $D_R = \{z, |z| < R\}$, *where* $R = \delta^{-1}. \max\{|\text{zeros}| \text{ of } Q\}$ *and all the zeros of* Q *are poles of* f.

For each $k \in N$ we denote by

$$D_{R_k} = D_{R_k}(f)$$

the largest disk where the function f admits a continuation as a meromorphic function with not more than k poles. Since $f \in H(\bar{D})$, we have $R_k > 1$, $k \in N$. We set

$$\Gamma_{R_m} = \partial D_{R_m} .$$

It follows from Theorem 1 that $R_m \geq \delta^{-1}. \max\{|\text{zeros}| \text{ of } Q\}$.

THEOREM 2. *Let* $f \in H(\bar{D})$ *and* $\omega \in K_\lambda(\bar{D})$. *Suppose there exists a polynomial* Q *of the form*

$$Q(z) = \sum_{k=1}^{\kappa} (z - \gamma_k^*)^{\tau_k}, \quad \tau_1 + \ldots + \tau_\kappa = m, \quad \gamma_i^* \neq \gamma_j^*, \quad i \neq j,$$

$$i, j = 1, \ldots, \kappa, \quad 1 < |\gamma_1^*| \leq \cdots \leq |\gamma_\mu^*| < |\gamma_{\mu+1}^*| = \ldots = |\gamma_\kappa^*|$$

such that

$$\lim_{n \in N} Q_{n,m} \to Q .$$

For each $k = 1, \ldots, \kappa$ *we denote by* $\gamma_{n,\ell}^{(k)}$, $\ell = 1, \ldots, \tau_k$ *those zeros of* $Q_{n,m}$ *which approach* γ_k^* *as* $n \in N$. *Suppose*

$$\limsup_{n \in N} |\gamma_{n,\ell}^{(k)} - \gamma_k^*|^{1/n} = \delta < 1, \quad k = 1, \ldots, \mu, \ \ell = 1, \ldots, \tau_k. \quad (3)$$

Let $m_1 = \tau_1 + \ldots + \tau_\mu$.

Then $R_{m_1} = |\gamma_\kappa^*|$ *and the function* F, *given by*

$$F(z) = f(z) . \prod_{k=1}^{\mu} (z - \gamma_k^*)$$

is holomorphic in $D_{|\gamma_\kappa^*|}$.

THEOREM 3. *Let* $f \in H(\bar{D})$ *and* $\omega \in K_1(\bar{D})$. *If there exists a polynomial* Q: $Q(z) = z - \gamma$, $|\gamma| > 1$, *such that*

$$Q_{n,1} \to Q \quad \textit{as} \quad n \in N,$$

then $R_0 = |\gamma|$ *and* f *has a singularity at* γ.

In the conditions of Theorem 3 the function f has a simple pole at γ, if $Q_{n,1}$ tend to Q as $n \in N$ with a speed of a geometric progression. In this case $R_1 \geq |\gamma| . (\limsup_{n\in N} \|Q_n - Q\|^{1/n})^{-1}$.

2. The next theorem is the basis to our later considerations.

THEOREM 4. *Let* $\phi \in H(\bar{D})$, $\phi(z) = \Sigma_{\nu\in N} \phi_\nu z^\nu$. *Suppose*

$$R_0(\phi) = ((\limsup_{\nu\in N} |\phi_\nu|^{1/\nu})^{-1}) < \infty.$$

Suppose also that the functions ψ_n, $n \in N$, *are holomorphic in a neighbourhood* U *of* Γ_{R_0} *and*

$$\psi_n \to \psi \textit{ as } n \in N, \textit{ uniformly on } \bar{U}, \psi(z) \neq 0 \textit{ for } z \in \bar{U}.$$

We set

$$(\phi\psi_n)(z) = \sum_{\nu=-\infty}^{\infty} c_\nu^{(n)} (z/R_0)^\nu .$$

Then

$$0 < \limsup \left|\frac{c_n^{(n)}}{\mu_n}\right| < \infty \quad ,$$

where $\{\mu_n\}_{n\in N} = \{\mu_n(\phi)\}_{n\in N}$ *is a monotone sequence such that* $\mu_{n+1}\mu_n^{-1} \to 1$ *as* $n \in N$.

Theorem 4 is due to Kakehashi (see [2]). Gonchar proved (unpublished) that in the conditions of Theorem 4

$$\limsup_{n\in N} \left|\frac{1}{2\pi i} \int_{\Gamma_\rho} \phi(z)\psi_n(z).z^{-n}\, dz\right|^{1/n} = (R_0(\phi))^{-1}, \quad 1 < \rho < R_0(\phi).$$

As a consequence of the last equality, we obtain

$$\limsup_{n\in N} |I_n(f)|^{1/n} = (R_0(f))^{-1}, \tag{4}$$

where

$$I_n(f) = \frac{1}{2\pi i} \int_{\Gamma_\rho} f(z).\omega_{n+1}^{-1}(z)\, dz, \quad n \in N, \quad 1 < \rho < R_0(f).$$

3. Proof of Theorem 1. It follows from the conditions of the theorem that f is not a rational function with less than m finite poles in $\bar{C}$. Since $f \in H(\bar{D})$, we have $R_m > 1$. Suppose that the number of poles of f in D_{R_m} is less than m. It follows from the results of [3] that all the poles of f are situated among the zeros of the polynomial Q. Let $Q(\gamma_k) = 0$, $k = 1, \ldots, m$ (some of the poles may

coincide) and let $\gamma_{k_1}, \dots, \gamma_{k_s}$, $s < m$, be the poles of f (in D_{R_m}). We set

$$\phi(z) = \prod_{i=1}^{s} (z - \gamma_{k_i}) \quad \text{and} \quad F = f \cdot \phi .$$

Obviously,

$$R_o(F) = R_m(f) = R_o(F \cdot Q) \quad (= R_m^{-1}).$$

We obtain from (2) that

$$I_{n+m}(F \cdot Q) = I_{n+m}(F \cdot (Q - Q_{n,m})), \qquad n \geq n_o .$$

The last equality and the conditions of theorem imply that

$$\limsup_{n \in N} |I_n(F \cdot Q)|^{1/n} \leq R_m^{-1} \delta .$$

The last inequality is possible only in the case when $R_m = \infty$, i.e. when F is an entire function.

Suppose that F is an entire function. We shall prove that such an assumption contradicts the conditions of the theorem. The main idea we shall use is contained, in essence, in [4].

Let θ and ρ' be positive numbers such that $\exp \theta < \rho'$. Now fix a number ρ, $\rho > \rho'$. We introduce the notation

$$F(z) = \sum_{\nu \geq 0} \Phi_\nu \cdot (z/\rho)^\nu . \tag{5}$$

We have following our assumption $(F \in H(\mathbb{C}))$:

$$|\Phi_\nu|^{1/\nu} \to 0 \quad \text{as} \quad \nu \in N.$$

Let $\{\Phi_n\}$, $n \in \Lambda$, be the "maximal" coefficients, namely, let $\{\Phi_n\}_{n \in \Lambda} = \{\Phi_n,\ n \in N,\ \Phi_n \neq 0$ and $|\Phi_n| \geq |\Phi_\nu|$ for all $\nu \geq n$, $\nu \in N\}$. Since F is not a polynomial, the sequence Λ is infinite.

For each $k = 0, \dots, m$ and $n \in \Lambda$ we consider the functions $\phi_{n,k}$, given by

$$\phi_{n,k}(z) = \sum_{\nu \geq n+m-k} (\Phi_\nu / \Phi_n) \cdot (z/\rho)^{\nu - n} \cdot z^{k-m} .$$

For each $k = 0, \dots, m$ the functions $\phi_{n,k}$ are holomorphic in D_ρ and uniformly bounded inside it. Let ρ_1 be fixed, $1 < \rho_1 < \rho$. There exists a subsequence Λ_1, $\Lambda_1 \subset \Lambda$ such that the sequences $\phi_{n,k}$, $k = 0, \dots, m$ converge as $n \in \Lambda_1$. The limit functions ϕ_k are holomorphic on $\overline{D}_{\rho_1}$, $k = 0, \dots, m$. Let

$$\phi_k(z) = \phi_o^{(k)} + \phi_1^{(k)} \cdot z + \phi_2^{(k)} \cdot z^2 + \dots, \qquad k = 0, \dots, m.$$

Obviously,

$$|\phi_\nu^{(k)}| \leq \rho^{-(m-k+\nu)}, \qquad \nu \in N, \quad k = 0, \dots, m; \tag{6}$$

$$\phi_o^{(m)} = 1. \tag{6'}$$

Let

$$Q(z) = c_o z^m + c_1 z^{m-1} + \ldots + c_m, \quad c_o = 1.$$

It follows from the conditions of the theorem that for all $n \in N$, $n \geq n'$, $\deg Q_{n,m} = m$. We set for $n \geq n'$

$$Q_n(z) = Q_{n,m}(z) = c_{n,o} z^m + c_{n,1} z^{m-1} + \ldots + c_{n,m}, \quad c_{n,o} = 1.$$

Let C_1 be a positive number such that

$$C_1 > \max\{|c_k|, \quad k = 0, \ldots, m\}.$$

We get from the conditions of the theorem that

$$|c_{n,k}| \leq C_1, \quad n \geq n_1, \quad n \in N, \quad k = 0, \ldots, m. \tag{7}$$

Let $n \in \Lambda_1$. Since $\Phi_n \neq 0$, we obtain from (2) that

$$I_{n+m}(F \cdot Q_n / F_n) = 0.$$

This implies that

$$- I_{n+m}(\phi_{n,m}(z) \cdot z^{n+m}) = c_{n,1} I_{n+m}(\phi_{n,m-1}(z) \cdot z^{n+m}) + \ldots + \\ + c_{n,m} I_{n+m}(\phi_{n,0}(z) \cdot z^{n+m}). \tag{8}$$

For each $k = 0, \ldots, m$ we have

$$I_{n+m}(\phi_{n,k}(z) \cdot z^{n+m}) \to \frac{1}{2\pi i} \int_{\Gamma_{\rho_1}} \phi_k(z) \frac{1}{\lambda(z)} \frac{dz}{z}.$$

We set

$$\frac{1}{2\pi i} \int_{\Gamma_{\rho_1}} \phi_k \frac{1}{\lambda(z)} \frac{dz}{z} = a_k, \quad k = 0, \ldots, m.$$

We obtain from (8) and from the conditions of the theorem that

$$- a_m = a_{m-1} \cdot c_1 + \ldots + a_o \cdot c_m. \tag{9}$$

Consider a_k, $k = 0, \ldots, m$. Since $\lambda(z) \neq 0$ in D^c, $\lambda \in H(D^c)$, we have $\lambda^{-1} \in H(D^c)$. Let

$$\lambda^{-1}(z) = \lambda_o + \lambda_1 . z^{-1} + \lambda_2 . z^{-2} + \ldots .$$

We have

$$\lambda_o = 1$$

and

$$|\lambda_\nu| \leq \exp(\nu \theta) \quad \text{for all} \quad \nu, \ \nu \geq n_2. \tag{10}$$

Using (6), (7) and the last inequality, we obtain

$$|a_k| \le C_2 \cdot \rho^{k-m}, \quad k = 0, \ldots, m-1, \tag{10'}$$

where C_2 is a positive constant which does not depend on ρ, $C_2 = C_2(\theta, \rho')$. On the other hand, we get for a_m, from (10) and (6),

$$|a_m| = |1 + \ldots| \ge 1 - C_2 \cdot \rho^{-1} \cdot e^Q.$$

If the number ρ from (5) is taken in the way that $\rho > m C^2 + C \exp\theta$, where $C = \max(C_1, C_2)$, then the equality (10') and the last one contradict (9). Consequently the function F is a polynomial, i.e. the sequence Λ is finite. Yet such an assumption contradicts the conditions of the theorem. Thus the function f has exactly m poles in D_{R_m}. It follows from [3] that each point $\gamma_1, \ldots, \gamma_m$ is a pole of f; i.e. $\gamma_k \in D_{R_m}$, $k = 1, \ldots, m$.

Now it easy to complete the proof of the theorem. We renumber the zeros of Q so that $|\gamma_1| \le \ldots \le |\gamma_m|$. Suppose $|\gamma_1| \le \ldots \le |\gamma_\mu| < |\gamma_{\mu+1}| = \ldots = |\gamma_m|$ and consider the function F, given by $F(z) = f(z) \cdot \prod_{k=1}^{\mu} (z - \gamma_k)$. We have

$$R_o(F) = |\gamma_m|$$

and

$$R_o(F.Q) = R_m. \tag{11}$$

On the other hand, we get from (2) and from the conditions of the theorem that

$$I_{n+m}(F.Q) = I_{n+m}(F \cdot (Q - Q_n)), \quad n \ge n_o.$$

Using Theorem 4 and (11), we get

$$R_m \ge \delta^{-1} \cdot |\gamma_m|.$$

This completes the proof of the theorem.

Theorem 1 has been proved at the seminar guided by prof. Gonchar. The idea presented in this work is a new one.

4. Proof of Theorem 2. It follows from Theorem 4 that under the conditions of the theorem either the function f is an entire one or it has singularities which lie on γ_k^*, $k = 1, \ldots, \kappa$. As in the proof of Theorem 1, we can prove that the first statement is not correct.

We set $\phi(z) = \prod_{k=1}^{\mu} (z - \phi_k^*)^{\tau_k}$ and $F = f.\phi$. Now we are going to prove that $R_o(F) = |\gamma_k^*|$.

Suppose that it is not the case. In the same way as in Theorem 1, we can show that $R_o(F) < \infty$. Then it follows from our assumption that $R_o(F) < |\gamma_k^*|$.

We notice now that if f has only polar singularities in $D_{|\gamma_k^*|}$, they lie among the points $\gamma_1^*, \ldots, \gamma_\mu^*$ and their orders do not exceed the multiplicity of the zero of Q at the corresponding point (see [3]). This means that $F \in H(D_{|\gamma_k^*|})$. Since $R_0(F) < |\gamma_k^*|$, there is at least one nonpolar singularity in $D_{|\gamma_k^*|}$ and it lies on some $\Gamma_{|\gamma_k^*|}$, $k = 1, \ldots, \mu$. Let $\Gamma_{|\gamma_\nu^*|}$ be the circle of the smallest radius which lies on a nonpolar singularity. We set

$$q_n(z) = \prod_{|\gamma_k^*| \le |\gamma_\nu^*|} \prod_{\ell=1}^{\tau_k} (z - \gamma_{n,k,\ell})$$

and

$$Q_n^* = Q_{n,m}/q_n.$$

It follows from the conditions of the theorem that

$$\limsup_{n \in N} \|q_n - q\|^{1/n} \le \delta < 1 \tag{12}$$

and

$$Q_n^* \to Q^* \quad \text{as} \quad n \in N,$$

where $q(z) = \prod_{|\gamma_k^*| \le |\gamma_\nu^*|} (z - \gamma_k^*)^{\tau_k}$ and $Q^* = Q/q$; $Q^*(z) \ne 0$ for $z \in \bar{\Gamma}_{|\gamma_\nu^*|}$.

Applying Theorem 4 and the notion of Gonchar to the function $F \cdot q$ with respect to the sequence $Q_n^* \cdot \omega_{n+m+1}^{-1}$, we obtain

$$\limsup_{n \in N} |I_{n+m}(F \cdot q \cdot Q_n^*)|^{1/n} = |\gamma_\nu^*|^{-1}. \tag{13}$$

On the other hand, we get from (2) and from the conditions of the theorem

$$I_{n+m}(F \cdot q \cdot Q_n^*) = I_{n+m}(F \cdot (q - q_n) \cdot Q_n^*).$$

The last equality implies that

$$\limsup_{n \in N} I_{n+m}(F \cdot q \cdot Q_n^*)^{1/n} \le \delta |\gamma_\nu^*|^{-1}.$$

This contradicts (13). Thus $R_0(F) = |\gamma_\kappa^*|$.

Now it is easy to show that $R_{m_1} = |\gamma_\kappa^*|$. Indeed, denote by ν_ρ the numbers of the poles of f in the disk D_ρ, $\rho > 1$. We have proved that

$$\nu(D_{|\gamma_\kappa^*|}) \le m_1, \qquad (m_1 = \tau_1 + \ldots + \tau_\kappa). \tag{14}$$

The assumption $R_{m_1} > |\gamma_\kappa^*|$ yields $R_{m_1} = \infty$. This is impossible (see the proof of Theorem 1). Suppose that $R_{m_1} < |\gamma_\kappa^*|$. Since f admits only polar singularities in $D_{\gamma_\kappa^*}$, then we obtain that $\nu(D_{|\gamma_\kappa^*|}) \ge m_1 + 1$. Yet this contradicts (14).

5. Proof of Theorem 3. Let $f(z)=\sum_{\nu=0}^{\infty} f_\nu z^\nu$. It follows from Theorem 4 (see the notion of Gonchar) that a) either the function f is entire or b) it has singularities on $\Gamma_{|\gamma|}$. As we have seen, a) is impossible. Then we obtain

$$\limsup_{\nu\in N} |f_\nu|^{1/\nu} = |\gamma|^{-1}.$$

Select a number ρ such that $1<\rho<|\gamma|$. We set

$$f(z) = \sum_{\nu\geq 0} F_\nu (z/\rho)^\nu.$$

Obviously,

$$\limsup |F_\nu|^{1/\nu} = \rho\cdot|\gamma|^{-1}.$$

Let $\{F_n\}$, $n\in\Lambda$, be the sequence of the "maximal" coefficients: $\{F_n\}_{n\in\Lambda}=\{F_n,\ F_n\neq 0$ and $|F_n|\geq|F_\nu|$ for all $\nu\in N,\ \nu\geq N\}$. Since f is not a polynomial (see the conditions of the theorem), the sequence Λ is infinite. We define the functions

$$\phi_n(z) = \sum_{\nu\geq n} (F_\nu/F_n)(z/\rho)^{\nu-n}, \qquad n\in\Lambda.$$

We fix a number ρ_1, $1<\rho_1<\rho$. As we have seen in the proof of Theorem 1 there is a subsequence Λ_1, $\Lambda_1\subset\Lambda$, such that $\phi_n\to\phi$ as $n\in\Lambda_1$, uniformly on D_{ρ_1}. We expand the function ϕ in a Taylor series

$$\phi(z) = 1+\lim_{n\in\Lambda_1}\frac{F_{n+1}}{F_n}\frac{z}{\rho}+\ldots. \tag{15}$$

Now, for $n>n'$, we set

$$Q_{n,1}(z) = z-\gamma_n.$$

It follows from the conditions of the theorem that

$$\gamma_n\to\gamma \qquad \text{as } n\in N. \tag{16}$$

We obtain from (2) that

$$\gamma_n = I_{n+1}(z\cdot f(z))\cdot I_{n+1}^{-1}(f), \qquad n\geq n_3,\quad n\in N.$$

Let $n\in\Lambda_1$. We write γ_{n-1} in the form

$$\gamma_{n-1} = \frac{I_n(z\cdot f(z))}{f_n}\cdot\left(\frac{I_n(f)}{f_n}\right)^{-1}.$$

It is easy to prove that

$$\gamma_{n-1} = (I_n(z^{n+1}\cdot\phi_n(z))+f_{n-1}\cdot f_n^{-1}).(I_n(z^n\cdot\phi_n(z))^{-1}, \qquad n\in\Lambda_1. \tag{17}$$

Using (1) and (15), we obtain

$$\lim I_n(z^{n+1}.\phi_n(z)) \to 0 \quad \text{as} \quad n \in \Lambda$$

and

$$\lim I_n(z^n.\phi_n(z)) \to 1 \quad \text{as} \quad n \in \Lambda_1.$$

Combining (16), (17) and the last equalities, we get

$$f_{n-1}.f_n \to \gamma \quad \text{as} \quad n \in \Lambda_1. \tag{18}$$

It is easy to see now that $\Lambda \equiv N$, $n \geq n'$. Indeed, if that is not the case there is a subsequence Λ', $\Lambda' \subset \Lambda$, such that $|F_{n-1}| < |F_n|$ for $n \in \Lambda'$. Yet, as we have shown, there exists Λ_1', $\Lambda_1' \subset \Lambda'$, such that (18) is valid for $n \in \Lambda_1'$. Relation (18) yields

$$F_{n-1} . F_n \to \gamma . \zeta^{-1} \quad \text{as} \quad n \in \Lambda_1.$$

Since $\zeta < |\gamma|$, we get a contradiction to the definition of Λ'.

Now it is not difficult to establish that

$$f_{n-1} . f_n^{-1} \to \gamma \quad \text{as} \quad n \in N. \tag{19}$$

Indeed, if that is not the case, we will obtain a contradiction to (18). Then, following the theorem of Fabry, we get that f has a singularity at γ.

References

[1] SAFF, E.B.: 'An extension of Montessus de Balore's theorem on the convergence of interpolating rational functions', *J. Approximation Theory* 6 (1972), 63-67.

[2] KAKEHASHI, T.: 'On interpolation of analytic functions I', *Proc. Japan Acad.* 10 (1956), 707-712.

[3] GONCHAR, A.A.: 'On the convergence of generalized Padé approximants of meromorphic functions', *Mat. Sb.* 98 (140) (1975), 546-577.

[4] SUETIN, S.P.: 'On the poles of the m-th row of a Padé table', *Mat. Sb.* 120 (162) (1983), 500-504.

[5] BUSLAEV, V.I.: 'Relations for the coefficients and singular points of functions', *Mat. Sb.* 131 (1986), 317-384.

THREE REMARKS ABOUT THE CARATHÉODORY DISTANCE

Marek Jarnicki
Institute of Mathematics
Jagiellonian University
PL-30-059 Kraków, Poland

and

Peter Pflug
Universität Osnabrück
Abteilung Vechta, Mathematik
D-2848 Vechta 1, BRD

ABSTRACT and INTRODUCTION. Let D be a domain in $\mathbb{C}^n$. The Carathéodory pseudodistance c_D on D is defined by

$$c_D(z', z'') = 1/2 \log \frac{1 + c_D^*(z', z'')}{1 - c_D^*(z', z'')},$$

where c_D^* denotes the Möbius function

$$c_D^*(z', z'') = \sup\{|f(z'')| : f : D \to E \text{ holomorphic}, f(z') = 0\};$$

here E is the unit disc in the complex plane. The Carathéodory pseudodistance is a very useful tool in complex analysis. For its standard properties we refer, for example, to the book by T. Frazoni-E. Vesentini [4].

The aim of this paper is to contribute to partial results about the product property of the Carathéodory distance and to provide explicite formulae about how to calculate the Carathéodory distance for a class of Reinhardt domains. The third remark deals with counterexamples to a conjecture of J. Burbea [1].

1. THE PRODUCT-FORMULA FOR THE CARATHÉODORY-DISTANCE

It is well known that for domains $D_1 \subset \mathbb{C}^{n_1}$ and $D_2 \subset \mathbb{C}^{n_2}$ we have the following formula for the Kobayashi pseudodistance $k_{D_1 \times D_2}$ on $D_1 \times D_2$:

$$k_{D_1 \times D_2}((z', w'), (z'', w'')) = \max(k_{D_1}(z', z''), k_{D_2}(w', w'')),$$

whenever $(z', w'), (z'', w'') \in D_1 \times D_2$.

If D_1 and D_2 are convex bounded domains then, following the work by L. Lempert [8], the Carathéodory distance and the Kobayashi distance coincide. Hence, there is the following formula for convex bounded domains D_1 and D_2:

J. Ławrynowicz (ed.), Deformations of Mathematical Structures, 161–170.

$$c_{D_1 \times D_2}((z', w'), (z'', w'') = \max (c_{D_1}(z', z''), c_{D_2}(w', w'')). \qquad (*)$$

In his survey article [7], S. Kobayashi claimed that the product-formula (*) holds for any product-domain $D_1 \times D_2$. Yet, so far there has been neither a proof of (*) nor a counterexample to (*).

In our paper we shall show that (*) is true for product-domains $D_1 \times D_2$, where one factor is a bounded convex domain. Before we state our results we claim the following

LEMMA 1. *Let* $D = \{z \in \mathbb{C}^n : h(z) < 1\}$ *be a bounded balanced domain of holomorphy with its Minkowski functional* h, *then the following holds true:*

a) $c_D^*(0,z) \leqq h(z)$ *for any* $z \in D$;

b) $c_D^*(0,z^o) = h(z^o)$ *for* $z^o \in D$, *iff there exists a homogeneous polynomial* Q *of order* 1 *with* $|Q| \leqq 1$ *on* D *and* $|Q(z^o)| = h(z^o)$;

c) $c_D^*(0, \ ) \equiv h$ *on* D, *iff* h *is a norm and iff* D *is convex.*

Proof. The claim in a) is a simple consequence of the Schwarz lemma. To see b) we assume $c_D^*(0,z^o) = h(z^o) = f(z^o)$, where $f : D \to E$ is a holomorphic function with $f(0) = 0$. Then we have the expansion of f in homogeneous polynomials

$$f(z) = \sum_{\nu \geqq 1} Q_\nu(z)$$

with $|Q_\nu| \leqq 1$ on D. The assumption and Schwarz Lemma lead to the equation

$$f(\lambda \frac{z^o}{h(z^o)}) = e^{i\Theta}\lambda = \sum_{\nu=1}^{\infty} \frac{Q_\nu(z^o)}{h^\nu(z^o)} \lambda^\nu$$

on the unit disc E. Hence $|Q_1(z^o)| = h_1(z^o)$, which proves b). The last assertion c) is a simple consequence of b) and the Hahn-Banach theorem.

Now we can formulate our first result.

THEOREM 1. *Let* $D_1 := \{z \in \mathbb{C}^{n_1} : h(z) < 1\}$ *be a convex homogeneous bounded balanced domain of holomorphy with its Minkowski functional* h *and let* $D_2 \subset \mathbb{C}^{n_2}$ *be an arbitrary domain. Then the product-formula* (*) *is true.*

Proof. By the assumptions and the distance decreasing property we only have to prove

$$c^*_{D_1 \times D_2}((0,w'), (z'', w'') \leqq \max (c^*_{D_1}(0,z''), c^*_{D_2}(w', w'')),$$

whenever $z'' \in D_1$ and $w', w'' \in D_2$. We can assume $w' \neq w''$ and $z'' \neq 0$. Let z'' satisfy $h(z'') \leq c^*_{D_2}(w', w'')$ and define $\Phi : D_2 \to D_1 \times D_2$ by

$$\Phi(w) := (z'' \cdot \frac{f(w)}{c^*_{D_2}(w', w'')}, w);$$

here $f : D_2 \to E$ is a holomorphic function with $f(w') = 0$ and $f(w'') = c^*_{D_2}(w', w'')$. Since Φ is holomorphic we receive

$$c^*_{D_1 \times D_2}((0,w'), (z'', w'')) \leq c^*_{D_2}(w',w'') = \max(c^*_{D_1}(0,z''), c^*_{D_2}(w',w'')).$$

In the second case we assume that z'' satisfies the inequality

$$h(z'') > c^*_{D_2}(w', w'').$$

We regard the following function:

$$E \setminus \{0\} \ni \lambda \to c^*_{D_1 \times D_2}((0,w'), (\lambda \frac{z''}{h(z'')}, w'')) \cdot \frac{1}{|\lambda|} =: g(\lambda).$$

It is obvious that $\log g$ is subharmonic on $E \setminus \{0\}$ and that g satisfies the following inequalities:

$$\log g(\lambda) = 0, \quad \text{if} \quad |\lambda| = c^*_D(w', w''),$$

and

$$\overline{\lim_{|\lambda| \to 1}} \log g(\lambda) \leq 0.$$

By the maximum principle, we conclude that

$$c^*_{D_1 \times D_2}((0,w'), (z'',w'') \leq h(z'') = c^*_{D_1}(0,z'').$$

The last equality is due to the lemma. Hence the theorem is completely verified.

When using the existence of complex geodesics in bounded convex domains (cf. [11]) we can improve our knowledge about the product-property; we obtain

THEOREM 2. *Let* D_1 *be a bounded convex domain in* $\mathbb{C}^{n_1}$ *and let* D_2 *be an arbitrary domain in* $\mathbb{C}^{n_2}$, *then the product-property holds true for* $D_1 \times D_2$.

Proof. Let $z', z'' \in D1$ and $w', w'' \in D_2$. According to Vigué's result there is a holomorphic map $\phi : E \to D_1$ with $\phi(\xi') = z'$, $\phi(\xi'') = z''$ $(\xi', \xi'' \in E)$ and $c_E(\xi', \xi'') = c_{D_1}(z', z'')$; E denotes the unit disc in the complex plane. Then we get

$$\max(c_{D_1}(z', z''), c_{D_2}(w', w'')) = \max(c_E(\xi', \xi''), c_{D_2}(w', w''))$$
$$= c_{E \times D_2}((\xi', w'), (\xi'', w'') \qquad \text{(use Theorem 1)}$$

$$\geq c_{D_1\times D_2}((z',w'),(z'',w'')) \qquad \text{(use the holomorphic map } \phi\times \mathrm{id}: E\times D_2 \to D_1\times D_2)$$
$$\geq \max(c_{D_1}(z',z''), c_{D_2}(w',w''))$$

which implies the product property.

Remark 1. a) The arguments show that the existence of a complex geodesic for at least one of the pairs (z',z''), (w',w'') is sufficient to conclude the product-property. b) It remains still open whether (*) is true without the hypothesis that D_1 is convex and bounded. c) It seems interesting to prove the product formula for the product of convex domains without using Lempert's result. d) In his paper [2], H. Cartan has already discussed some aspects of the product formula. We would like to thank Prof. P. Vigué for showing us the paper by H. Cartan.

2. AN EFFECTIVE FORMULA FOR THE CARATHÉODORY PSEUDODISTANCE

In this section we consider special Reinhardt domains of holomorphy

$$G=\{z\in\mathbb{C}^n : |z_1|^{\alpha_1^j}\cdots|z_n|^{\alpha_n^j}<c_j,\ 1\leq j\leq r\},$$

where $\alpha^1,\ldots,\alpha^r\in(\mathbb{R}_+)^n_*$, $c_1>0,\ldots,c_r>0$, such that

$$S=S(\alpha^1,\ldots,\alpha^r):=(\mathbb{Z})^n_*\cap(\mathbb{R}_+\alpha^1+\ldots+\mathbb{R}_+\alpha^r)\neq\emptyset. \qquad (**)$$

Remark 2. In view of Prop. 3.18 [6] we have the following situation:

$$S(\alpha^1,\ldots,\alpha^r)=\emptyset \quad \text{iff} \quad H^\infty(G)\cong\mathbb{C} \quad \text{and iff} \quad c_G^*\equiv 0.$$

We introduce the following notation:

Definition. Let $\alpha^1,\ldots,\alpha^r\in(\mathbb{R}_+)^n_*$ with $S=S(\alpha^1,\ldots,\alpha^r)\neq\emptyset$. An element $\alpha\in S$ is called *irreducible* if α does not belong to $S+S$. By $B=B(\alpha^1,\ldots,\alpha^r)$ we denote the set of all irreducible elements of S.

Remark 3. For every $\alpha\in S$, there exist $k\in\mathbb{N}$, $\beta^1,\ldots,\beta^k\in B$ such that $\alpha=\beta^1+\ldots+\beta^k$.

The following example shows that, in general, this decomposition is not unique.

Example. Let $n=r=2$ and $\alpha^1:=(1,0)$, $\alpha^2:=(2,7)$. Then $B(\alpha^1,\alpha^2)=\{(1,0),(1,1),(1,2),(1,3),(2,7)\}$ and $(2,4)=(1,1)+(1,3)=2\cdot(1,2)$.

The set B satisfies the following properties:

PROPOSITION 1. *a) If* $\alpha^1,\dots,\alpha^r \in \mathbb{R}_{>0}(\mathbb{Q}_+)^n_*$, *then the set* $B(\alpha^1,\dots,\alpha^r)$ *is finite. b) If* $\alpha^1,\dots,\alpha^r \in (\mathbb{R}_+)^n_*$ *satisfy rank* $(\alpha^1,\dots,\alpha^r) = r$ *and if the space* $\mathbb{R}\alpha^1 + \dots + \mathbb{R}\alpha^r$ *is of rational type (compare Def. 3.2 in [6]), then the following conditions are equivalent:*

α) $\alpha^1,\dots,\alpha^r \in \mathbb{R}_{>0}(\mathbb{Q}_+)^n_*$,

β) $B(\alpha^1,\dots,\alpha^r)$ *is finite.*

Proof. a) Let $\alpha^j = \sigma_j \cdot q^j$ with $\sigma_j > 0$ and $q^j \in (\mathbb{Q}_+)^n_*$. We write $q^j = \tilde{q}^j \cdot \frac{1}{q}$ with $\tilde{q}^j \in (\mathbb{Z}_+)^n_*$ and $q \in \mathbb{N}$. Then $\tilde{q}^1,\dots,\tilde{q}^r \in S(\alpha^1,\dots,\alpha^r) = S$. Let

$$F := S \cap ([0,1]\cdot\tilde{q}^1 + \dots + [0,1]\cdot\tilde{q}^r).$$

Since F is finite, it suffices to see that $B \subset F$. Let $\alpha \in B$ possess a representation

$$\alpha = t_1\cdot\alpha^1 + \dots + t_r\cdot\alpha^r = \tilde{t}_1\cdot\tilde{q}^1 + \dots + \tilde{t}_r\cdot\tilde{q}^r \quad (t_j,\ \tilde{t}_j \geq 0),$$

then we have

$$\begin{aligned}\alpha &= \sum_{j=1}^r [\tilde{t}_j]\,\tilde{q}^j + \sum_{j=1}^r (\tilde{t}_j - [\tilde{t}_j])\tilde{q}^j \\ &=: \quad \beta' \quad + \quad \beta''.\end{aligned}$$

Since β' and β'' belong to $S \cup \{0\}$ and since α is irreducible we conclude that either $\beta' = 0$ or $\beta'' = 0$. If $\beta' = 0$, it is obvious that $\alpha \in F$. If $\beta'' = 0$ then $\alpha = \beta'$, which implies that for some $j_o (1 \leq j_o \leq r)$

$$\alpha = 1 \cdot \tilde{q}^{j_o} \in F.$$

b) Remark: The assumptions in b) imply that $S(\alpha^1,\dots,\alpha^r) \neq \emptyset$. In view of a) we only have to verify that β) implies α). Assume that $B = B(\alpha^1,\dots,\alpha^r)$ is finite and define

$$D := \{z \in \mathbb{C}^n : |z_1|^{\alpha_1^j} \dots |z_n|^{\alpha_n^j} < 1 \quad \text{for} \quad 1 \leq j \leq r\}.$$

Because of Theorem 3.9 in [6], D is the domain of existence of a bounded holomorphic function. With

$$R(z) := \max\{|z^\alpha| : \alpha \in B\},$$

it is clear that $D \subset \{z \in \mathbb{C}^n : R(z) < 1\}$. Using Prop. 3.18 in [6] one can prove, moreover, the equality which implies that $\alpha^1,\dots,\alpha^r \in \mathbb{R}_{>0}\cdot B$. Thus Proposition 1 is verified.

The following example shows that, in general, if the space

$\mathbb{R}\alpha^1 + \ldots + \mathbb{R}\alpha^r$ is not of rational type, then $B(\alpha^1,\ldots,\alpha^r)$ may also be finite in the case where $\alpha^1,\ldots,\alpha^r \notin \mathbb{R}_{>0}\cdot(\mathbb{Q}_+)^n_*$.

Example 1 (comp. Example 3.20 in [6]). Let $n = 3$, $r = 2$, $\alpha^1 = (\alpha_1, \alpha_2, \alpha_3)$, $\alpha^2 = (1-\alpha_1, 1-\alpha_2, 1-\alpha_3)$, where $0 < \alpha_1, \alpha_2, \alpha_3 < 1$, $\alpha_1 \neq \alpha_2$ and $(\alpha_1 - \alpha_3)\cdot(\alpha_1-\alpha_2)^{-1} \notin \mathbb{Q}$. Then $B(\alpha^1, \alpha^2) = \{(1,1,1)\}$. Let

$$\alpha^1,\ldots,\alpha^r \in (\mathbb{R}_+)^n_* \quad \text{with} \quad S(\alpha^1,\ldots,\alpha^r) \neq \emptyset$$

and put

$$G := \{z \in \mathbb{C}^n : |z_1|^{\alpha^j_1}\cdots|z_n|^{\alpha^j_n} < c_j \quad \text{for} \quad 1 \leq j \leq r\} \qquad (c_j > 0).$$

Then all the functions f with $f_\alpha(z) = z^\alpha$, $\alpha \in S$, belong to $H^\infty(G)$. Therefore one has the following inequality:

$$c^*_G(0,z) \geq \sup\{\frac{|z^\alpha|}{\|z^\alpha\|_G} : \alpha \in S\} \geq \sup\{\frac{|z^\alpha|}{\|z^\alpha\|_G} : \alpha \in B\}, \quad z \in G.$$

Here $\|\ \|_G$ denotes the supremum norm on G.

The following example shows that, in general, both of the above inequalities need not to be identities.

Example 2. Let $G := \{z \in \mathbb{C}^2 : |z_1| < 1,\ |z_2| < 1,\ |z_1\cdot z_2| < 1/2$, i.e.

$$n = 2,\ r = 3,\ \alpha^1 = (0,1),\ \alpha^2 = (0,1),\ \alpha^3 = (1,1),\ c_1 = c_2 = 1,\ c_3 = 1/2.$$

It is clear that $S(\alpha^1, \alpha^2, \alpha^3) = (\mathbb{Z}_+)^2$ and $B(\alpha^1, \alpha^2, \alpha^3) = \{\alpha^1, \alpha^2\}$. Observe that for $z \in G$

$$\sup\{\frac{z^\alpha}{\|z^\alpha\|_G} : \alpha \in (\mathbb{Z}_+)^2\} = \max\{|z_1|, |z_2|, 2|z_1z_2|\}$$

$$\neq \max\{|z_1|, |z_2|\} = \sup\{\frac{z^\alpha}{\|z^\alpha\|_G} : \alpha \in B\}$$

On the other hand, we have $G \subset \{z \in \mathbb{C}^2 : |z_1| + |z_2| < 3/2\} =: D$; therefore we obtain for $z \in G$

$$c^*_G(0,z) \geq c^*_D(0,z) = \tfrac{2}{3}(|z_1| + |z_2|).$$

Here we have applied Lemma 1. In particular, we get for $0 < t < 2/3$

$$c^*_G(0, (t,t)) \geq 4/3 \cdot t > \max\{t, 2t^2\}.$$

What is still not known is whether the following formula is true:

$$c^*_G(0,z) = \max\{|z_1|, |z_2|, 2|z_1|\cdot|z_2|, \tfrac{2}{3}(|z_1| + |z_2|)\}.$$

On the other hand, if we assume the linear independency of the

$\alpha^1,\dots,\alpha^r$, we have the following theorem:

THEOREM 3. *Let* $\alpha^1,\dots,\alpha^r \in (\mathbb{R}_+)^n_*$ *and* $c_1,\dots,c_r \in \mathbb{R}_{>0}$ *and assume that*

a) $S = S(\alpha^1,\dots,\alpha^r) \neq \emptyset$

b) $\mathrm{rank}(\alpha^1,\dots,\alpha^r) = r.$

Set

$$G := \{z \in \mathbb{C}^n : |z_1|^{\alpha_1^j} \dots |z_n|^{\alpha_n^j} < c_j \quad \text{for} \quad 1 \leq j \leq r\},$$

then, for $z \in G$, *the following equalities hold:*

$$c_G^*(0,z) = \sup\{\frac{|z^\alpha|}{\|z^\alpha\|_G} : \alpha \in S\} = \sup\{\frac{|z^\alpha|}{\|z^\alpha\|_G} : \alpha \in B\}.$$

Proof. Without loss of generality we assume that $c_1 = \dots = c_r = 1$. Let $R(z) := \sup\{|z^\alpha| : \alpha \in B\}$, $z \in G$. Since the family $\{z^\alpha : \alpha \in B\}$ is equicontinuous on G, the function R is continuous. Therefore it suffices to find a set $T \subset G$ such that $\log T$ is dense in

$$\log G =: X \quad (\log A := \{x \in \mathbb{R}^n : (e^{x_1},\dots,e^{x_n}) \in A\},\ A \subset \mathbb{C}^n)$$

and that, for $z \in T$, we have

$$c_G^*(0,z) \leq R(z).$$

Define

$$T := \{z(\nu,\lambda) := (\lambda^{-\nu_1},\dots,\lambda^{-\nu_n}) : \nu \in X \cap Z^n,\ \lambda \in E \setminus \{0\}\}.$$

It is easy to see that $T \subset G$ and $X \cap Q^n \subset \log T$, thus $\log T$ is dense in X.

Let $f : G \to E$ be a holomorphic function with $f(0) = 0$. We write $f(z) = \Sigma\, a_\alpha z^\alpha$ and mention that $|a_\alpha z^\alpha| < 1$ on G, $\alpha \in (\mathbb{Z}_+)^n$. In particular, for those α with $a_\alpha \neq 0$, we obtain

$$X \subset \{x \in \mathbb{R}^n : \langle x,\alpha\rangle < \text{const}\};$$

here $\langle\, ,\, \rangle$ denotes the standard scalar product in $\mathbb{R}^n$. Because of the proof of Prop. 3.18 in [6] we find $a_\alpha = 0$ for $\alpha \notin S$.

In particular, for $\nu \in X \cap \mathbb{Z}^n$ and $\lambda \in E \setminus \{0\}$, we have the series representation

$$f(z(\nu,\lambda)) = \sum_{a \in S} a_\alpha \lambda^{\langle -\nu,\alpha\rangle}$$

which yields

$$|f(z(\nu,\lambda))| \leq |\lambda|^{P(\nu)} \qquad (\lambda \in E\setminus\{0\}),$$

where $P(\nu) := \inf\{\langle -\nu, \alpha\rangle : \alpha \in S\}$. On the other hand, we find that

$$R(z(\nu,\lambda)) = |\lambda|^{Q(\nu)} \qquad (\lambda \in E\setminus\{0\})$$

with $Q(\nu) := \inf\{\langle -\nu, \alpha\rangle : \alpha \in B\}$. The observation that $P(\nu) = Q(\nu)$ completes the proof of Theorem 3.

We conclude with an example which illustrates Theorem 3.

Example 3. Let $G := \{z \in \mathbb{C}^2 : |z_1| < 1,\ |z_1^{345} \cdot z_2^{128}| < 1\}$. Then

$$c_G^*(0,z) = \max\{|z_1|,\ |z_1^{3}\cdot z_2|,\ |z_1^{11}\cdot z_2^{4}|,\ |z_1^{19}\cdot z_2^{7}|,$$
$$|z_1^{27}\cdot z_2^{10}|,\ |z_1^{62}\cdot z_2^{23}|,\ |z_1^{345}\cdot z_2^{128}|\}.$$

Remark 4. If $G_1 \in \mathbb{C}^{n_1}$ and $G_2 \in \mathbb{C}^{n_2}$ are Reinhardt domains of the type in Theorem 3, then we have

$$c^*_{G_1\times G_2}((0,0),\ (z',w')) = \max(c^*_{G_1}(0,z'),\ c^*_{G_2}(0,w')),$$

whenever $z' \in G_1$ and $w' \in G_2$.

3. COMPARISON BETWEEN THE CARATHÉODORY-REIFFEN-METRIC AND THE BERGMAN METRIC

Let D be a bounded domain in $\mathbb{C}^n$, let $C_D(z:v)$ be the Carathéodory -Reiffen-metric on $D \times \mathbb{C}^n$ and let $B_D(z:v)$ be the classical Bergman metric on $D \times \mathbb{C}^n$. For the definitions compare [1, 4]. Then the following inequality between C_D and B_D is well known:

$$C_D(z:v) \leq B_D(z:v) \quad \text{on} \quad D\times\mathbb{C}^n$$

(compare [1, 5]). In his paper [1], J. Burbea conjectured that

$$B_D(z:v) \leq k(D)C_D(z:v) \tag{***}$$

on $D\times\mathbb{C}^n$, where $k(D)$ is a positive constant depending only on the bounded domain D.

According to the comparison results due to K. Diederich, J.E. Fornaess and G. Herbort, this conjecture is not true. Here we give another counterexample which shows that there is no comparison between the Carathéodory distance and the Bergman distance.

Let $D_o := \{(z,w) \in \mathbb{C}^2 : |z| < 1,\ |w| < e^{-V(z)}\}$ be that bounded domain of holomorphy which was introduced by N. Sibony in [10]. Recall that D_o is not an $H^\infty(D_o)$-domain of holomorphy [10] and that D_o is complete w.r.t. the Bergman distance b_D [9].

We denote by c_D^i the associated inner distance to c_D. Then we have the following simple lemma:

LEMMA 2. *Let* D *be a bounded domain in* $\mathbb{C}^n$, *which is complete w.r.t.* c_D^i, *then* D *is an* $H^\infty(D)$*-domain of holomorphy.*

Proof. Assume that D is not an $H^\infty(D)$-domain of holomorphy. This implies that there exist a point $z^* \in D$ and a number $R' > R := \mathrm{dist}(z^*, \partial D)$ such that the following holds: For any bounded holomorphic function f on D there is a holomorphic function F on the ball $U' := U(z^*, R')$ around z^* with radius R' such that

$$F_{|U} \equiv f_{|U}, \quad U := U(z^*, R).$$

Hence we obtain $c_D(z', z'') \leq c_{U'}(z', z'')$ $(z', z'' \in U)$.

We choose $\hat{z} \in \partial D \cap \partial U$ and a sequence $\{z^\nu\} \subset U$, $z^\nu \in \overline{z^* \hat{z}}$, with $z^\nu \to \hat{z}$. By $\gamma : [0,1] \to U$ we denote the straight line from z^ν to z^μ. Then the following inequalities are obvious:

$$c_D^i(z^\nu, z^\mu) \leq L_{c_D}(\gamma) \leq L_{c_{U'}}(\gamma) = c_{U'}(z^\nu, z^\mu);$$

here L_{c_D} $(L_{c_{U'}})$ denotes the length w.r.t. c_D $(c_{U'})$. Hence $\{z^\nu\}$ is a c_D^i-Cauchy-sequence tending to the boundary point $\hat{z}$ of D. This contradicts the c_D^i-completeness and so Lemma is proved.

When using Lemma 2 and the quoted properties of the domain D_o we have the following result:

THEOREM 4. *There is no constant* $k > 0$ *such that* $b_{D_o} \leq k \cdot c_{D_o}^i$ *on* $D_o \times D_o$.

Remark 5. 1) From Theorem 4, it is clear that also (***) is not fulfilled on $D_o \times D_o$.

2) It is not clear whether the example in [3] does not satisfy an inequality $b_D \leq k \cdot c_D^i$.

We would like to conclude this paper by posing the problem of finding geometrical conditions which ensure the validity of Burbea's conjecture.

References

[1] BURBEA, J.: 'Inequalities between intrinsic metrics', *Proc. Am. Math. Soc.* 67 (1977), 50-54.

[2] CARTAN, H.: 'Sur les fonctions de n variables complexes: Les transformation du produit topologique de deux domaines bornés', *Bull. Soc. Math. France* 64 (1936), 37-48.

[3] DIEDERICH, K., FORNAESS, J.E., and HERBORT, G.: 'Boundary behaviour of the Bergman metric. Complex Analysis of Several Variables', *Proc. Symposia in Pure Mathematics* 41 (1984), 59-67.

[4] FRANZONI, T. and VESENTINI, E.: *Holomorphic maps and invariant distances*, North Holland, Amsterdam 1980.

[5] HAHN, K.T.: 'On completeness of the Bergman metric and its subordinate metrics II, *Pacific Journal of Math.* 68 (1977), 437-446.

[6] JARNICKI, M. and PFLUG, P.: 'Existence domains of holomorphic functions of restricted growth' (Preprint 1986).

[7] KOBAYASHI, S.: 'Intrinsic distances, measures and geometric function theory', *Bull. Am. Math. Soc.* 82 (1976), 357-416.

[8] LEMPERT, L.: 'La métrique de Kobayashi et la représentation des domaines sur la boule', *Bull. Soc. Math. France* 109 (1981), 427-474.

[9] PFLUG, P.: 'Applications of the existence of well growing holomorphic functions', in *Analytic functions, Błażejewko 1982, Proceedings*, ed. by J. Ławrynowicz (Lecture Notes in Math. 1039), Springer, Berlin-Heidelberg-New York-Tokyo 1983, pp. 376-388.

[10] SIBONY, N.: 'Prolongement des fonctions holomorphes bornées et métrique de Carathéodory', *Inventiones math.* 29 (1975), 205-230.

[11] VIGUÉ, J.P.: 'Points fixes d'applications holomorphes dans un domaine borné', *T.A.M.S.* 289 (1985), 345-355.

ON THE CONVEXITY OF THE KOBAYASHI INDICATRIX

Giorgio Patrizio
Dipartimento di Matematica
II Università di Roma
I-00173 Roma, Italy

ABSTRACT. It is shown that the Kobayashi indicatrix of a strictly convex domain $D \subset \mathbb{C}^n$ is strictly convex at every point $p \in D$. As a consequence, it follows that a strictly pseudoconvex complete domain, which is not strictly convex, cannot be biholomorphic to a strictly convex domain. Some condition for the convexity of the Kobayashi metric in more general domains is also given.

Let U be the unit disk in $\mathbb{C}$ and M be a complex manifold. Denote by $H(U,M)$ the set of all holomorphic maps from U to M. The Kobayashi metric of M is defined by

$$K_M(p,v) = \inf\{|u| \mid u \in \mathbb{C} \text{ and } df(0)(u) = v \text{ for some } f \in H(u,M) \text{ with } f(0) = p\} \tag{0.1}$$

for all $p \in M$ and $v \in T_p(M)$. Here we identify $T_0(U)$ with C and $|\ |$ is the euclidean norm. The *Kobayashi indicatrix* of M at p is the complete circular domain defined by

$$I_p(M) = \{v \quad T_p(M) \mid K_M(p,v) < 1\}. \tag{0.2}$$

The Kobayashi indicatrix is intrinsic to the geometry of the underlying manifold. In fact, if $\phi\colon M \to N$ is a biholomorphic map between complex manifolds, then, for any $p \in M$, the differential $d\phi(p)$ is a linear biholomorphic map between $I_p(M)$ and $I_{\phi(p)}(N)$. Therefore it is of great interest to study the properties of the indicatrix and, in particular, it is important to determine when it is strictly convex (see [4], Problem A.1).

1. It is known that, if $D \subset \mathbb{C}^n$ is a bounded convex domain, then, given any point $p \in D$, the Kobayashi indicatrix is convex ([3], Exercise 13). Let $D \subset \mathbb{C}^n$ be a *strictly convex domain*, i.e. a smoothly bounded domain with a definining function ρ whose real Hessian H_ρ is strictly positive definite in a neighbourhood of ∂D. The main

J. Ławrynowicz (ed.), Deformations of Mathematical Structures, 171–176.

result of this note is the following:

THEOREM 1. *For any* $p \in D$, *the Kobayashi indicatrix* $I_p(D)$ *is a strictly convex domain.*

In order to prove Theorem 1, we need to recall some known facts (see [5]). Given $p \in D$, there exists a C^∞ surjective map $F: U \times S \to D$, where $S \subset \mathbb{C}^n$ is the unit sphere, such that for all $b \in S$ the map $F(.,b): U \to D$ is a proper holomorphic embedding with $F(0,b) = p$ and $F'(0,b) = dF(0,b)(1) = \|F'(0,b)\| b$ (here we identify $T_p(0)$ with $\mathbb{C}^n$ and $\| \ \|$ is the euclidean norm). Moreover for every $b \in S$, $F(.,b)$ is the unique element in $H(U,D)$ such that $K(p,b) = K_D(p,b) = (\|F'(0,b)\|)^{-1}$, i.e. $F(.,b)$ is the unique extremal map for the Kobayashi metric at p in the direction b. For $\lambda \in \partial U$ one has $F(z,\lambda b) = F(\lambda z, b)$ and if $b_1, b_2 \in S$ and $L_j = F(.,b)(U)$, $j = 1, 2$, then either $L_1 \cap L_2 = \{p\}$ or $L_1 = L_2$ and $b_2 = \mu b_1$ for some $\mu \in \partial U$. A strictly convex exhaustion $\tau: D \to [0,1)$, called the *Lempert exhaustion* of D at p, is well defined by:

$$\tau(F(z, b)) = |z|^2. \tag{1.1}$$

The pair (D,τ) is, using the terminology of [7], a *manifold of circular type*. This means that he exhaustion τ satisfies the following properties (see [5], [6]):

$$\tau \in C^0(D) \cap C^\infty(D_*), \quad \text{where} \quad D_* = D - \{p\}; \tag{1.2}$$

$$dd^c\tau = \frac{i}{4\pi}\,\partial\bar\partial\tau > 0 \quad \text{and} \quad dd^c \log \tau \geq 0 \quad \text{on} \quad D_*; \tag{1.3}$$

$$(dd^c \log \tau)^n = 0 \quad \text{on} \quad D_*; \tag{1.4}$$

with respect to coordinates centered at p, one has $C\|Z\|^2 \geq \tau(Z) \geq K\|Z\|^2$ for some $C, K > 0$; (1.5)

if $\pi: D \to \tilde{D}$ is the blow up of D at p, then $\tau \circ \pi \in C^\infty(\tilde{D})$. (1.6)

We define a function $\sigma: \mathbb{C}^n \to [0,+\infty)$ by $\sigma(0) = 0$ and

$$\sigma(Z) = \frac{\|Z\|^2}{\|F'(0,Z/\|Z\|)\|^2} \quad \text{if} \quad Z \neq 0. \tag{1.7}$$

Then the Kobayashi indicatrix $I_p(D)$ of D at p can be described as follows:

$$\begin{aligned} I_p(D) &= \{Z \in \mathbb{C}^n \mid K(p, Z) < 1\} \\ &= \{Z \in \mathbb{C}^n - \{0\} \mid \|Z\|\, K(p, Z/\|Z\|) < 1\} \cup \{0\} \\ &= \{Z \in \mathbb{C}^n \mid \sigma(Z) < 1\}. \end{aligned} \tag{1.8}$$

Clearly $\sigma \in C^{\infty}(\mathbb{C}^n - \{0\})$. Thus, to complete the proof of the theorem, it is enough to show that the function σ is strictly convex on $\mathbb{C}^n - \{0\}$. From (1.7) it is immediate that, if $\lambda \in \mathbb{C}$ and $Z \in \mathbb{C}^n$, then

$$\sigma(\lambda Z) = |\lambda|^2 \sigma(Z). \tag{1.9}$$

Without loss of generality we can assume $p = 0$. It is known (see [7], proof of Proposition 4.1, p. 359) that, for some ε, $\varepsilon_o > 0$ if

$$A = \{(t, W) \in R \times C^n \mid |t| < \varepsilon, \ 0 < \|W\| < \varepsilon_o\},$$

then for $(t, W) \in A$ one has:

$$\sigma(t W) = \tau(t W) + t^3 L(t, W) \tag{1.10}$$

where $L \in C^{\infty}(A)$. Let $W \in \mathbb{C}^n$ with $0 < \|W\| < \varepsilon_o$ with $\|X\| = 1$. Because of (1.6), the function h_W: $[-\varepsilon/2, \varepsilon/2] \to (0, +\infty)$ defined by

$$h_W(t) = H\ (t W)(X, X) = \frac{\partial^2 \tau(t W)}{\partial x^i \partial x^j} x^i x^j$$

if of class C^{∞}. Let

$$M_W = \min h_W(t) \quad \text{for} \quad t \in [-\varepsilon/2, \varepsilon/2]$$

and

$$N_W = \max \left| \frac{\partial^2 L(t, w)}{\partial x^i \partial x^j} x^i x^j \right| \quad \text{for} \quad t \in [-\varepsilon/2, \varepsilon/2].$$

Let $Z \in \mathbb{C}^n - \{0\}$ and $W_o = \frac{\varepsilon_o Z}{2\|Z\|}$. Then $0 < \|W\| \leq \varepsilon_o/2 < \varepsilon_o$ and, using (1.9) and (1.10), one computes, choosing $0 < t < \min\{M/N, \varepsilon/2\}$,

$$\begin{aligned} H_\sigma(Z)(X, X) &= H_\sigma(W_o)(X, X) = H_\sigma(t W_o)(X, X) \\ &= H_\tau(t W_o)(X, X) + t \frac{\partial^2 L(t, W_o)}{\partial x^i \partial x^j} x^i x^j \\ &> M_{W_o} - t N_{W_o} > 0. \end{aligned}$$

Since $X \in \mathbb{R}^{2n}$ with $\|X\| = 1$ was arbitrary, it follows that σ is strictly convex on $\mathbb{C}^n - \{0\}$ and thus the proof of Theorem 1 is complete.

2. It is well-known that, given a strictly pseudoconvex domain D in $\mathbb{C}^n$ and a point $p \in D$, one can find holomorphic coordinates around p so that, with respect to these coordinates, D is strictly convex. It is also true [2] that there exists a strictly convex domain $\mathfrak{D}$ in $\mathbb{C}^N$, $N \geq n$, such that D can be embedded in $\mathfrak{D}$ via a proper holomorphic map. On the other hand there is no characterization of the strictly pseudoconvex domains which are biholomorphic to a strictly convex one.

In fact using Theorem 1 it can be shown that there are many strictly pseudoconvex domains which fulfill the obvious topological conditions but are not biholomorphic to any strictly convex domain. Using Theorem 1 and the fact that $I_0(G) = G$ (see [1]), in analogy with [1], we have

THEOREM 2. *If a strictly pseudoconvex, complete circular domain* G *in* $\mathbb{C}^n$ *is biholomorphic to a strictly convex domain, then* G *is strictly convex.*

3. It is possible to give a version of Theorem 1 in the general setting of manifolds of circular type. For simplicity we shall outline it in the case of a domain in $\mathbb{C}^n$.

Let $D \subset \mathbb{C}^n$ be a strictly pseudoconvex domain with $0 \in D$. We assume that there exists an exhaustion $\tau: D \to [0,1)$ with the properties (1.2) - (1.6) so that, in the terminology of [7], the pair (D,τ) is a manifold of circular type with radius 1 and center 0. The assumptions (1.3) and (1.4) imply that on $D_* = D - \{0\}$ the form $\omega = dd^c \log \tau$ has rank $n-1$. Thus ω defines on D_* a distribution Σ the complex rank 1 which is integrable since $d\omega = 0$. The maximal integral manifolds of Σ are Riemann surfaces and provide a foliation of D_* which is known as the *Monge-Ampère foliation* associated to τ. A simple calculation shows that at every point the distribution Σ is generated by the vector field X defined by

$$X = X^{\mu} \frac{\partial}{\partial z^{\mu}} = \tau^{\bar{\nu}\mu} \tau_{\bar{\nu}} \frac{\partial}{\partial z^{\mu}} \neq 0, \tag{3.1}$$

where $(\tau^{\bar{\nu}\mu}) = (\tau_{\mu\bar{\nu}})^{-1}$. Hence the leaves of the Monge-Ampère foliation one just the complex integral curves of X (for the details see [9] for example). In [7] it was given the following description of the Monge-Ampere foliation. There exists uniquely a C^{∞} surjective map $F: U \times S \to D$ such that

$$F(0,b) = p \quad \text{for all} \quad b \in S; \tag{3.2}$$

$$\tau(F(z,b)) = |z|^2 \quad \text{for all} \quad (z,b) \in U \times S; \tag{3.3}$$

$$F(z,\lambda b) = F(\lambda z, b) \quad \text{for all} \quad (z,b) \in U \times S \quad \text{and} \quad \lambda \in \partial U; \tag{3.4}$$

$$F(.,b): U \to D \ \text{is proper, injective and holomorphic for all} \ b \in S; \tag{3.5}$$

$$X(F(z,b)) = z F'(z,b) \quad \text{for all} \quad z \in U - \{0\} \quad \text{and} \quad b \in S; \tag{3.6}$$

$$F'(0,b) = \|F'(0,b)\| \neq 0 \quad \text{for all} \quad b \in S. \tag{3.7}$$

From this it follows that every leaf of the Monge-Ampère foliation is parametrized by $F(.,b)$ for some $b \in S$ and that the leaf space of the foliation is $\mathbb{P}_{n-1}$. Clearly there is a strong similarity between the Monge-Ampere foliation and the foliation in extremal disks of a strictly convex domain. In fact, using properties (3.2)-(3.7), in [8] it was shown the following

THEOREM 3. *For every* $b \in S$ *we have:*

(i) $F(.,b): U \to D$ *is the unique extremal map for the Kobayashi metric with respect to* 0 *and* b.

(ii) $F(.,b): U \to D$ *is the unique extremal map for the Kobayashi distance with respect to* 0 *and* $F(w,b)$ *for all* $w \in U$.

As a consequence, if we define $\sigma: \mathbb{C}^n \to [0,\infty)$ by $\sigma(0)=0$ and

$$\sigma(Z) = \frac{\|Z\|^2}{\|F'(0, Z/\|Z\|)\|^2} \qquad \text{if} \quad Z \neq 0, \tag{3.8}$$

then the Kobayashi indicatrix of D at 0 is given by

$$I_0(D) = \{Z \in \mathbb{C}^n \mid \sigma(Z) < 1\}. \tag{3.9}$$

It should also remarked that in [7] was proved that near 0 we have

$$\tau(Z) = \sigma(Z) + L(Z)\,\|Z\|^3, \tag{3.10}$$

where L is a suitable bounded function.

Finally, if δ is the Kobayashi distance from 0, then part (ii) of Theorem 2 and (3.3) imply that

$$\tau(Z) = (\text{tgh}\,\delta(Z))^2 \tag{3.11}$$

for every $Z \in D$. This shows in particular that the pseudoballs defined by τ are exactly the pseudoballs of the Kobayashi distance with center at 0.

We have the following

THEOREM 4. *The Kobayashi indicatrix* $I_0(D)$ *of* D *at* 0 *is strictly convex if either one of following statements are verified:*

(i) *there exists a neighbourhood* U *of* 0 *such that* τ *is strictly convex on* $U - \{0\}$,

(ii) *there exists* $R > 0$ *such that if* $0 < r < R$ *then the Kobayashi pseudoballs* $K(r) = \{Z \in D \mid \delta(Z) < r\}$ *are strictly convex.*

Proof. Using (3.10), the same proof of Theorem 1 shows that (i) implies the claim. Furthermore (i) and (ii) are equivalent. That (i) implies (ii) is obvious. To prove the other implication it is enough to note that if $r \in (0,R)$ and $t = (\text{tgh}\,r)^2$, then $t^{-1}\tau$ is the Lempert exhaustion of the strictly convex domain $K(r)$ and thus it is strictly convex.

References

[1] BARTH, T.: 'The Kobayashi indicatrix at the center of a circular domain', *Proc. A.M.S.* 88 (1983), 527-530.

[2] FORNAESS, J.E.: ' Embedding strictly pseudoconvex domains in convex domains', *Am. J. Math.* 98 (1976), 529-569.

[3] HARRIS, L.A.: 'Schwarz-Pick systems of pseudometrics for domains is normed linear spaces', in: *Advances in Holomorphy*, North-Holland Math. Studies 34, North-Holland, Amsterdam and New York]979, 345-406.

[4] KOBAYASHI, S.: 'Intrinsic distances, measures and geometric function theory', *Bull. A.M.S.* 82 (1976), 357-416.

[5] LEMPERT, L.: 'La métrique de Kobayashi et la represéntation des domains sur al boule', *Bull. Soc. Math. Fr.* 109 (1981), 427-474.

[6] PATRIZIO, G.: 'Parabolic exhaustions for strictly convex domains', *Manuscr. Math.* 47 (1984), 271-309.

[7] ——: 'A characterization of complex manifolds biholomorphic to a circular domain', *Math. Z.* 189 (1985), 343-363.

[8] ——: 'The Kobayashi metric and the homogeneous complex Monge--Ampère equation', in: *Complex analysis and applications, Varna 1985, Proceedings*, Edited by L. Iliev, I. Ramadanov, and T. Tonev, Publ. House of the Bulgarian Acad. of Sciences, Sofia 1986, pp. 515-523.

[9] WONG, P.M.: 'Geometry of the homogeneous complex Monge-Ampère equation', *Invent. Math.* 67 (1982), 261-274.

BOUNDARY REGULARITY OF THE SOLUTION OF THE $\bar{\partial}$-EQUATION IN THE POLYDISC

Piotr Jakóbczak
Mathematical Institute
Jagiellonian University
PL-30-059 Kraków, Poland

ABSTRACT. Henkin and Sergeev [4] introduced the solution operator for the $\bar{\partial}$-equation with uniform estimates for strictly pseudoconvex polyhedra (satisfying some additional assumption). We prove that in the unit polydisc in $\mathbb{C}^n$ (which is one of the simplest examples of strictly pseudoconvex polyhedra) the operator admits C^k-estimates for (0,1)-forms (Theorem 1).

0. NOTATIONS

Let $F(D)$ be a function space on an open subset D of $\mathbb{C}^n$. By $F_{oq}(D)$ we denote the space of differential forms of type $(0,q)$ on D, with coefficients in $F(D)$.

Given a domain D in C^n, $\|\ \|_D$ and $\|\ \|_{D,k}$ denotes the usual uniform and C^k-norm on D.

If $\alpha = (\alpha_1, \ldots, \alpha_n) \in \mathbb{Z}^n_+$ is a multiindex, we set $|\alpha| = \alpha_1 + \ldots + \alpha_n$, and for $\alpha, \beta \in Z^n_+$, $\alpha < \beta$ means that for every $i = 1, \ldots, n$, $\alpha_i \leq \beta_i$, and $|\alpha| < |\beta|$.

The differential operators $\partial/\partial z_j$ and $\partial/\partial \overline{z_j}$ in $\mathbb{C}^n$ will be denoted by D_j and $D_{\bar{j}}$, $j = i, \ldots, n$, respectively, For $\alpha \in Z^n_+$, we set

$$D^{\alpha} = \frac{\partial^{|\alpha|}}{\partial z_i^{\alpha_1} \ldots \partial z_n^{\alpha_n}}, \quad D^{\bar{\alpha}} = \frac{\partial^{|\alpha|}}{\partial \bar{z}_1^{\alpha_1} \ldots \partial \bar{z}_n^{\alpha_n}}, \quad D^{\alpha\bar{\beta}} = D^{\alpha} D^{\bar{\beta}}.$$

If the differentiated function depends on two variables, $\zeta, z \in \mathbb{C}^n$, we denote the respective differentation by D_{ζ_j}, $D_{\bar{z}_j}$, $D^{\alpha\bar{\beta}}$, etc.

1. INTRODUCTION

Henkin [3] as well as Grauert and Lieb [2] introduced the solution operator for the $\bar{\partial}$-equation with uniform estimates in strictly pseudoconvex domains. More exactly, they proved that if D is a strictly pseudoconvex domain in $\mathbb{C}^n$ with C^2-boundary and f is a $\bar{\partial}$-closed differential form of type $(0,q)$ with coefficients which are smooth in D

J. Ławrynowicz (ed.), Deformations of Mathematical Structures, 177–189.

and continuous up to ∂D, $q = 1, \ldots, n$, then there exists a $(0, q-1)$-form $T_q f$ with coefficients smooth in D and continuous up to ∂D, such that $\bar{\partial} T_q f = f$, the operator $f \to T_q f$ is linear, and there exists $c > 0$ independent of f, such that $\|T_q f\|_D \leq c\|f\|_D$. This solution operator is given by an integral formula.

Siu has shown in [6] that if D has sufficiently smooth boundary, then the operator T_1 admits also C^k-estimates, i.e. if $f \in C^k_{01}(\bar{D}) \cap C^\infty_{01}(D)$ and $\bar{\partial} f = 0$, then the solution $T_1 f$ satisfies the estimate $\|T_1 f\|_{D,k} \leq c\|f\|_{D,k}$ with some c indenendent of f.

On the other hand, Henkin and Sergeev introduced in [4] the solution operator for the $\bar{\partial}$-equation with uniform estimates for a wide class of domains, called strictly pseudoconvex polyhedra (satisfying some additional assumption). This operator is also given by an integral formula, which is a generalization of the formula for the solution in strictly pseudoconvex domains.
In this paper we prove that in the unit polydisc $U^n = \{z = (z_1, \ldots, z_n) \in \mathbb{C}^n \mid |z_i| < 1, i = 1. \ldots, n\}$ in $\mathbb{C}^n$ (which is one of the simplest examples of strictly pseudoconvex polyhedra), Henkin-Sergeev's operator admits C^k-estimates for $(0,1)$-forms:

<u>THEOREM 1</u>. *Let* f *be a* $\bar{\partial}$*-closed* $(0,1)$*-form in* U^n *with coefficients which are smooth and bounded with derivatives up to order* k *in* U^n. *Then the function* $T_1 f$ *- Henkin-Sergeev's solution of the equation* $f = \bar{\partial} T_1 f$, *is in* $C^\infty(U^n) \cap L^{\infty,k}(U^n)$, *and there exists* $c > 0$ *independent of* f, *such that*

$$\|T_1 f\|_{U^n,k} \leq c\|f\|_{U^n,k}. \tag{1.1}$$

Moreover, if f *has coefficients which are* C^k *up to* U^n, *so has* $T_1 f$.

In both papers [6] and [4], it is necessary to reformulate the original integral formula for the solution operator, in order to obtain a new formula, which admits already the desired estimates. Our method of proving it is a combination of the techniques of Henkin and Sergeev, and of some variation of Siu's procedure of the proof of C^k-estimates. It turned out, that the essential ingredient of our proof is the fact that Henkin-Lieb's kernel for the unit disc $U \subset \mathbb{C}$ is holomorphic in both variables (it is simply the Cauchy kernel). This is in contrast with the several dimensional case, where the kernel has not this property, and our method fails. E.g. we do not know, how to prove Theorem 1 in the case of the product $D_1 \times \ldots \times D_n$, where every D_i is a strictly pseudoconvex domain in C^{n_i}, $n_i > 1$.

Landucci proved in [5] that if f is a $\bar{\partial}$-closed $(0,1)$-form in U^2 with coefficients smooth in U^2 and C^k up to ∂U^2, then the canonical solution u of the equation $\bar{\partial} u = f$ (i.e. the unique function u in U^2 such that

$$\int_{U^2} f(x)\overline{u(x)}\,dm(x) = 0$$

for every $f \in L^2(U^2)$ and holomorphic in U^2 - m is a Lebesgue measure in C^2) is also C^k up to ∂U^2.

2. THE FORMULA FOR THE SOLUTION OF $\bar\partial$ WITH C^k-ESTIMATES IN THE POLYDISC

Given $i \in \{1, \ldots, n\}$ and $\zeta, z \in \mathbb{C}^n$, set

$$\eta^i = \eta^i(\zeta,z) = (0, \ldots, (\zeta_i - z_i)^{-1}, \ldots, 0)^t$$

$(\zeta_i - z_i)^{-1}$ on the i-th place). Set also

$$\eta^0 = (\bar\zeta_1 - \bar z_1)/|\zeta - z|^2, \ldots, (\bar\zeta_n - \bar z_n)/|\zeta - z|^2)^t.$$

For $I = \{i_1, \ldots, i_s\} \subset \{1, \ldots, n\}$ with $s < n$ and $1 \le i_1 < \ldots < i_s \le n$, let

$$M_I = M_I(\zeta,z) = \det(\eta^{i_1}, \ldots, \eta^{i_s}, \eta^o, \bar\partial_\zeta\eta^o, \ldots, \bar\partial_\zeta\eta^o),$$

where $\bar\partial_\zeta\eta^o$ appears n-s-1 times. Set $L = L(\zeta,z) = \det(\eta^o, \bar\partial_\zeta\eta^o, \ldots, \bar\partial_\zeta\eta^o)$ ($\bar\partial_\zeta\eta^o$ appears $n-1$ times), and let $d\zeta = d\zeta_1 \wedge \ldots \wedge d\zeta_n$.

It follows from [4], that the solution $u = T_1 f$ of the equation $\bar\partial u = f$, where $f \in C^k_{01}(\bar U^n) \cap C^\infty_{01}(U^n)$ is $\bar\partial$-closed, has the form

$$T_1 f(z) = \sum_I c_I \int_{S_I} M_I \wedge d\zeta \wedge f(\zeta) + c_n \int_{U^n} L \wedge d\zeta \wedge f(\zeta), \tag{2.1}$$

where the summation is extended over all non-void subsets $I = \{i_1, \ldots, i_s\}$ of $\{1, \ldots, n\}$ with $|I| = s < n$, C_I and c_n are constants, and $S_I = \{z = (z_1, \ldots, z_n) \in \bar U^n \mid |z_i| = i, i = i_1, \ldots, i_s\}$.

In this chapter we derive the formula for $D^\alpha T_1 f$ with $|\alpha| \le k$ (see (2.5)). The procedure which leads to (2.5) is based on the method used by Siu in [6]. The formula (2.5) allows then to apply the techniques of Henkin and Sergeev from [4] in order to obtain some new formula for $D^\alpha T_1 f$, which admits uniform estimates; this will be done in the next chapter.

As in [6], p. 173, set for $g \in C^k(\bar U^n)$ and $\ell \le k$,

$$g^{(\ell)}(\zeta,z) = g(\zeta) - \sum_{|\beta|+|\beta'| \le 1} \frac{1}{\beta!} \frac{1}{(\beta')!} (D^{\beta\bar\beta'} g)(z)(\zeta - z)^\beta (\bar\zeta - \bar z)^{\beta'}, \tag{2.2}$$

$$g^{(-1)}(\zeta,z) = g(\zeta),$$

and for $f \in C^k_{01}(\bar U^n)$, $f = \sum_{i=1}^n f_i \, d\bar\zeta_i$, put $f^{(\ell)}(\zeta,z) = \sum_{i=1}^n f_i^{(\ell)}(\zeta,z) d\bar\zeta_i$, $\ell = -1, 0, 1, \ldots, k$. Note that by [6], p. 173,

$$D_{\zeta_j} g^{(\ell)}(\zeta,z) = (D_j g)^{(\ell-i)}(\zeta,z), \quad D_{\bar\zeta_j} g^{(\ell)}(\zeta,z) = (D_{\bar j} g)^{(\ell-i)}(\zeta,z), \tag{2.3}$$

and

$$(D_{z_j} + D_{\zeta_j}) g^{(\ell)}(\zeta,z) = (D_j g)^{(\ell)}(\zeta,z), \quad \ell = 0, \ldots, k. \tag{2.4}$$

Moreover, if f is a differential form with sufficiently smooth coefficients, by $D^{\alpha\beta}f$ we will denote the form, obtained from f by applying the differential operator $D^{\alpha\bar{\beta}}$ to the coefficients of f.

PROPOSITION 2.1. *Let f be $\bar\partial$-closed $(0,1)$-form in U^n with coefficients which are smooth in U^n and C^k up to boundary. Then for every $\alpha = (\alpha_1, \ldots, \alpha_n) \in Z^n_+$ with $|\alpha| \leq k$,*

$$\begin{aligned} D^\alpha T_1 f(z) = &\sum_{\beta < \alpha} c_{\alpha\beta} D^\beta T_1 (D^{\alpha-\beta} f)(z) + \\ &+ \sum_{i=1}^{n} \sum_{|\beta|+|\beta'|<|\alpha|} a_{\alpha\beta\beta' i}(z)\, D^{\beta\bar{\beta}'} f_i(z) \\ &+ \sum_{\beta\leq\alpha} d_{\alpha\beta} \int_{U^n} D^\beta_\zeta L \wedge d\zeta \wedge (D^{\alpha-\beta} f)^{(|\beta|-1)}(\zeta,z) \\ &+ \sum_{|I|<n} \sum_{\beta\leq\alpha} \sum_{\gamma\leq\beta_I} \int_{S_I} b^I_{\alpha\beta\gamma}(\zeta)\, D^{\beta'_I} \bar{I}^{\gamma} M_I \wedge d\zeta \wedge \\ &\quad \overline{(D^{(\alpha-\beta)'_I(\alpha-\beta)_I} f)^{(|\beta|-1)}}(\zeta,z), \end{aligned} \tag{2.5}$$

where $c_{\alpha\beta}$, $d_{\alpha\beta}$ are constants, $a_{\alpha\beta\beta' i} \in C^\infty(\mathbb{C}^n)$, $b^I_{\alpha\beta\gamma}(\zeta)$ is a polynomial in $\bar\zeta_i$, where $i \in I$ are such that $\alpha_i > 0$, and for $J \subset \{1, \ldots, n\}$ and $\delta = (\delta_1, \ldots, \delta_n) \in Z^n_+$, $\delta_J = ((\delta_J)_1, \ldots, (\delta_J)_n)$, where $(\delta_J)_i = {}_i$ if $i \in J$ and 0 otherwise, and $\delta'_J = \delta - \delta_J$.

In order to prove Proposition 2.1 we need first the following

PROPOSITION 2.2. *Given $\beta, \gamma \in Z^n_+$, the function $T_1((\zeta-z)^\beta \bar\partial_\zeta (\bar\zeta - \bar z)^\gamma)$ is smooth in $\mathbb{C}^n$.*

Proof. Since $(\zeta-z)^\beta \bar\partial_\zeta (\bar\zeta - \bar z)^\gamma = \bar\partial_\zeta ((\zeta-z)^\beta (\bar\zeta - \bar z)^\gamma)$, it follows from [4], p. 529, that $(\zeta-z)^\beta (\bar\zeta - \bar z)^\gamma = \bar\partial T_0((\zeta-z)^\beta(\bar\zeta-\bar z)^\gamma) + T_1((\zeta-z)^\beta \bar\partial_\zeta(\bar\zeta-\bar z)^\gamma)$. Therefore, it is enough to prove that $T_0((\zeta-z)^\beta(\bar\zeta-\bar z)^\gamma)$ is smooth in $\mathbb{C}^n$. Yet by [4],

$$(T_0 u)(z) = \int_{(\partial U)^n} \frac{u(\zeta)}{(\zeta_1 - z_1) \ldots (\zeta_n - z_n)}\, d\zeta.$$

It is simple to prove that this integral is smooth in $\mathbb{C}^n$ for $u = (\zeta - z)^\beta (\bar\zeta - \bar z)^\gamma$. Therefore we are done.

Proof of Proposition 2.1. Consider first the case $k = 1$, and let $D^\alpha = D_j$. We have

$$T_1 f(z) = \sum_{i=1}^{n} f_i(z) T_i(d\bar\zeta_i)(z) + \int_{U^n} L \wedge d\zeta \wedge f^{(0)}(\zeta,z) + \tag{2.6}$$

$$+ \sum_{|I|<n} \int_{S_I} M_I \wedge d\zeta \wedge f^{(0)}(\zeta,z).$$

Applying the operator D_j to both sides of (2.6), using the identity $D_j = (D_{z_j} + D_{\zeta_j}) - D_{\zeta_j}$, (2.4), and the fact that $(D_{z_j} + D_{\zeta_j})L = (D_{z_j} + D_{\zeta_j})M_I = 0$ (since both depend only on $\zeta_i - z_i$ and $\bar{\zeta}_i - \bar{z}_i$, $i = 1, \ldots, n$), we obtain

$$\begin{aligned} D_jT_1f(z) &= \sum_{i=1}^{n} (D_jf_i)(z)T_1(d\bar{\zeta}_i)(z) + \sum_{i=1}^{n} f_i(z)D_jT_1(d\bar{\zeta}_i)(z) \\ &\quad + \int_{U^n} (D_{z_j} + D_{\zeta_j})(L \wedge d\zeta \wedge f^{(0)}(\zeta,z)) \\ &\quad - \int_{U^n} D_{\zeta_j}(L \wedge d\zeta \wedge f^{(0)}(\zeta,z)) + \sum_{|I|<n} \int_{S_I} (D_{z_j} + D_{\zeta_j})(M_I \wedge \\ &\quad d\zeta \wedge f^{(0)}(\zeta,z)) - \sum_{|I|<n} \int_{S_I} D_{\zeta_j}(M_I \wedge d\zeta \wedge f^{(0)}(\zeta,z)) \\ &= \sum_{i=1}^{n} (D_jf_i)(z)T_1(d\bar{\zeta}_i)(z) + \sum_{i=1}^{n} f_i(z)D_jT_1(d\bar{\zeta}_i)(z) \\ &\quad + \int_{U^n} L \wedge d\zeta \wedge (D_jf)^{(0)}(\zeta,z) - \int_{U^n} D_{\zeta_j}(L \wedge d\zeta \wedge f^{(0)}(\zeta,z)) \\ &\quad + \sum_{|I|<n} \int_{S_I} M_I \wedge d\zeta \wedge (D_jf)^{(0)}(\zeta,z) \\ &\quad - \sum_{|I|<n} \int_{S_I} D_{\zeta_j}(M_I \wedge d\zeta \wedge f^{(0)}(\zeta,z)). \end{aligned} \tag{2.7}$$

The formula (2.6) for D_jf has the form

$$\begin{aligned} T_1(D_jf)(z) &= \sum_{i=1}^{n} (D_jf_i)(z)T_1(d\bar{\zeta}_i)(z) + \int_{U^n} L \wedge d\zeta \wedge (D_jf)^{(0)}(\zeta,z) \\ &\quad + \sum_{|I|<n} \int_{S_I} M_I \wedge d\zeta \wedge (D_jf)^{(0)}(\zeta,z). \end{aligned} \tag{2.8}$$

Insert (2.8) to (2.7). Then

$$\begin{aligned} D_jT_1f(z) &= T_1(d_jf)(z) + \sum_{i=1}^{n} f_i(z)\,D_jT_1(d\bar{\zeta}_i)(z) \\ &\quad - \int_{U^n} D_{\zeta_j}(L \wedge d\zeta \wedge f^{(0)}(\zeta,z)) - \sum_{|I|<n} \int_{S_I} D_{\zeta_j}(M_I \wedge d\zeta \wedge f^{(0)}(\zeta,z)). \end{aligned} \tag{2.9}$$

Consider the term $\int_{S_I} D_{\zeta_j}(M_I \wedge d\zeta \wedge f^{(0)}(\zeta,z))$. If $j \notin I$, this is equal, by (2.2), to

$$\int_{S_I} (D_{\zeta_j} M_I) \wedge d\zeta \wedge f^{(0)}(\zeta,z) + \int_{S_I} M_I \wedge d\zeta \wedge (D_jf)(\zeta). \tag{2.10}$$

If $j \in I$, write $\int_{S_I} D_{\zeta_j}(M_I \wedge d\zeta \wedge f^{(0)}(\zeta,z))$ as

$$\int_{S_{I-j}} \Big(\int_{\zeta_j \in \partial U} D_{\zeta_j} (M_I \wedge d\zeta \wedge \sum_{\substack{i=1 \\ i\neq j}}^{n} f_i^{(0)}(\zeta,z)d\bar{\zeta}_i)\Big), \tag{2.11}$$

where $I-j = I \setminus \{j\}$ (we use the fact that $d\zeta_j \wedge d\bar{\zeta}_j = 0$ on $\{\zeta_j \in \partial U\}$).

As in [6], p. 167, apply Stokes' theorem to the form $M_I \wedge d\zeta\,[j] \wedge \sum_{\substack{i=1 \\ i\neq j}}^{n} f_i^{(0)}(\zeta,z)d\bar{\zeta}_i$ in the interior integral (here $d\zeta[j] = d\zeta_1 \wedge \ldots \wedge d\zeta_{j-1} \wedge d\zeta_{j+1} \wedge \ldots \wedge d\zeta_n$).

We can see that (2.11) is equal to

$$c_1 \int_{S_{I-j}} \Big(\int_{\zeta_j \in \partial U} D_{\bar{\zeta}_j} (M_I \wedge d\bar{\zeta}_j \wedge d\zeta[j] \wedge \sum_{i\neq j} f_i^{(0)}(\zeta,z)d\bar{\zeta}_i)\Big) =$$

$$c_2 \int_{S_{I-j}} \Big(\int_{\zeta_j \in \partial U} \bar{\zeta}_j^{2} D_{\bar{\zeta}_j} (M_I \wedge d\zeta \wedge \sum_{i\neq j} f_i^{(0)}(\zeta,z)\, d\bar{\zeta}_i)\Big) =$$

$$c_3 \int_{S_I} \bar{\zeta}_j^{2} D_{\bar{\zeta}_j} (M_I \wedge d\zeta \wedge f^{(0)}(\zeta,z)),$$

where c_i are constants, because $d\bar{\zeta}_j = -\bar{\zeta}_j^{2}\, d\zeta_j$ and $d\zeta_j \wedge d\bar{\zeta}_j = 0$ on $\{\zeta_j \in \partial U\}$. This last integral is equal, by (2.2), to

$$c_4 \int_{S_I} \bar{\zeta}_j^{2} D_{\bar{\zeta}_j} M_I \wedge d\zeta \wedge f^{(0)}(\zeta,z) + c_5 \int_{S_I} \bar{\zeta}_j^{2} M_I \wedge d\zeta \wedge (D_{\bar{j}}f)(\zeta). \tag{2.12}$$

Moreover, by (2.2),

$$\begin{aligned} &\int_{U^n} D_{\zeta_j} (L \wedge d\zeta \wedge f^{(0)}(\zeta,z)) = \\ &\quad = \int_{U^n} D_{\zeta_j} L \wedge d\zeta \wedge f^{(0)}(\zeta,z) + \int_{U^n} L \wedge d\zeta \wedge (D_j f)(\zeta) \end{aligned} \tag{2.13}$$

Inserting the integrals from (2.10), (2.12) and (2.13) to (2.9), and using the fact that $D_j T_1 (d\bar{\zeta}_i) \in C^{\infty}(\mathbb{C}^n)$ by Proposition 2.2, we can see that $D_j T_1 f$ consists of terms, which satisfy (2.5). Therefore, Proposition 2.1 is valid for $k=1$.

Proceeding inductively, suppose the validity of (2.5) for $|\alpha| \leq k$. Consider the differential operator $D_j D^{\alpha}$ with $j \in \{1,\ldots,n\}$ and $|\alpha| = k$. As in [6], p. 174, define for $\gamma \in Z_+^n$ with $|\gamma| > 0$, $J_\gamma = \{(\delta,i)\,|\,\delta \in Z_+^n,\ i=1,\ldots,n,$ and $D^{\delta} D_i = D^{\gamma}\}$, and let

$$f_\gamma(z) = D^{\bar{\delta}} f_i \tag{2.14}$$

for $(\delta,i) \in J_\gamma$ (this definition is independent of the choice of (δ,i) because $\bar{\partial} f = 0$). Similarly as in [6], p. 174, we obtain

$$f^{(k)}(\zeta,z) = f(\zeta) - \sum_{|\beta|+|\gamma|\le k+1} \frac{1}{\beta!}\frac{1}{\gamma!}(\zeta-z)^{\beta}(D^{\beta}f_{\gamma})(z)\bar{\partial}_{\zeta}(\bar{\zeta}-\bar{z})^{\gamma}. \tag{2.15}$$

By the induction assumption, (2.5) holds for DT_1f. Replace f by $f^{(k)}(\zeta,z')$ in (2.5), using the expression (2.15), and then set $z'=z$. Then

$$\begin{aligned}
D^{\alpha}T_1 f(z) = &\sum_{|\beta|+|\gamma|\le k+1} \frac{1}{\beta!}\,\frac{1}{\gamma!}(D^{\beta}f_{\gamma})(z)T_1((\zeta,z)^{\beta}\bar{\partial}_{\zeta}(\bar{\zeta}-\bar{z}')^{\gamma})\Big|_{z'=z} \\
&+ \sum_{\beta<\alpha} c_{\alpha\beta}D^{\beta}T_1(D^{\alpha-\beta}f^{k}_{z'})(z)\Big|_{z'=z} \\
&+ \sum_{i=1}^{n} \sum_{|\beta|+|\beta'|<|\alpha|} a_{\beta\beta' i}(z)\, D^{\beta\bar{\beta}'}(f_i^{(k)}(\zeta,z'))\Big|_{\zeta=z'=z} \\
&+ \sum_{\beta<\alpha} \int_{U^n} d_{\alpha\beta}D^{\beta}_{\zeta}\, L\wedge d\zeta\wedge(D^{\alpha-\beta}f^{k}_{z'})^{(|\beta|-1)}(\zeta,z)\Big|_{z'=z} \\
&+ \sum_{|I|<n}\ \sum_{\beta\le\alpha}\ \sum_{\gamma\le\beta_I} \int_{S_I} b^{I}_{\alpha\beta\gamma}(\zeta)D_{\zeta}^{\beta'_I\bar{\gamma}}\, M_I\wedge d\zeta\wedge \\
&\qquad (D^{(\alpha-\beta)'_I\overline{(\alpha-\beta)}_I} f^{k}_{z'})^{(|\beta|-1)}(\zeta,z)\Big|_{z'=z},
\end{aligned}$$

where $f^{k}_{z'}(\zeta) = f^{(k)}(\zeta,z')$. Note that

$$g^{(\ell)}(z,z) = 0 \quad \text{for} \quad \ell \ge 0 \tag{2.16}$$

and

$$(f^{k}_{z'})^{(\ell)}(\zeta,z) = f^{(k)}(\zeta,z) \quad \text{for} \quad z'=z \quad \text{and} \quad \ell\le k. \tag{2.17}$$

It follows from (2.3), (2.16) and (2.17), and from the fact that $|\alpha| = k$, that

$$(D^{\alpha-\beta}f^{k}_{z'})(\zeta) = (D^{\alpha-\beta}f)^{(k-|\alpha|-|\beta|)}(\zeta,z\ ) = (D^{\alpha-\beta}f)^{(|\beta|)}(\zeta,z'),$$

$$D_{\zeta}^{\beta\bar{\beta}'}(f_i^{(k)}(\zeta,z'))\Big|_{\zeta=z'=z} = (D^{\beta\bar{\beta}'}f_i)^{(k-|\beta|-|\beta'|)}(z,z) = 0,$$

and similarly

$$(D^{\alpha-\beta}f^{k}_{z'})^{(|\beta|-1)}(\zeta,z)\Big|_{z'=z} = (D^{\alpha-\beta}f)^{(|\beta|)}(\zeta,z)$$

and

$$(D^{(\alpha-\beta)'_I\overline{(\alpha-\beta)}_I} f^{k}_{z'})^{(|\beta|-1)}(\zeta,z)\Big|_{z'=z} = (D^{(\alpha-\beta)'_I\overline{(\alpha-\beta)}_I} f^{(|\beta|)}(\zeta,z).$$

Moreover, by (2.15)

$$\begin{aligned}
T_1((D^{\alpha-\beta}f)^{(|\beta|)}(\zeta,z'))(z)\Big|_{z'=z} &= T_1(D^{\alpha-\beta}f)(z) - \\
&- \sum_{|\eta|+|\delta|\le|\beta|+1} \frac{1}{\eta!}\,\frac{1}{\delta!}(D^{\eta}(D^{\alpha-\beta}f)_{\delta})(z)\, T_1((\zeta-z)^{\eta}\,\bar{\partial}_{\zeta}(\bar{\zeta}-\bar{z})^{\delta}),
\end{aligned}$$

where $(D^{\alpha-\beta}f)$ is defined as in (2.14). Therefore, using Proposition

2.2, we can see that

$$
\begin{aligned}
D^{\alpha}T_1(z) &= \sum_{\beta<\alpha} c_{\alpha\beta} D^{\beta}T_1(D^{\alpha-\beta}f)(z) \\
&+ \sum_{i=1}^{n} \sum_{|\beta|+|\beta'|\leq|\alpha|} \tilde{a}_{\alpha\beta\beta' i}(z) D^{\beta}\bar{D}^{\beta'} f_i(z) \\
&+ \sum_{\beta\leq\alpha} d_{\alpha\beta} \int_{U^n} D_{\zeta}^{\beta} L \wedge d\zeta \wedge (D^{\alpha-\beta} f)^{(|\beta|)}(\zeta,z) \qquad (2.18) \\
&+ \sum_{|I|<n} \sum_{\beta\leq\alpha} \sum_{\gamma\leq\beta_I} \int_{S_I} b^{I}_{\alpha\beta\gamma}(\zeta) D^{\beta}\, {}_I\acute{D}^{\gamma} M_I \wedge d\zeta \wedge \\
&\qquad (D^{(\alpha-\beta)}{}_I\acute{\overline{D}}^{(\alpha-\beta)_I} f)^{(|\beta|)}(\zeta,z),
\end{aligned}
$$

where $\tilde{a}_{\alpha\beta\beta' i} \in C^{\infty}(\mathbb{C}^n)$.

Differentiating both sides of (2.18) with respect to D_j and proceeding as in the case $k=1$, we obtain

$$
\begin{aligned}
D_j D^{\alpha}T_1 f(z) &= \sum_{\beta<\alpha} c_{\alpha\beta} D_j D^{\beta}T_1(D^{\alpha-\beta}f)(z) \\
&+ \sum_{i=1}^{n} \sum_{|\beta|+|\beta'|\leq|\alpha|} D_j \tilde{a}_{\alpha\beta\beta' i}(z) D^{\beta}\bar{D}^{\beta'} f_i(z) \\
&+ \sum_{i=1}^{n} \sum_{|\beta|+|\beta'|\leq|\alpha|} \tilde{a}_{\alpha\beta\beta' i}(z) D^{\beta}\bar{D}^{\beta'} (D_j f)(z) \\
&+ \sum_{\beta\leq\alpha} d_{\alpha\beta} \int_{U^n} D_{\zeta}^{\beta} L \wedge d\zeta \wedge (D^{\alpha-\beta} D_j f)^{(|\beta|)}(\zeta,z) \\
&- \sum_{\beta\leq\alpha} d_{\alpha\beta} \int_{U^n} D_{\zeta_j} (D_{\zeta}^{\beta} L \wedge d\zeta \wedge (D^{\alpha-\beta} f)^{(|\beta|)}(\zeta,z)) \\
&+ \sum_{|I|<n} \sum_{\beta\leq\alpha} \sum_{\gamma\leq\beta_I} \int_{S_I} b^{I}_{\alpha\beta\gamma}(\zeta) D_{\zeta}^{\beta}\, {}_I\acute{\bar{D}}^{\gamma} M_I \wedge d\zeta \wedge \\
&\qquad (D^{(\alpha-\beta)}{}_I\acute{\overline{D}}^{(\alpha-\beta)_I} D_j f)^{(|\beta|)}(\zeta,z) \\
&- \sum_{|I|<n} \sum_{\beta\leq\alpha} \sum_{\gamma\leq\beta_I} \int_{S_I} D_{\zeta_j} (b^{I}_{\alpha\beta\gamma}(\zeta) D_{\zeta}^{\beta}\, {}_I\acute{\bar{D}}^{\gamma} M_I \wedge d\zeta \wedge \\
&\qquad (D^{(\alpha-\beta)}{}_I\acute{\overline{D}}^{(\alpha-\beta)_I} f)^{(|\beta|)}(\zeta,z)).
\end{aligned}
$$

Writing (2.18) for $D_j f$, inserting it into (2.19), performing then with the terms in the last integral in (2.19) the same procedure as in the case $k=1$ (depending on whether $j \notin I$ or $j \in I$), and differentiating with respect to D_{ζ_j} in the second integral in the right-hand side of (2.19), we obtain the desired form of $D_j D^{\alpha}T_1 f$, as asserted in (2.5). This ends the proof of Proposition 2.1.

3. ESTIMATES

We prove first, that if f has coefficients smooth and bounded with derivatives up to order k in U^n, then all derivatives $D^\alpha T_1 f$ with $|\alpha| \leq k$ are also bounded in U^n. Consider the formula (2.5) from Proposition 2.1. Proceeding inductively, we can assume that for $\beta < \alpha$, $T_1(D^{\alpha-\beta}f)$ have coefficients bounded in U^n with all derivatives of order $\leq |\beta|$, therefore $D^\beta T_1(D^{\alpha-\beta}f)$ is bounded in U^n. Similarly, the terms in the second sum on the right-hand side of (2.5) are bounded in U^n. Hence it rests to prove, that all the integrals, appearing in (2.5), define functions, which are bounded in U^n.

It is well-known that the derivatives of the Bochner-Martinelli kernel, $D^\beta_\zeta L(\zeta,z)$, have singularity of order $1/|\zeta-z|^{2n-1+|\beta|}$. Since $|(D^{\alpha-\beta}f)^{(|\beta|-1)}(\zeta,z)| \leq c\|f\|_{U^n,k}|\zeta-z|^{|\beta|}$, where c is independent of β, z and f, we see, that the terms of type

$$\int_{U^n} D^\beta_\zeta L \wedge d\zeta \wedge (D^{\alpha-\beta}f)^{(|\beta|-1)}(\zeta,z)$$

are bounded in U^n (they are even Hölder continuous in U^n with every exponent $\alpha < 1$).

Consider now the integrals over S_I from the last sum in (2.5). Those terms admit no direct estimate. Therefore, similarly as in [4], it is necessary to perform some transformation using Stokes theorem and Bochner-Martinelli formula, in order to represent every such integral as a sum of terms, which can already be estimated.

Given $I = (i_1, \ldots, i_s)$, set

$$\eta^I = (\eta^{i_1}, \ldots, \eta^{i_s}), \qquad \eta^{(0)} = (\eta^0, \bar\partial_\zeta \eta^0, \ldots, \bar\partial_\zeta \eta^0)$$

($\bar\partial_\zeta \eta^0$ is taken $n-s-1$ times). Then $M_I = \det(\eta^I, \eta^{(0)})$, and for $\beta \leq \alpha$ and $\gamma \leq \beta_I$, we have

$$D^{\beta_{\acute{I}}\bar\gamma}_\zeta M_I = \det(\eta^I, D^{\beta_{\acute{I}}\bar\gamma}_\zeta \eta^{(0)}). \tag{3.1}$$

The fact, that η^I is actually not differentiated and remains unchanged in the determinant in the right-hand side of (3.1), is essential for the estimates, as was already mentioned in the introduction.

It is easy to verify, that the method of transformation of terms

$$\int_{S_I} \det(\eta^I, \eta^{(0)}) \wedge d\zeta \wedge f,$$

with the help of Stokes Theorem and Bochner-Martinelli formula, described in [4] (see also [1]), and used in the proof of C^0-estimates, can be applied without essential changes, also for the integrals of the form

$$\int_{S_I} b^I_{\alpha\beta\gamma}(\zeta)\det(\eta^I, D^{\beta_{\acute{I}}\bar\gamma}_\zeta \eta^{(0)}) \wedge d\zeta \wedge (D^{(\alpha-\beta)_{\acute{I}}\overline{(\alpha-\beta)}_I} f)^{(|\beta|-1)}(\zeta,z), \tag{3.2}$$

which appear in (2.5). Proceeding as in [1] or in [4], one can prove the following proposition, which gives another formula for the integrals (3.2). The terms appearing in the formula admit already the direct estimate. In order to formulate this proposition, we need some notations:

Given $K \subset \{1, \dots, n\}$, $K = \{k_1, \dots, k_r\}$ with $1 \le k_1 < \dots < k_r \le n$, set

$$\zeta_K = (\zeta_{k_1}, \dots, \zeta_{k_r}), \qquad z_k = (z_{k_1}, \dots, z_{k_r}).$$

For $l = 1, \dots, n$, let

$$(\eta^{(K)})_\ell(\zeta,z) = (\bar{\zeta}_{k_i} - \bar{z}_{k_i})/|\zeta_K - z_K|^2 \quad \text{if } \ell = k_i \text{ for some } i = 1, \dots, r,$$

and $(\eta^{(K)})_\ell = 0$ otherwise, $\eta^{(K)} = ((\eta^{(K)})_1, \dots, (\eta^{(K)})_n)^t$. Set

$$C_\ell^{(K)} = (\eta^{(K)}, \bar{\partial}_\zeta \eta^{(K)}, \dots, \bar{\partial}_\zeta \eta^{(K)}) \tag{3.3}$$

($\bar{\partial}_\zeta \eta^{(K)}$ is taken ℓ-1 times), $\ell = 1, \dots, n$. Also, let $K^c = \{1, \dots, n\} \setminus K$. Given $j_1, \dots, j_\ell \in \{1, \dots, n\}$, $j_p \ne j_q$ for $p \ne q$, set $S_{j_1 \dots j_\ell} = S_{\{j_1, \dots, j_\ell\}}$ (the indices $j_1, \dots, j_\ell$ need not be ordered increasingly). Set also $D_K = \{(\zeta_{k_1}, \dots, \zeta_{k_r}) \in \mathbb{C}^r \mid |\zeta_{k_i}| \le 1,\ i = 1, \dots, r\}$. For simplicity, denote

$$(D^{(\alpha-\beta)} \acute{I}\, \overline{(\alpha-\beta)}\, I f)^{(|\beta|-1)}(\zeta,z) = h(\zeta,z).$$

PROPOSITION 3.1. *Given* I $\{1, \dots, n\}$, $I = (i_1, \dots, i_s)$, *the integral* (3.2) *is the sum of terms of the following form:*

$$\int_{S_{j_{\tau(1)} \dots j_{\tau(r)}}} \det(C_{d_t}^{(I_t)}, C_{d_{\sigma(0)}}^{(I_{\sigma(0)})}, \dots, C_{d_{\sigma(r-1)}}^{(I_{\sigma(r-1)})}, \bar{\partial}_\zeta C_{d_{\sigma(r)}}^{(I_{\sigma(r)})}, \dots, \bar{\partial}_\zeta C_{d_{\sigma(t-1)}}^{(I_{\sigma(t-1)})} \wedge d\zeta \wedge b_{\alpha\beta\gamma}^I(\zeta) h(\zeta,z) \tag{3.4}$$

$$\int_{S_{j_{\tau(1)} \dots j_{\tau(r)}}} \det(C_{d_t}^{(I_t)}, C_{d_{\sigma(0)}}^{(I_{\sigma(0)})}, \dots, C_{d_{\sigma(r)}}^{(I_{\sigma(r)})}, \bar{\partial}_\zeta C_{d_{\sigma(r+1)}}^{(I_{\sigma(r+1)})}, \dots, \bar{\partial}_\zeta C_{d_{\sigma(t-1)}}^{(I_{\sigma(t-1)})} \wedge d\zeta \wedge \bar{\partial}_\zeta (b_{\alpha\beta\gamma}^I(\zeta) h(\zeta,z) \tag{3.5}$$

$$\int_{D_{(I_{t+1}^c)}} \det\nolimits_{S^{I_{r+1}}_{j_{\tau(1)} \dots j_{\tau(r)}}} (C_{d_t}^{(I_t)}, C_{d_{\sigma(0)}}^{(I_{\sigma(0)})}, \dots, C_{d_{\sigma(r-1)}}^{(I_{\sigma(r-1)})}, \bar{\partial}_\zeta C_{d_{\sigma(r)}}^{(I_{\sigma(r)})}, \dots,$$

$$\bar{\partial}_\zeta C_{d_{\sigma(t-1)}}^{(I_{\sigma(t-1)})}) \times [z_{I_{t+1}}, \zeta_{I_{t+1}^c}] \wedge d\zeta_{I_{t+1}^c} \tag{3.6}$$

$$\wedge\, b_{\alpha\beta\gamma}^I(\zeta) h(\zeta,z)) \big|_{\zeta_{I_{t+1}} = z_{I_{t+1}}}$$

and

$$\int_{D_{I^c_{t+1}} \cap S_{j_{\tau(1)}\cdots j_{\tau(r)}}} \widetilde{\det}_{I_{r+1}}(C_{d_t}^{(I_t)}, C_{d_{\sigma(0)}}^{(I_{\sigma(0)})}, \ldots, C_{d_{\sigma(r)}}^{(I_{\sigma(r)})}, \bar\partial_\zeta C_{d_{\sigma(r+1)}}^{(I_{\sigma(r+1)})}, \ldots,$$

$$\bar\partial_\zeta C_{d_{\sigma(t-1)}}^{(I_{\sigma(t-1)})}) \times [z_{I_{t+1}}, \zeta_{I^c_{t+1}}] \wedge d\zeta_{I^c_{t+1}} \tag{3.7}$$

$$\wedge \bar\partial_\zeta (b^I_{\alpha\beta\gamma}(\zeta) h(\zeta,z)) \Big|_{\zeta_{I_{t+1}} = z_{I_{t+1}}}$$

where: $t \in \{1, \ldots, s\}$, $I_0, I_1, \ldots, I_{t+1}$ *are the subsets of* $\{1, \ldots, n\}$ *such that* $I_0 = \{1, \ldots, n\}$, $I_1 = I$, $I_1 \; I_2 \; \ldots \; I_t \; I_{t+1}$, *and* $I_{t+1} = \emptyset$, $d_\ell = |I_\ell| - |I_{\ell+1}|$ *for* $\ell = 0, \ldots, t$, $j \notin I$ *for* $\ell = 1, \ldots, t$, σ *is a permutation of* $\{0, 1, \ldots, t-1\}$, τ *is a permutation of* $\{1, \ldots, t\}$, $r = 0, 1, \ldots, t$ *in* (3.4) *and* (3.6), *and* $r = 0, \ldots, t-1$ *in* (3.5) *and* (3.7), *for* $r = 0$, $S_{j_{\tau(1)}\cdots j_{\tau(r)}}$ *should be replaced by* U^n, $\widetilde{\det}_{I_{t+1}}(\ldots)[z_{I_{t+1}}, I^c_{t+1}]$ *in* (3.6) *and* (3.7) *means the algebraic complement of* η^I *in*

$$\det(\eta^{I_{t+1}}, C_{d_t}^{(I_t)}, \ldots),$$

restricted to $D_{(I_{t+1})^c}$, *and* $C_{d_\ell}^{(I_\ell)}$ *are defined by* (3.3), *except for* $\ell = 0$: *we set namely*

$$C_{d_o}^{(I_o)} = D^\beta_\zeta \acute{I} \bar{\gamma} \eta^{(0)} = D^\beta_\zeta \acute{I} \bar{\gamma}(\eta^0, \bar\partial_\zeta \eta^0, \ldots, \bar\partial_\zeta \eta^0)$$

($\bar\partial_\zeta \eta^0$ *taken* s-1 *times*).

In order to prove that integral of the form (3.4)-(3.7) can be estimated by $c\|f\|_{U^n,k}$, where c is independent of f, one can use the same considerations, as described in [4] and especially in [1], p. 75-86. The only difference in comparison to [4] and [1] is, that in our case $\eta^{(0)}$ and f are replaced by $D^\beta \acute{I} \gamma \eta^{(0)}$ and $h(\zeta,z)$, respectively; but the estimate $|h(\zeta,z)| \le c\|f\|_{U^n,k}|\zeta-z|^{|\beta|}$, where c is independent of f, cancels the additional singularity (in comparison to the singularity of $\eta^{(0)}$), which occurs in $D^\beta \acute{I} \gamma \eta^{(0)}$; one can easily verify, that this additional singularity is at most of type $1/|\zeta-z|^{|\beta|}$, since $|\beta \acute{I}| + |\bar\gamma| \le |\beta|$. As to the integrals (3.6) and (3.7), one should use the fact, that

$$|h(\zeta,z)|_{z_{I_{t+1}} = \zeta_{I_{t+1}}}| \le c\|f\|_{U^n,k} |\zeta_{I^c_{t+1}} - z_{I^c_{t+1}}|^{|\beta|},$$

where c is independent of f; this cancels the additional singularity from $D^\beta \acute{I} \bar\gamma \eta^{(0)}$ in $\widetilde{\det}_{I_{t+1}}(\ldots)[z_{I_{t+1}}, \zeta_{I^c_{t+1}}]$ in (3.6) and (3.7). This ends the $L^{\infty,k}$-part of Theorem 1.

To prove that for $f \in C^k_{01}(\bar{U}_n)$, we have $u \in C^k(\bar{U}^n)$, one proceeds as follows: Consider the forms $f_r(z) = f(rz)$, $r < 1$. Then every f_r is smooth in $\bar{U}^n$, and hence $T_1 f_r$ is also smooth in $\bar{U}^n$. By (1.1),

$$\|T_1 f_r - T_1 f\|_{\bar{U}^n,k} \le c\|f_r - f\|_{\bar{U}^n,k},$$

and since

$$\|f_r - f\|_{\bar{U}^n,k} \to 0 \quad \text{as} \quad r \to i,$$

it follows that $T_1 f_r$ converges to $T_1 f$ uniformly in $\bar{U}^n$ with all derivatives of order $\le k$. Hence $T_1 f \in C^k(\bar{U}^n)$. This ends the proof of Theorem 1.

Remark. Given $t > 0$, $t \notin N$, let

$$\Lambda_t(\bar{U}^n) = \{f \in C^{[t]}(\bar{U}^n) \mid \quad \text{for every } \alpha \in Z^n_+ \text{ with } |\alpha| = [t],$$

$$D^\alpha f \text{ is } t-[t]\text{-Hölder continuous in } \bar{U}^n\}.$$

One can prove the following theorem:

THEOREM 2. *Let* f *be a* $\bar{\partial}$*-closed* $(0,1)$*-form in* U^n *such that the coefficients of* f *are in* $C^\infty(U^n) \cap \Lambda_t(\bar{U}^n)$, *where* t *is as above. Then* $T_1 f \in C^\infty(U^n) \cap \Lambda_t(\bar{U}^n)$.

In the proof one checks the $t-[t]$-Hölder continuity of all terms appearing in (2.5) and (3.4)-(3.7). The details are tedious and will appear elsewhere.

Remark. It was proved in [4], that if $D \subseteq \mathbb{C}^n$ is a strictly pseudoconvex polyhedron, satisfying conditions (C) and (CR) (for definition see [4], p. 523), then for every $\bar{\partial}$-closed $(0,q)$-form f with coefficients smooth in D and continuous up to ∂D, the $(0,q-1)$-form $u = T_q f$, defined in [4], p. 528, which is a solution of the $\bar{\partial}$-equation $\bar{\partial} u = f$, satisfies the estimate $\|u\|_D \le c\|f\|_D$ with some c independent of f, and has coefficients which are continuous up to D. It is not true that for *every* strictly pseudoconvex polyhedron satisfying the conditions (C) and (CR), if f has coefficients which are C^k up to ∂D, so is $T_q f$. The simple counterexample can be found even in $\mathbb{C}$. Let D be an upper semidisc $\{z \in \mathbb{C} \mid |z| < 1, \operatorname{Im} z > 0\}$ in $\mathbb{C}$. One can check that D is a strictly pseudoconvex polyhedron, satisfying the conditions (C) and (CR), and that for $f \in C_{01}(\bar{D})$,

$$T_1 f(z) = \frac{1}{2\pi i} \int_D \frac{d\zeta}{\zeta - z} \wedge f(\zeta).$$

Let $f(z) = z\,d\bar{z}$. Then f is $\bar{\partial}$-closed and is C^∞ up to ∂D. Since $f = \bar{\partial}|z|^2$, by the Cauchy-Green formula,

$$T_1 f(z) = |z|^2 - \frac{1}{2\pi i} \int_{\partial D} \frac{|\zeta|^2 d\zeta}{\zeta - z}, \quad z \in D.$$

$T_1 f$ is continuous up to ∂D (this follows from the general result of Henkin and Sergeev, but in this simple case can be checked directly). Let L denote the upper boundary of D (oriented positively with respect to D). Then for $z \in D$,

$$D_z \int_{\partial D} \frac{|\zeta|^2 d\zeta}{\zeta - z} = -\int_{\partial D} \frac{|\zeta|^2 d\zeta}{(\zeta - z)^2} = \int_{-1}^{+1} \frac{x^2 dx}{(x-z)^2} + \int_L \frac{d\zeta}{(\zeta - z)^2} = \int_{-1}^{+1} \frac{(x^2+1)dx}{(x-z)^2}$$

by residue theorem. One can check that this last integral is not bounded, as $z \to 1$, $z \in D$. Therefore, $T_1 f \notin C^1(\bar{D})$, although $f \in C^{\infty}_{01}(\bar{D})$.

Therefore, to solve the $\bar{\partial}$-equation with C^k-estimates in arbitrary strictly pseudoconvex polyhedra, one should look for another solution operator. The validity of Theorem 1 shows, in comparison to the above example, that the polydisc is a very special case of strictly pseudoconvex polyhedron.

References

[1] BERTRAMS, J.: *Das $\bar{\partial}$-Problem auf pseudokonvexen Polyedern*, Bonn University, Thesis, 1983.

[2] GRAUERT, H. and I. LIEB: 'Das Ramirezsche Integral und die Lösung der Gleichung $\bar{\partial}f = \alpha$ im Bereich der beschränkten Formen', *Rice Univ. Study* 56,2 (1970), 29-50.

[3] HENKIN, G.M.: 'Integral representation of functions in strictly pseudoconvex domains and applications to the $\bar{\partial}$-problem', *Mat. Sbornik* 82 (1970), 300-308 (in Russian).

[4] —— and A.G. SERGEEV: 'Uniform estimates of the solutions of the $\bar{\partial}$-equation in pseudoconvex polyhedra', *Mat. Sbornik* 112 (1980), 523-567 (in Russian).

[5] LANDUCCI, M.: 'Uniform bounds on derivatives for the $\bar{\partial}$-problem in the polydisc', *Proc. Symp. Pure Math.* 50 (1977), 177-180.

[6] SIU, Y.-T.: 'The $\bar{\partial}$-problem with uniform bounds on derivatives', *Math. Ann.* 207 (1974), 163-176.

Added in proof. The result similar to that contained in Theorem 1 has been proved independently, for a wider class of domains, in:

[7] BERTRAMS, J.: 'Randregularität von Lösungen der $\bar{\partial}$-Gleichung auf dem Polyzylinder und zweidimensionalen analytischen Polyedern', *Bonner Math. Schriften* 176 (1986), IV + 164 pp.

Editor's note. It is perhaps worth-while to quote, in addition to (or instead of) [1], the paper:

[8] BERTRAMS, J.: 'Das $\bar{\partial}$-Problem auf pseudokonvexen Polyedern nach Sergeev und Henkin, *Bonner Math. Schriften* 167 (1985), IV + 166 pp.

HOLOMORPHIC CHAINS AND EXTENDABILITY OF HOLOMORPHIC MAPPINGS

Pierre Dolbeault and Julian Ławrynowicz
Mathématiques, L.A. 213 du C.N.R.S. Université Pierre et Marie Curie F-75252 Paris Cédex 05, France
Institute of Mathematics Polish Academy of Sciences PL-90-136 Łódź, Poland

ABSTRACT. The authors introduce a Dirichlet integral-type biholomorphic-invariant pseudodistance connected with bordered holomorphic chains of dimension one whose regular part is treated as a Riemann surface. The condition for a complex manifold that the pseudodistance on it is a distance defines a class of hyperbolic-like manifolds which have an important property of extendability of holomorphic mappings, analogous to the hyperbolic manifolds, Stein spaces, and complex spaces with a Stein covering.

1. INTRODUCTION

In 1960 Andreotti and Stoll [2] published the following theorem: *Let* $f: X \setminus A \to Y$ *be a holomorphic mapping of an open subset of a normal complex space* X *into a Stein space* Y. *Let further* A *be a thin set. If* A *has topological codimension* ≥ 3, *then* f *can be extended to a holomorphic mapping of* X *into* Y.

We recall some definitions. Let X be a complex space as given in Serre [18]. We denote by H the sheaf of germs of holomorphic functions on X and by H_x the stalk of H at $x \in X$. If H_x is an integral domain (ring), the space X is called *irreducible at* x. If X is irreducible at every point, we say that X is *locally irreducible*. If H_x is an integral domain integrally closed in its quotient field, X is called *normal at* x. If X is normal at every point, we say that X is *normal*.

A complex space Y is said to be *K-complete* if for every point $y_o \in Y$ there exists a finite number of holomorphic functions $f_1,\dots,f_p$ on Y such that y_o is an isolated point of the set

$$\bigcap_{j=1}^{p} \{y \in Y:\ f_j(y) = f_j(y_o)\}.$$

A K-complete space is said to be a *Stein space* if for every sequence (y_k) of points in Y without accumulation points we can find a holomorphic function f on Y such that $|f(y_k)|$ is unbounded.

J. Ławrynowicz (ed.), Deformations of Mathematical Structures, 191–204.

The proof given by Andreotti and Stoll is surprisingly simple. The space X can be assumed as being of pure dimension n. If g is a holomorphic function on Y, then $g \circ f$ is holomorphic on $Y \setminus A$. By a theorem of Thullen [20] the singular points of $g \circ f$ form an analytic set S_g which is of complex dimension $n-1$ or is empty. Since $S_f \subset A$, the topological dimension of S_g is $\leq 2n-3$. This implies that S_f is empty, so $g \circ f$ has a holomorphic extension h_g to all of X.

It remains to show that if Y is a Stein space, then f has no hole. Actually Andreotti and Stoll prove that f is gapless. Namely, let (x_k) be a sequence in $X \setminus A$ converging to a point $x \in A$. Assume, if possible, that $(f(x_k))$ has no accumulation point in Y. Then we can find a holomorphic function g on Y such that $|g \circ f(x_k)|$ is unbounded. Yet

$$h_g(x_k) := g \circ f(x_k) \to h_g(x) \neq \infty \quad \text{as} \quad k \to \infty,$$

so $\{g \circ f(x_k)\}$ is bounded. This contradiction shows that $\{f(x_k)\}$ has got to have an accumulation point, i.e. (cf. Stoll [19], Definition 1.4 and Theorem 1.2), f has to be gapless. This completes the proof.

Actually, in [2] Andreotti and Stoll formulated also the following result: *Let* $f: X \setminus A \to Y$ *be a holomorphic mapping of an open subset* $X \setminus A$ *of a complex manifold* X *into a space* Y *which has a covering space* (W, π) *such that* W *is a Stein space. Let further* A *be a thin set. If* A *has topological codimension* ≥ 3, *then* f *can be extended to a holomorphic mapping of* X *into* Y.

Kobayashi [11-13] obtained several results of a similar type by another method, employing his concept of a hyperbolic manifold. One of his theorems (Theorem 5.2 in [12], p. 90) states the following: *Let* X *be a complex manifold and* A *its subset which is nowhere dense in an analytic subset, say* B, *of* X, *with topological codimension* ≥ 1. *Let further* Y *be a complete hyperbolic manifold. Then every holomorphic mapping* $f: X \setminus A \to Y$ *can be extended to a holomorphic mapping of* X *into* Y.

We recall the definitions of a hyperbolic manifold and a complete hyperbolic manifold. Let X be a complex manifold (without or with boundary) and $F(x)$ the family of all holomorphic functions $f: X \to \Delta = \{z \in \mathbb{C}: |z| < 1\}$. We call an arbitrary sequence of points $z_0, \ldots, z_p \in X$, where $z_p = z$, a *holomorphic chain* joining z_0 to z if there are points $a_1, \ldots, a_p, b_1, \ldots, b_p \in \Delta$ and holomorphic functions $f_1, \ldots, f_p: \Delta \to X$ such that $f_j(a_j) = z_{j-1}$ and $f_j(b_j) = z_j$ for $j = 1, \ldots, p$. With each of such chains we associate the number $\rho(a_1, b_1) + \ldots + \rho(a_p, b_p)$, ρ denoting the hyperbolic distance, and call the infimum $d_X(z_0, z)$, $z_0, z \in X$, of these numbers the *Kobayashi pseudodistance* of $z_0, z \in X$. It is easy to verify that d_X is a pseudodistance indeed. If d_X is a distance, i.e., $d_X(z_0, z) > 0$ for $z_0 \neq z$, then X is called a *hyperbolic manifold*. A hyperbolic manifold X is said to be *complete* if it is complete with respect to d_X.

The quoted theorems of Andreotti-Stoll and of Kobayashi show that neither Stein spaces nor complex spaces with a Stein covering nor hyperbolic manifolds are the most general spaces with the above formulated properties of extendability of holomorphic mappings since in each class one can find an example of a space which does not belong to at least one of the other two classes. Therefore in this paper, following our previous research [1, 3, 5-8, 14-16], we introduce a Dirichlet integral--type biholomorphic-invariant pseudodistance connected with bordered holomorphic chains of dimension one whose regular part is treated as a Riemann surface. The condition for a complex manifold that the pseudodistance on it is a distance defines a class of hyperbolic-like manifolds which have an important property of extendability of holomorphic mappings. Namely, our main result reads: *Let* X *be a complex manifold and* A *its subset which is nowhere dense in an analytic subset, say* B, *of* X, *with topological codimension* ≥ 1. *Let further* Y *be a complete* $(\alpha, \mathcal{U})$-*hyperbolic-like manifold. Then every holomorphic mapping* $f: X \setminus A \to Y$ *can be extended to a holomorphic mapping of* X *into* Y.

The effective formulae for our pseudodistance, even in simple cases, as well as the comparison of hyperbolic-like and hyperbolic manifolds, Stein spaces, and complex spaces with a Stein covering will be published in a separate paper.

2. AN ANALOGUE OF THE HYPERBOLIC PSEUDODISTANCE

Let X be a complex manifold of complex dimension n. Consider a compact connected C^1-cycle γ of (real) dimension one on X. Following [6], under the C^1-*cycle* we mean a cycle of the class C^1 except perhaps for a closed subset of the one-dimensional Hausdorff measure zero. Suppose that Γ is an irreducible complex analytic subvariety of complex dimension one of $U = X \setminus \operatorname{spt}\gamma$, with support $\operatorname{spt}\Gamma$ relatively compact on X.

By an *elementary bordered holomorphic chain* $[\Gamma]^{\sim}$ *related to* Γ or, for short, by an *elementary chain related to* Γ we understand the current of integration $[\Gamma]^{\sim}$ defined by Γ with the following properties:

(i) there exists a C^1-cycle γ of dimension one with support $\operatorname{spt}\gamma$ relatively compact on X,

(ii) Γ is as before,

(iii) the current of integration $[\Gamma]$ defined by Γ in U admits a simple extension $[\Gamma]^{\sim}$ to X such that $d[\Gamma]^{\sim} = \gamma$ (cf. also [9, 10, 17].

For the sake of simplicity, without ambiguity, we can denote $[\Gamma]^{\sim}$ also by Γ. For its *regular part* $\operatorname{Reg}\Gamma$ (cf. [6], p. 183), we have

LEMMA 1. *If* Γ *is an elementary chain, then its regular part* $\operatorname{Reg}\Gamma$ *is a complex one-dimensional submanifold of the subvariety* Γ *and - more generally - of an open set of* X. *Thus* $\operatorname{Reg}\Gamma$ *is the image of a connected Riemann surface* S *of* X *under a biholomorphic mapping* f.

A *bordered holomorphic chain* Γ passing through points z_0, z of X is defined as a finite sum $\Sigma_{j\in I}\Gamma_j$ of elementary chains Γ_j such that $\operatorname{spt}\Gamma$ is connected and contains z_0 and z_1.

Consider now a locally finite open covering $\mathcal{U}=\{U_j: j\in I\}$ of X and denote by $F[\mathcal{U}]\equiv \operatorname{adm}(X,\mathcal{U})$ the family of all pluriharmonic C^2-functions u_j in U_j, defined in each member of the covering, which satisfy the following conditions (cf. [4], in particular pp. 126 and 134):

(a) the oscillation of u_j in U_j is less then one,

(b) $d u_j = d u_k$ in $U_j \cap U_k \neq \emptyset$.

$F[\mathcal{U}]$ will be called the *admissible family for* X *with respect to* $\mathcal{U}$.

Remark 1. As in the general case of the Chern-Levine-Nirenberg seminorms we may also consider the case of globally defined pluriharmonic C^2-functions with $|u|<1$, but in the case of compact X the resulting pseudodistance will be zero.

Condition (b) describes a closed real one-form on X. Similarly, $d^c u_j$ is also well defined on X. Without ambiguity, we can denote them omitting the indices. Thus, considering the differential form $du\wedge d^c u = 2i\, d'u\wedge d''u$ (some authors use the notation ∂ and $\bar\partial$ instead of d' and d'', respectively) for an elementary chain Γ, we have, by Lemma 1,

$$\int_\Gamma du\wedge d^c u = \int_{\operatorname{Reg}\Gamma} du\wedge d^c u = \int_S f^*(du\wedge d^c u) = \int_S dv\wedge d^c v.$$

If (U, z) is a chart on S, we get

$$dv\wedge d^c v = 2i\, d'v\wedge d''v = 2i(\partial/\partial z)v(\partial/\partial\bar z)v\, dz\wedge d\bar z. \tag{1}$$

Now we equip X with an hermitian metric h so that, in local coordinates, we have the following associated geodesic distance:

$$ds^2 = h_{j\bar k}\, dz_j \otimes d\bar z_{\bar k} \quad \text{(with the Eistein summation convention).} \tag{2}$$

The induced hermitian metric on S can be expressed, via the associated geodesic distance, as

$$f^* ds^2 = g\, dz\otimes d\bar z,$$

so that, by (2), relation (1) becomes

$$\int_\Gamma du\wedge d^c u = \int_S |(\partial/\partial z)v|^2\, dz\wedge d\bar z. \tag{3}$$

Let

$$g(d^c v, d^c v) \equiv \|\nabla v\|^2_{ds^2} := g^{1\bar 1}(\partial/\partial z)v(\partial/\partial\bar z)v.$$

In analogy to [15] and [8], motivated by Theorem 8 in [15], we prefer to consider instead of (3), starting from the very beginning, a more general integral

$$2i \int_S \hat{g}^{\alpha} |(\partial/\partial z)v|^2 \, dz \wedge d\bar{z}, \quad \hat{g} := g(d^c v, d^c v), \quad \alpha \geq 0, \tag{4}$$

where, eventually, we may take a sufficiently small α.

Next, given an elementary chain Γ, we consider the related complex-analytic variety Γ which is the image of a connected Riemann surface Σ under a holomorphic mapping ϕ:

$$\phi: \Sigma \to \Gamma \subset X \setminus \operatorname{spt} \gamma \tag{5}$$

such that

$$S \subset \Sigma \quad \text{and} \quad \phi | S = f.$$

Hence we get

LEMMA 2. *The mapping* (5) *is biholomorphic except perhaps for a discrete set of points.*

Remark 2. Within Γ there exist compact connected C^1-cycles of dimension one.

Let Γ' be a compact connected C^1-cycle of complex dimension one within Γ. Assume that the length $|\gamma'|$ of the border γ' of Γ' (denoted by $\gamma' = d\Gamma'$) is uniformly bounded in Γ. Let Δ' be the inverse image of Γ' by ϕ:

$$\Delta' := \phi^{-1}[\Gamma']. \tag{6}$$

It is an elementary chain with the border

$$\delta' := \phi^{-1}[\gamma'], \tag{7}$$

lying on the Riemann surface Σ.

Now we can consider an arbitrary chain the image under the mapping (5) of an elementary chain with border on the Riemann surface Σ; the mapping (5) has the property given in Lemma 2.

Given a bordered holomorphic chain passing through distinct points z_0, z of U, we can consider it the sum of elementary chains passing through distinct points z_{j-1}, z_j of U, $j = 1, \ldots, p$, so that z_0 is the first given point, while z is the last one: $z_p = z$.

For each elementary chain Γ_j' passing through the points z_{j-1}, z_j with Γ_j' contained in a fixed elementary chain Γ_j, we consider a holomorphic mapping

$$\phi_j: \Sigma_j \to \Gamma_j \subset X \setminus \operatorname{spt} \gamma_j.$$

Then, by Lemma 2 applied to ϕ_j, we get

$$\inf_{\Gamma_j' \subset \Gamma_j} \left\{ \frac{|\gamma_j'|}{|\Gamma_j'|} \left| \int_{\Gamma_j'} (\phi_j^* \hat{g})^\alpha du \wedge d^c u \right| \right\} = \inf_{\Delta_j' \subset \Delta_j} \left\{ \frac{|\delta_j'|}{|\Delta_j'|} \left| \int_{\Delta_j'} \hat{g}^\alpha dv \wedge d^c v \right| \right\}$$

with g^α as in (4), where $|\Gamma_j'|$ denotes the volume of Γ_j', while Δ_j' and Δ_j denote the inverse images of Γ_j' and Γ_j by ϕ_j, respectively:

$$\Delta_j' := \phi_j^{-1}[\Gamma_j'], \qquad \Delta_j := \phi_j^{-1}[\Gamma_j].$$

Consequently, with any bordered holomorphic chain Γ passing through points z_0, z of X, such that the length $|\gamma'|$ of $\gamma' = d\Gamma'$ is uniformly bounded in Γ, we may associate the expression

$$\mu_\Gamma^\alpha(z_0, z)[u] := \sum_{j \in I} \inf_{\Gamma_j'} \mu_{\Gamma_j'}^\alpha[u], \tag{8}$$

where

$$\mu_{\Gamma_j'}^\alpha[u] := (|\delta_j'|/|\Delta_j'|) \left| \int_{\Delta_j'} \hat{g}^\alpha dv \wedge d^c v \right|,$$

so that we obtain

PROPOSITION 1. *Let* z_0, z *be arbitrary points of a complex manifold* X. *Suppose that* Γ *is a bordered holomorphic chain passing through* z_0, z, *such that the length* $|\gamma'|$ *of* $\gamma' = d\Gamma'$ *is uniformly bounded in* Γ. *Finally, let* $u \in F[U]$. *Then the expression*

$$\mu_X^\alpha(z_0, z)[u, U] := \inf\{\mu_\Gamma^\alpha(z_0, z)[u] : \Gamma \text{ passing through } z_0, z \subset U\} \tag{9}$$

is well defined.

In particular, we get

PROPOSITION 2. *Let* z_0, z_1 *be points of the same coordinate neighbourhood* U *of a complex manifold* X *of complex dimension* n *identified through a chart with an open set in* $\mathbb{C}^n$. *Suppose that* z_0, z_1 *are sufficiently near to each other so that the segment* $[z_0, z_1]$ *is contained in* U. *Let* L *be the complex line in* $\mathbb{C}^n$ *containing* z_0, z_1. *Consider the set*

$$\Gamma_\varepsilon := \{z \in L : \operatorname{dist}(z, [z_0, z_1]) < \varepsilon\}, \qquad \varepsilon > 0.$$

Suppose that ε *is so small that the closure of* Γ_ε *with respect to* X *is contained in* U *and therefore it gives rise to an elementary chain in* X, *also denoted by* Γ_ε. *Then the expression*

$$\rho_X^\alpha(z_0, z)[U] := \sup\{\mu_X^\alpha(z_0, z)[u, U] : u \in F[U]\} \tag{10}$$

is well defined for every $z_0, z \in X$.

3. THE EXPRESSION $\rho_\alpha(z_o, z)$ AS A HYPERBOLIC-LIKE PSEUDODISTANCE

The expression $\rho_X^\alpha(z_o, z)[\mathcal{U}]$ defined by (10) resembles the hyperbolic pseudodistance $d_X(z_o, z)$ due to Kobayashi [11-13]. We are going to prove that it is a pseudodistance and a biholomorphic invariant. We start with

LEMMA 3. *Suppose that* $f: Y \to X$ *is a holomorphic mapping;* X *and* Y *being complex-analytic manifolds. Then, for every pluriharmonic function* u *on* X*, the function* $f^*u = u \circ f$ *is pluriharmonic as well.*

Proof. We have $dd^c u = 0$, i.e. $d'd''u = 0$. Let (z_k) and (z_j') be local co-ordinates on X and Y, respectively. Then we get

$$d'' f^*u = \frac{\partial}{\partial \bar{z}_j'} f^*u \, d\bar{z}_j' = (f^* \frac{\partial u}{\partial \bar{z}_k}) \frac{\partial \bar{f}_k}{\partial \bar{z}_j'}$$

and hence

$$d'd'' f^*u = \frac{\partial^2 u}{\partial z_m \partial \bar{z}_k} \frac{\partial f_m}{\partial z_\ell} \frac{\partial \bar{f}_k}{\partial \bar{z}_j} dz_\ell' \wedge d\bar{z}_j + \frac{\partial u}{\partial \bar{z}_k} \frac{\partial^2 \bar{f}_k}{\partial z_\ell' \partial \bar{z}_j'} .$$

Since u is harmonic and all $\bar{f}_k$ are antiholomorphic, then

$$(\partial^2/\partial z_m \partial \bar{z}_k) u = 0 \quad \text{and} \quad (\partial^2/\partial z_\ell' \partial \bar{z}_j') \bar{f}_k = 0,$$

so $d'd'' f^*u = 0$, as desired.

Next we prove

LEMMA 4. *Let* z_0, z *be points of a complex manifold* X. *Then, for any number* $\alpha \geq 0$ *and any locally finite open covering* $\mathcal{U}$ *of* X, *we have*

$$\rho_X^\alpha(z_o, z)[\mathcal{U}] < +\infty.$$

Proof. Without any loss of generality we may suppose that Γ is an elementary chain passing through z_0, z_1. Since the closure $cl_X \operatorname{spt} \Gamma$ is compact, then it is contained in the union of a finite number of open sets $U_1, \ldots, U_q$ in $\mathcal{U}$; the dependence on any function $u \in F[\mathcal{U}] \equiv \operatorname{adm}(X, \mathcal{U})$ being restricted to $du, d^c u$ only. We may suppose that every u of $F[\mathcal{U}]$ has its modulus less than one in U_1, and it is well defined in the union $U_1 \cup \ldots \cup U_q$, having its modulus less than q there.

It is well known [4] that the derivatives of the functions of F, being pluriharmonic and of the class C^2, are uniformly bounded on $cl_X \operatorname{spt} \Gamma$. By Lemma 3 the same statement is true for the pluriharmonic functions of the form $v = f^*u$, u $F[\mathcal{U}]$, on the Riemann surface S. Then there exists a number $A_\Delta > 0$ such that on S we have

$$\hat{g}^\alpha |(\partial/\partial z) v|^2 < A_\Delta,$$

where

$$\Delta = \phi^{-1}[\Gamma];$$

ϕ being as in (5). Then, passing to (6), we get

$$(1/|\Delta'|)\left|z \int_{\Delta'} \hat{g}^{\alpha} |(\partial/\partial z)v|^2 \, dz \wedge d\bar{z}\right| < B_{\Delta}, \tag{11}$$

where B_{Δ} is a positive number. Since, by assumption, the length $|\gamma'|$ of $\gamma' = d\Gamma'$ is uniformly bounded in Γ, the product of the expression on the left-hand side of (11) and the length of (7) is estimated by a number $C_{\Delta} > 0$.

Finally, $\mu_{\Gamma}^{\alpha}(z_0, z)[u, \mathcal{U}]$ is estimated by a finite number indendent of $u \in F[\mathcal{U}]$ and the same holds true for (10), as desired.

R e m a r k 3. It is obvious that, under the hypotheses of Lemma 4, $\rho_X^{\alpha}(z_0, z) \geq 0$ and $\rho_X^{\alpha}(z, z_0) = \rho_X^{\alpha}(z_0, z)$. If $z = z_0$, then the length of (7) can be as small as we desire, so $\rho_X^{\alpha}(z_0, z) = 0$.

Besides we have

LEMMA 5. *Let* z_0, z_1, *and* z_2 *be points of a complex manifold* X. *Then, for any number* $\alpha \geq 0$ *and for any locally finite open covering* $\mathcal{U}$ *of* X, *we have*

$$\rho_X^{\alpha}(z_0, z_2) \leq \rho_X^{\alpha}(z_0, z_1) + \rho_X^{\alpha}(z_1, z_2). \tag{12}$$

P r o o f. Let Γ_o, Γ_1, and Γ_2 be bordered holomorphic chains passing through z_0, z_1; z_1, z_2; z_0, z_2, respectively. Then $\Gamma_o + \Gamma_1$ is also a bordered holomorphic chain passing through z_0, z_2 and everywhere in $F[\mathcal{U}]$ we have

$$\mu_{\Gamma_1+\Gamma_2}^{\alpha}(z_0, z_2) \leq \mu_{\Gamma_1}^{\alpha}(z_0, z_1) + \mu_{\Gamma_2}^{\alpha}(z_1, z_2).$$

Hence

$$\begin{aligned}\mu_X^{\alpha}(z_0, z_2) &= \inf_{\Gamma} \mu_{\Gamma}^{\alpha}(z_0, z_2) \leq \inf_{\Gamma_1+\Gamma_2} \mu_{\Gamma_1+\Gamma_2}^{\alpha}(z_0, z_2) \\ &\leq \inf_{\Gamma_1,\Gamma_2} [\mu_{\Gamma_1}^{\alpha}(z_0, z_1) + \mu_{\Gamma_2}^{\alpha}(z_0, z_2)] \\ &= \inf_{\Gamma_1} \mu_{\Gamma_1}^{\alpha}(z_0, z_1) + \inf_{\Gamma_2} \mu_{\Gamma_2}^{\alpha}(z_1, z_2) \\ &= \mu_X^{\alpha}(z_0, z_1) + \mu_X^{\alpha}(z_1, z_2)\end{aligned}$$

and, consequently, we get (12).

From Lemma 4, by Remark 3 we infer

PROPOSITION 3. *Let* X *be a complex manifold and* α *a positive number. Then, for any locally finite open covering* $\mathcal{U}$ *of* X, *the corresponding expression* ρ_X^{α} *given by* (10) *is a continuous pseudodistance.*

Besides, we have

PROPOSITION 4. *Let* X *be a complex manifold,* M *a covering manifold of* X *with covering projection* $\pi: M \to X$, *and* α *a positive number. Let* $z_0, z \in X$ *and* $s_o, s \in M$ *so that* $\pi(s_o) = z_0$ *and* $\pi(s) = z$. *Then*

$$\rho_X^\alpha(z_0, z) = \inf\{\rho_M^\alpha(s_o, s): s \in M, \quad \pi(s_o) = z_0 \quad \textit{and} \quad \pi(s) = z\}.$$

Proof. Since $\pi: M \to X$ is distance-decreasing, we have $\rho_X^\alpha(z_0, z) \leq \inf_s \rho_M^\alpha(s_o, s)$. Assuming the strict inequality, let

$$\rho_X^\alpha(z_0, z) + \varepsilon < \inf_s \rho_M^\alpha(s_o, s), \; \varepsilon \text{ being a positive number.} \quad (13)$$

By the definition of d_X^α there exist points ζ_{j-1}, ζ_j of Σ_j, $j = 1, \ldots, p$, where ζ_o is the first given point, while ζ is the last one: $\zeta_p = \zeta$, and

$$z_0 = \phi_1(\zeta_o), \quad \phi_1(\zeta_1) = \phi_2(\zeta_1), \ldots, \phi_{p-1}(\zeta_{p-1}) = \phi_p(\zeta_{p-1}),$$
$$\phi_p(\zeta_p) = z$$

and

$$\rho_X^\alpha(z_0, z) + \varepsilon > \Sigma_{j=1}^p \rho_{\Sigma_j}^\alpha(\zeta_{j-1}, \zeta_j).$$

Then we can lift ϕ_j to holomorphic mappings $f_j: \Sigma_j \to M$, $j = 1, \ldots, p$, so that

$$s_o = f_1(\zeta_o), \quad f_1(\zeta_1) = f_2(\zeta_1), \ldots, f_{p-1}(\zeta_{p-1}) = f_p(\zeta_{p-1}),$$
$$\pi \circ f_p = \phi_p.$$

If we set $s = f_p(z)$, then $\pi(s) = z$ and

$$\rho_M^\alpha(s_o, s) \leq \Sigma_{j=1}^p \rho_{\bar{z}_j}^\alpha(\zeta_{j-1}, \zeta_j).$$

Hence $\rho_M^\alpha(s_o, s) < \rho_X^\alpha(z_0, z) + \varepsilon$ which contradicts (13).

PORPOSITION 5. *Let* X *and* X' *be complex manifolds, and* α *a positive number. Then*

$$\rho_X^\alpha(z_0, z) + \rho_{X'}^\alpha(z_0', z') \geq \rho_{X \times X'}^\alpha((z_0, z_0'), (z, z'))$$
$$\geq \max\{\rho_X^\alpha(z_0, z), \rho_{X'}^\alpha(z_0', z)\}$$

for $z_0, z \in X$ *and* $z_0', z' \in X'$.

Proof. We have

$$\rho_X^\alpha(z_0, z) + \rho_{X'}^\alpha(z_0', z') \geq \rho_{X \times X'}((z_0, z_0'), (z, z_0'))$$
$$+ \rho_{X \times X'}((z, z_0'), (z, z')) \geq \rho_{X \times X'}((z_0, z_0'), (z, z')).$$

Indeed, the first inequality holds since the mappings $f\colon X \to X \times X'$ and $f'\colon X' \to X \times X'$, defined by $f(s) = (s, z_o')$ and $f'(s') = (z, s')$, respectively, are distance-decreasing, while the other inequality is a consequence of Lemma 5. The inequality

$$\rho^{\alpha}_{X \times X'}((z_o, z_o'), (z, z')) \geq \max\{\rho^{\alpha}_{X}(z_o, z), \rho^{\alpha}_{X'}(z_o', z')\}$$

holds since the projections $X \times X' \to X$ and $X \times X' \to X'$ are both distance-decreasing.

Finally, by Proposition 3, we trivially get

PROPOSITION 6. *Let* X *and* Y *be complex manifolds,* α *a positive number, and* $f\colon X \to Y$ *a holomorphic mapping. Then* *)

$$\rho^{\alpha}_{X}(z_o, z) \geq \rho^{\alpha}_{Y}(f(z_o), f(z)) \quad \textit{for} \quad z_o, z \in X.$$

In particular, every biholomorphic mapping $f\colon X \to Y$ *is a* ρ^{α}*-isometry:*

$$\rho^{\alpha}_{X}(z_o, z) = \rho^{\alpha}_{Y}(f(z_o), f(z)) \quad \textit{for} \quad z_o, z \in X.$$

Propositions 3-6 motivate the following definition. Let X be a complex manifold, $\mathcal{U}$ a locally finite open covering of X, and α a positive number. If $\rho^{\alpha}_{X}(\ ,\ )[\mathcal{U}]$ is a distance, i.e., $\rho^{\alpha}_{X}(z_o, z)[\mathcal{U}] > 0$ for $z_o \neq z$, then X is called an *$(\alpha, \mathcal{U})$-hyperbolic-like manifold.* Examples of such manifolds, different from hyperbolic manifolds, will be given in a subsequent paper. An $(\alpha, \mathcal{U})$-hyperbolic-like manifold X is said to be *complete* if it is complete with respect to $\rho^{\alpha}_{X}(\ ,\ )[\mathcal{U}]$.

4. MAPPINGS FROM THE PUNCTURED DISC INTO AN $(\alpha, \mathcal{U})$-HYPERBOLIC-LIKE MANIFOLD

We are going to prove (cf. [12], pp. 83-86):

THEOREM 1. *Let* Δ^* *denote the punctured unit disc and* Y *be an* $(\alpha, \mathcal{U})$*-hyperbolic-like manifold. Let further* $f\colon \Delta^* \to Y$ *be a holomorphic mapping such that, for a suitable sequence of points* $z_k \in \Delta^*$ *converging to the origin,* $f(z_k)$ *converges to a point* $s_o \in Y$*. Then* f *extends to a holomorphic mapping of the unit disc* Δ *into* Y.

Proof. Let γ_k be the image of the circle $\{z\colon |z| = r_k\}$ by f. Let further U be a neighbourhood of s_o in Y with a local coordinate system $(w^1, \ldots, w^n)$. Without any loss of generality we may assume that s_o is at the origin of $(w^1, \ldots, w^n)$. Let ε be a positive number and V the open neighbourhood of s_o defined by the conditions $|w^j| < \varepsilon$, $j = 1, \ldots, n$. Taking ε sufficiently small we may assume that $\mathrm{cl}\, V \subset U$. Let W be the neighbourhood of s_o defined by $|w^j| < \frac{1}{2}\varepsilon$, $j = 1, \ldots, n$. In order to prove the theorem it suffices to show that, for a suitable number $\delta > 0$, the small punctured disc $\Delta^*_{\delta} = \{z \in \Delta^*\colon |z| < \delta\}$ is mapped by f into U.

Let $L(r)$ be the arc length of $\{z: |z| = r < 1\}$ with respect to the invariant distance $\rho^{\alpha}_{D*}(\ ,\ )[U]$. We apply Proposition 3 (continuity) and Proposition 4 (covering projection into Δ). Taking into account that the Kobayashi pseudodistance $d_Y(z_0, z)$ is a limit case of $\rho^{\alpha}_Y(z_0, z)[U]$ and $L(r)$ for $d_{D*}(z_0, z)$ equals $2\pi/\log(1/r)$ ([12], p. 81), we infer that, also in the general case of $\rho^{\alpha}_{D*}(\ ,\ )[U]$,

$$L(r) \to 0 \quad \text{as} \quad r \to 0+. \tag{14}$$

Then, since the diameter of γ_k approaches zero as $k \to \infty$, all but a finite number of the curves γ_k are in W. Without any loss of generality we may assume that all the curves γ_k are in W. By taking, if necessary, a subsequence of (z_k), we may assume that (r_k) is monotone decreasing. Consider the set of integers k with the property that the image of $\{z: r_{k+1} < |z| < r_k\}$ by f is not entirely contained in W. If this set is finite, f maps a small punctured disc Δ^*_δ into $\mathrm{cl}\,W$. Assuming that the set is finite we shall obtain a contradiction.

Now, by taking a subsequence, we may assume that, for any k, the image of $\{z: r_{k+1} < |z| < r_k\}$ is not entirely contained in W. Let $\Delta_k = \{z \in \Delta^*: a_{k+1} < |z| < b_k\}$ be the largest open annulus such that $a_k < r_k < b_k$ and f maps Δ_k into W. Let σ_k and τ_k denote the inner and outer boundaries of Δ_k, respectively. Both $f[\sigma_k]$ and $f[\tau_k]$ are contained in $\mathrm{cl}\,W$ but not in W. By (14), the diameters $f[\sigma_k]$ and $f[\tau_k]$ approach zero as $k \to \infty$. By taking a subsequence, if necessary, we may assume that $(f[\sigma_k])$ and $(f[\tau_k])$ converge to points s and s' of $\mathrm{cl}\,W \setminus W$, respectively. Since s_o is in W and both s and s' are on the boundary of W, the points s and s' are different from s_o. By taking a new co-ordinate system around s_o, if necessary, we may assume that $w^1(s)$ and $w^1(s')$ are different from $w^1(s_o) = 0$.

Let $f(z) = (f^1(z), \ldots, f^n(z))$ be the local expression of f on $f^{-1}[U] \subset \Delta$. Then

$$f^1[\sigma_k] \to w^1(s), \quad f^1[\tau_k] \to w^1(s'), \text{ and } f^1(z_k) \to w^1(s_o) \quad \text{as } k \to \infty.$$

Hence, if k is sufficiently large, we have

$$f^1(z_k) \notin f^1[\sigma_k] \cup f^1[\tau_k].$$

Again, if k is sufficiently large, we can find a simply connected open neighbourhood G_k of $w^1(s)$ in C such that $f^1[\sigma_k] \subset G_k$ and $f^1(z_k) \notin G_k$. Applying Cauchy's theorem to the holomorphic function $w^1 \mapsto 1/[w^1 - f^1(z_k)]$ and the closed curve $f^1[\sigma_k]$ in G_k, we get

$$\int_{f^1[\sigma_k]} [w^1 - f^1(z_k)]^{-1}\, dw^1 = 0.$$

Hence

$$\int_{\sigma_k} f^{1\prime}(z)[f^1(z) - f^1(z_k)]^{-1}\, dz = 0, \quad \text{where} \quad f^{1\prime}(z) = (d/dz)f^1(z).$$

For a sufficiently large k,

$$\int_{\tau_k} f^{1'}(z)[f^1(z) - f^1(z_k)]^{-1}\,dz = 0$$

as well. On the other hand, by the principle of argument applied to $z \mapsto f^1(z) - f^1(z_k)$ in a neighbourhood of Δ_k which is bounded by the curve $\tau_k - b_k$, we obtain

$$\int_{\tau_k} \frac{f^{1'}(z)}{f^1(z) - f^1(z_k)}\,dz - \int_{\sigma_k} \frac{f^{1'}(z)}{f^1(z) - f^1(z_k)}\,dz = 2\pi i(N - P),$$

where N and P denote the numbers of zeros and poles of $z \mapsto f'(z) - f'(z_k)$ in Δ_k. Precisely, $P = 0$ and $N \geq 1$, so we have arrived at a contradiction, as desired.

COROLLARY 1. *If Y is a compact (α, U)-hyperbolic-like manifold, then every holomorphic mapping $f: \Delta^* \to Y$ extends to a holomorphic mapping of Δ into Y.*

5. HOLOMORPHIC MAPPINGS INTO (α, U)-HYPERBOLIC-LIKE MANIFOLDS

We proceed to generalize Theorem 1 to the cases where Δ^* is replaced by a complex manifold X with a suitable analytic subset excluded, while Y is supposed to be either compact or complete.

THEOREM 2. *Let X be a complex manifold and A any of its analytic subsets of topological codimension ≥ 1. Then every holomorphic mapping $f: X \setminus A \to Y$ can be extended to a holomorphic mapping of X into Y.*

Proof. If is analogous to the proof of Theorem VI.4.1 in [12], pp. 86-88, and based on Proposition 5 and Theorem 1.

If A is sufficiently small, then in Theorem 2 it suffices to assume that Y is complete (α, U)-hyperbolic-like. Firstly, we have

PROPOSITION 7. *Let $\Delta^n = \{(z^1, \ldots, z^n) \subset \mathbb{C}^n : |z^j| < 1$ for $j = 1, \ldots, n\}$ and let A be a subset of $\Delta^n = \Delta \times \Delta^{n-1}$ of the form $A = \{0\} \times A'$, where A' is nowhere dense in Δ^{n-1}. Then the distance $\rho^\alpha_X(\ ,\ )[U]$, where $X = \Delta^m - A$, is the restriction of the distance $\rho^\alpha_{\Delta^m}$ to X.*

In analogy to Proposition VI.5.1 in [12], pp. 88-90, the proof is based on the following lemma whose proof is essentially unchanged:

LEMMA 6. *Given points a, b, c in the unit disc Δ, for every $\varepsilon > 0$ there is a $\delta > 0$ such that for any $c' \in \Delta$ with $\rho^\alpha_\Delta(c, c')[U] < \delta$ there exists an automorphism $h: \Delta \to \Delta$ satisfying*

$$h(c) = c', \quad \rho^\alpha_\Delta(a, h(a)) < \varepsilon, \text{ and } \rho^\alpha_\Delta(b, h(b)) < \varepsilon.$$

As an application of Proposition 7, now we can prove our main

THEOREM 3. *Let* X *be a complex manifold and* A *its subset which is nowhere dense in an analytic subset, say* B, *of* X, *with topological codimension* ≥ 1. *Let further* Y *be a complete* $(\alpha, \mathcal{U})$*-hyperbolic-like manifold. Then every holomorphic mapping* $f: X\setminus A \to Y$ *can be extended to a holomorphic mapping of* X *into* Y.

Proof. As in the proof of Theorem 2, we can reduce the proof to the special case where $X = \Delta^n$, $n = \dim X$, and B is the subset defined by $z^1 = 0$, so that A is of the form $A = \{0\} \times A'$, where A' is nowhere dense in Δ^{n-1}. Since $f: \Delta^n \setminus A$ is distance-decreasing, f can be extended to a continuous mapping from the completion of the metric space $\Delta^n \setminus \Delta$ into Y. By Proposition 7, Δ^n is the completion of $\Delta^n \setminus \Delta$ with respect to the distance ρ_X^α, $X = \Delta^n \setminus A$. By the Riemann extension theorem, the extended continuous mapping $f: \Delta^n \to Y$ has to be holomorphic, as desired.

COROLLARY 2. *Let* X *be a complex manifold and* A *its analytic subset of codimension* ≥ 2. *Let further* Y *be a complete* $(\alpha, \mathcal{U})$ - *hyperbolic-like manifold. Then every holomorphic mapping* $f: X\setminus A \to Y$ *can be extended to a holomorphic mapping of* X *into* Y.

References

[1] ANDREOTTI, A. and J. ŁAWRYNOWICZ: 'On the generalized complex Monge-Ampère equation on complex manifolds and related questions', *Bull. Acad. Polon. Sci. Sér. Sci. Math. Astronom. Phys.* 25 (1977), 943-948.

[2] —— and W. STOLL: 'Extension of holomorphic maps', *Ann. of Math.* (2) 72 (1960), 312-349.

[3] BESNAULT, J. et P. DOLBEAULT: 'Sur les bords d'ensembles analytiques complexes dans $\mathbb{P}^n(C)$', *Sympos. Math.* 24 (1981), 205-213.

[4] CHERN, S.S., H.I. LEVINE, and L. NIRENBERG: 'Intrinsic norms on a complex manifold', in *Global analysis*, Papers in honor of K. Kodaira, ed. by D.C. Spencer and S. Iynaga, Univ. of Tokyo Press and Princeton Univ. Press, Tokyo 1969, pp. 119-139; reprinted in S.S. CHERN: *Selected papers*, Springer, New York-Heidelberg-Berlin 1978, pp. 371-391.

[5] DOLBEAULT, P.: 'On holomorphic chains with given boundary in $\mathbb{P}^n(\mathbb{C})$', in *Analytic functions, Błażejewko 1982, Proceedings*, ed. by J. Ławrynowicz (Lecture Notes in Math. 1039), Springer, Berlin-Heidelberg-New York-Tokyo 1983, pp. 118-129.

[6] ——: 'Sur les chaines maximalement complexes de bord donné', *Proc. Sympos. Pure Math.* 44 (1986), 171-205.

[7] GAVEAU, B. et J. ŁAWRYNOWICZ: 'Espaces de Dirichlet invariants bi-

holomorphes et capacités associées', *Bull. Acad. Polon. Sci. Sér. Sci. Math.* 30 (1982), 63-69.

[8] GAVEAU, B. et J. ŁAWRYNOWICZ: 'Intégrale de Dirichlet sur une variété complexe I', in *Séminaire Pierre Lelong - Henri Skoda (Analyse), Années 1980/81* (Lecture Notes in Math. 919), Springer, Berlin-Heidelberg-New York 1982, pp. 131-151 et 163-165.

[9] HARVEY, R.: 'Holomorphic chains and their boundaries', *Proc. Sympos. Pure Math.* 30, 1 (1977), 309-382.

[10] —— and B. LAWSON: 'On boundaries of complex analytic varieties I-II', *Ann. of Math.* (2) 102 (1975), 233-290 and (2) 106 (1977), 213-238.

[11] KOBAYASHI, S.: 'Invariant distances on complex manifolds and holomorphic mappings', *J. Math. Soc. Japan* 19 (1967), 460-480.

[12] ——: *Hyperbolic manifolds and holomorphic mappings*, Marcel Dekker, Inc., New York 1970.

[13] ——: 'Intrinsic distances, measures and geometric function theory', *Bull. Amer. Math. Soc.* 82 (1976), 357-416.

[14] ŁAWRYNOWICZ, J.: 'Condenser capacities and an extension of Schwarz's lemma for hermitian manifolds', *Bull. Acad. Polon. Sci. Sér. Sci. Math. Astronom. Phys.* 23 (1975), 839-844.

[15] ——: 'On a class of capacities on complex manifolds endowed with an hermitian structure and their relation to elliptic and hyperbolic quasiconformal mappings', (a) *Conf. Analytic Functions Abstracts*, Cracow 1974, pp. 33-34 (abstract), (b) *Ann. Polon. Math.* 33 (1976), 178 (abstract), (c) *Dissertationes Math.* 166 (1980), 48 pp. (in extenso).

[16] —— and M. OKADA: 'Canonical diffusion and foliation involving the complex hessian', (a) *Inst. of Math. Polish Acad. Sci. Preprint* no. 356 (1985), II + 10 pp., (b) *Bull. Polish Acad. Sci. Math.* 34 (1986), 661-667.

[17] SCHIFFMAN, B.: 'On the removal of singularities of analytic sets', *Michigan Math. J.* 15 (1968), 111-120.

[18] SERRE, J.-P.: 'Géométrie algébrique et géométrie analytique', *Ann. Inst. Fourier* 6 (1956), 1-41.

[19] STOLL, W.: 'Über meromorphe Abbildungen komplexer Räume I-II', *Math. Ann.* 136 (1958), 201-239 and 393-429.

[20] THULLEN, P.: 'Über die wesentlichen Singularitäten analytischer Funktionen und Flächen im Räume von n komplexen Veränderlichen', *Math. Ann.* 111 (1935), 137-157.

*) More precisely, one has to write $\rho_X^\alpha(z_0, z)[U, g; \Gamma_j': j \in I]$ instead of $\rho_X^\alpha(z_0, z)$, and $\rho_Y^\alpha(f(z_0), f(z))[f[U], f^*g, f_* \Gamma_j': j \in I]$ instead of $\rho_Y^\alpha(f(z_0), f(z))$.

REMARKS ON THE VERSAL FAMILIES OF DEFORMATIONS OF HOLOMORPHIC AND TRANSVERSELY HOLOMORPHIC FOLIATIONS

Xavier Gómez Mont and Duraiswami Sundararaman
Instituto de Matematicas
Universidad Nacional
Autonoma de México
04510 México, D.F.
México

Departamento de Matematicas
Centro de Investigacion y de
Estudios Avanzados
Apartado Postal 14-740
07000 México, D.F., México

ABSTRACT. After a brief discussion of holomorphic and transversely holomorphic nonsingular foliations the authors concentrate on constructing versal space of deformations of a holomorphic foliation with singularities in a compact manifold and describe a homological method to construct the tangent space to the versal space. Finally holomorphic foliations by curves are discussed.

1. HOLOMORPHIC AND TRANSVERSELY HOLOMORPHIC NONSINGULAR FOLIATIONS

1.1. We briefly outline first what Kodaira and Spencer did to construct versal spaces of deformations of a nonsingular holomorphic foliation F in a compact complex manifold X. The holomorphic structure of F is characterised by the sheaf $\mathcal{O}_X$ of germs of holomorphic functions on X and by the subsheaf $\mathcal{O}_F$ of germs of holomorphic functions constant on the leaves. Let I_X (I_F respectively) be the ideal generated by $d\mathcal{O}_X$ ($d\mathcal{O}_F$ respectively) in the algebra A of smooth complex valued differential forms on X. Let Θ_F be the sheaf of germs of holomorphic vector fields preserving F. Then the Kodaira--Spencer resolution of Θ_F is given by

$$0 \to \Theta_F \overset{\varepsilon}{\to} \underline{\phi}^0 \overset{D}{\to} \underline{\phi}^1 \overset{D}{\to} \underline{\phi}^2 \to \ldots,$$

where $\underline{\phi}^q$ is the sheaf of germs of real derivations of degree q of the algebra A mapping I_X into I_X and I_F into I_F. The operator D is defined as [d,] and ε associates to a vector field the corresponding Lie Derivative. We denote by ϕ^q the vector space of sections of $\underline{\phi}^q$. We get a differential graded algebra structure given by $[\phi^p, \phi^q] \subset \phi^{p+q}$. The above complex is elliptic and hence the cohomology groups $H^i(X, \Theta_F)$ are finite dimensional.

They show that there is a one to one correspondence between:

{Pairs of holomorphic foliations F, F' where F' is close to F with respect to a complex structure X' close to X together

J. Ławrynowicz (ed.), Deformations of Mathematical Structures, 205–213.

with a smooth automorphism of TX close to the identity which is a C-isomorphism of TX onto TX′ and $T\mathcal{F}$ onto $T\mathcal{F}'$}

and

{elements φ of the subspace $\mathcal{D}$ of ϕ^1 defined by $D\varphi - \frac{1}{2}[\varphi, \varphi] = 0$}.

If $\varphi(t)$ is a smooth path in $\mathcal{D}$ with $\varphi(0) = 0$, then the cohomology class of $\partial\phi/_{\partial t}(0) \in H^1(X, \Theta_{\mathcal{F}})$ is the infinitesimal deformation of the corresponding family $\mathcal{F}(t)$.

By arguments analogous to the case of complex structures, they showed [8] that if $H^2(X, \Theta_{\mathcal{F}}) = 0$, there exists a versal family of holomorphic foliations parametrized by a neighbourhood of 0 in $H^1(X, \Theta_{\mathcal{F}})$. They did not consider the difficult case of $H^2(X, \Theta_{\mathcal{F}}) \neq 0$, which is recently settled as a consequence of [5]. See the following section 1.2.

1.2. Now let us consider the case of a transversely holomorphic foliation $\mathcal{F}$ in a compact differentiable manifold M. Recall that such a foliation $\mathcal{F}$ of complex codimension q is given by an open covering $\{U_i\}$ of M and differentiable submersions $f_i : U_i \to \mathbb{C}^q$ such that, for each $i, j \in I$, there is a biholomorphism f_{ij} of $f_j(U_i \cap U_j)$ onto $f_i(U_i \cap U_j)$ such that

$$f_i = f_{ij} \circ f_j.$$

Hence every differentiable submanifold of M of real dimension 2q transversal to the leaves of $\mathcal{F}$ has a well defined complex manifold structure and the normal bundle to the foliation has a well defined complex structure.

Let $\hat{\Theta}_{\mathcal{F}}$ be the sheaf of germs of smooth vector fields preserving $\mathcal{F}$ and $\hat{\Theta}_{\mathcal{F}}$ be the quotient of $\Theta_{\mathcal{F}}$ by the subsheaf of germs of vector fields tangent to $\mathcal{F}$. Thus $\Theta_{\mathcal{F}}$ is the sheaf of germs of transversely holomorphic vector fields. Evidently we have $H^i(M, \hat{\Theta}_{\mathcal{F}}) = H^i(M, \Theta_{\mathcal{F}})$ for each $i > 0$. As above we have the resolution of $\hat{\Theta}_{\mathcal{F}}$:

$$0 \to \hat{\Theta} \hookrightarrow \underline{\phi}^0 \xrightarrow{D} \underline{\phi}^1 \to$$

where $\underline{\phi}^q$ is the sheaf of germs of derivations of $\mathcal{A}$ preserving the ideal $\mathcal{I}_{\mathcal{F}}$. The above complex is also elliptic except at 0.

Again there is a one to one correspondence between

{Pairs of transversely holomorphic foliations $\mathcal{F}$ and $\mathcal{F}'$ where $\mathcal{F}'$ is close to $\mathcal{F}$, together with a smooth automorphism of TM mapping $T\mathcal{F}$ onto $T\mathcal{F}'$ and $\mathbb{C}$-analytic on $TM/T\mathcal{F}$ and $TM/T\mathcal{F}'$.}

and

{Elements φ of the subspace $\mathcal{D}$ in ϕ^1 defined by $D\varphi - \frac{1}{2}[\varphi, \varphi] = 0$}.

We have also the analogous existence theorem: There exists a versal family of deformations of $\mathcal{F}$ parametrized by an open neighbourhood of 0 in $H^1(M, \Theta_F)$, when $H^2(M, \Theta_F) = 0$.

The general case (when $H^2(M, \Theta_F) \neq 0$) is settled in [5], also compare [4]. In a manner similar to that of Kuranishi [8], (also see Douady [2]), the existence of the versal space K is proved. K is the germ at 0 of the analytic subspace of ϕ^1 defined by

$$K = \{\varphi \in \phi^1;\quad D^*\varphi = 0 \quad \text{and} \quad D\varphi - \tfrac{1}{2}[\varphi, \varphi] = 0\},$$

where D^* is the adjoint of D with respect to a hermitian metric. The space K is the intersection of

$$\mathcal{D} = \{\varphi \in \phi^1:\quad D\varphi - \tfrac{1}{2}[\varphi, \varphi] = 0\}$$

with

$$\mathcal{H} = \{\varphi:\quad D^*\omega = 0 \quad \text{and} \quad D^*(D\varphi - \tfrac{1}{2}[\varphi, \varphi]) = 0\}.$$

An alternate description of $\mathcal{H}$ is

$$\mathcal{H} = \{\delta \in \phi^1,\quad \delta = H\delta + \tfrac{1}{2} D^*G[\delta, \delta]\},$$

where H is the orthogonal projection on the space of harmonic elements and G is the Green's operator defined by $\varphi = H\varphi + \Delta G\varphi$, where $\Delta = D^*D + DD^*$ and $G\delta$ is orthogonal to the harmonic elements. It is to be noted that $\mathcal{H}_1$ is finite dimensional whose tangent space at 0 is the space $H^1 (\approx H^1(M, \Theta))$ of harmonic elements in ϕ^1. For complete details see [5].

1.3. Example. Assume we have a holomorphic foliation F transverse to the fibres Y of a holomorphic fibre bundle $p: X \to B$, where X, B are compact connected complex manifolds. The holonomy homomorphism, $H: \pi_1(B) \to \mathrm{Aut}_{\mathbb{C}} Y$ completely determines the bundle and the foliation F. In fact $X = \tilde{B} \times_{\Pi_1(B)} Y$, where $\tilde{B}$ is the universal covering of B; and F is the quotient of the horizontal foliation on $\tilde{B} \times Y$. Let $(K_F, 0)$ be the versal (Kuranishi) space of deformations of the holomorphic foliation F. Let $(K_F^{tr}, 0)$ be the versal (Kuranishi) space of deformations of F considered as a transversely holomorphic foliation. Also let $(K_B, 0)$ be the Kuranishi space of the compact complex manifold. How are these spaces related ? It is shown in [5] that if K_B is non-singular and if the sheaf $\Theta_B \otimes H^1(M, \mathcal{O}_M)$ has no nontrivial sections, then $K_F \approx K_F^{tr} \times K_B$. Examples where these assumptions are valid and not valid are also given in [5].

Remark. The existence of versal space can also be shown for transversely holomorphic foliation in a manifold with boundary, if the foliation intersects transversely the boundary. This may be seen by constructing a manifold without boundary by the usual double construction, with the natural transversely holomorphic foliation. If K denotes the versal space of this foliation, it comes with a natural involution induced from the double construction. One may prove that the analytic space defined as the fixed point set of this involution is a versal space for the original foliation.

2. HOLOMORPHIC FOLIATIONS WITH SINGULARITIES

In this section we show how to construct versal space of deformations of a holomorphic foliation with singularities in a compact manifold and describe a homological method to construct the tangent space to the versal space.

Definition 2.1. Let M be a complex manifold. A *holomorphic foliation* with singularities in M is a coherent subsheaf F of the tangent sheaf Θ_M of M that is closed under Lie bracket. The *singular set of the foliation* F is the set of points where Θ_M/F is not locally free, and is denoted by $\operatorname{Sing} F$.

Recall that Douady has shown in [3] how to parametrize quotient subsheaves of a given sheaf in a compact manifold. Applying his construction to the tangent sheaf Θ_M we obtain the existence of a natural structure of a complex analytic space in the family of all coherent subsheaves of Θ_M. It is shown in [7] that the integrability conditions determine an analytic subvariety in the above space, where special care has to be taken since this space is not necessarily reduced. In this form we see that there is a universal space that parametrizes all foliations in M. The naturality of the tangent sheaf suggests a modifications of this universal space taking into consideration the following:

1. If we define two foliations in M as equivalent if there is a biholomorphism $\varphi: M \to M$ such that $F' = \varphi_* F$, where φ_* is the natural map induced by the Jacobian, then the equivalence class of foliations is represented by the orbit space of holomorphic foliations.

2. One should also let the complex structure of M to vary.

Using Kuranishi's theorem [9] on the existence of a versal space of deformations of complex structures of M, and a relative version of Douady's theorem due to Pourcin [10], one can push the construction sketched above to a satisfactory state for the very general case when M has no non-trivial holomorphic vector fields. In [7] it is shown how to take a suitable section of the above space to obtain a versal space of deformations when M has nontrivial automorphisms.

The essential fact that one is using above is the naturality of the tangent sheaf associated to a complex manifold (i.e. given a family $\{M_t\}_{t\in T}$, $\{\Theta_{M_t}\}$ is canonically defined) and the fact that an automorphism of M induces a canonical automorphism of Θ_M.

Since these facts also carry to the tensor sheaves $(\Theta_M^*)^{\otimes p} \otimes (\Theta_M)^{\otimes q}$ as well as to the alternating sheaves $\Lambda^p \Theta_M^* = \Omega_M^p$, we obtain similar versal spaces for subsheaves of these sheaves, where we require that the quotient sheaves are flat with respect to the parameter (which is a restriction only at points where the quotient sheaf is not locally free).

One may also find in [7] a homological method, which is a blend of Poincaré's method to linearise vector fields [1] and Kodaira-Spencer's expression of the infinitesimal deformation of a structure [8], to construct the tangent space of the versal space.

Definition 2.2. The *leaf complex* of a holomorphic foliation with singularities $F \subset \Theta_M$ is the complex of sheaves on M

$$\Theta_M \xrightarrow{D_o} \mathrm{Hom}_{\mathcal{O}_M}(F, \Theta_M|F) \xrightarrow{D_1} \mathrm{Hom}_{\mathcal{O}_M}(\Lambda^2 F, \Theta_M|F) \xrightarrow{D_2} \cdots$$

where

$$D_o X(Y) = \Pi[X, Y], \qquad X \in \Theta_M, \quad Y \in F,$$

and $\Pi: \Theta_M \to \Theta_M|F$ is the projection, the bracket is the Poisson bracket of vector fields, and for $p > 0$

$$(p+1)\, D_p w(X_1, \ldots, X_{p+1}) = \sum_{i=1}^{p+1} (-1)^i [X_i, w(\ldots, \hat{X}_i, \ldots)]$$

$$+ \sum_{i<j} (-1)^{i+j} w([X_i, X_j], \ldots, \hat{X}_i, \ldots, \hat{X}_j, \ldots).$$

THEOREM 2.3 ([7]). *Let* F *be a holomorphic foliation with singularities in the complex manifold* M; *then the space of infinitesimal deformations of* F *is represented by the first hypercohomology group* $\mathbb{H}^1_D(M, F)$ *of the leaf complex. If* M *is compact, then there is a versal family of deformations of* F *parametrized by a complex analytic space* V *and* $T_o(V) \simeq \mathbb{H}^1_D(M, F)$.

The integrability conditions imply that D_o vanishes on F, so the leaf complex gives rise to an exact sequence of complexes that we have called the *leaf diagram*

$$\begin{array}{ccccccc}
0 & & & & & & \\
\downarrow & & & & & & \\
F & & & & & & \\
\downarrow & & & & & & \\
\Theta_M & \xrightarrow{D_o} & \mathrm{Hom}(F, \Theta_M|F) & \xrightarrow{D_1} & \mathrm{Hom}(\Lambda^2 F, \Theta_M|F) & \xrightarrow{D_2} & \\
\downarrow & & & & & & \\
\Theta_M|F & \xrightarrow{d_o} & \mathrm{Hom}(F, \Theta_M|F) & \xrightarrow{d_1} & \mathrm{Hom}(\Lambda^2 F, \Theta_M|F) & \xrightarrow{d_2} & \\
\downarrow & & & & & & \\
0 & & & & & &
\end{array}$$

If we denote by $\mathbb{H}^q_d(M, F)$ the hypercohomology group of the lower complex, the long exact sequence of hypercohomology of the leaf diagram is

$$\cdots \to H^1(M, F) \to \mathbb{H}^1_D(M, F) \xrightarrow{\alpha} \mathbb{H}^1_d(M, F) \to H^2(M, F) \to \cdots \qquad (2.2)$$

where the terms $H^q(M, F)$ are the cohomology groups of the coherent sheaf F.

Let $M' = M - \mathrm{Sing}\,\mathcal{F}$ be the set of points where the foliation is nonsingular. Restricting the leaf diagram to M', we observe that the complexes are exact, forming resolutions of the kernel sheaves

$$0 \to \mathcal{F} \to \Theta_{\mathcal{F}}^{hol} \to \Theta_{\mathcal{F}}^{Tr} \to 0, \tag{2.3}$$

where $\Theta_{\mathcal{F}}^{hol}$ and $\Theta_{\mathcal{F}}^{Tr}$ are the sheaves of infinitesimal automorphism of the foliation $\mathcal{F}$ on M' and of the transversely holomorphic foliation $\mathcal{F}^{Tr}$ associated to $\mathcal{F}$ on M'. Restricting to M' we obtain that $\mathbb{H}_d^1(M', \mathcal{F})$ represents the infinitesimal deformations of F^{Tr} and that α' associates to an infinitesimal deformation of $\mathcal{F}$ on M' its induced transversal deformation. Applying a Mayer-Vietoris type argument to the leaf diagram, we may decompose α in (2.2) as α' plus a map which associates invariants concentrated at the singularities (see [7]). We will apply these results in the next section to foliations with leaf dimension one.

3. HOLOMORPHIC FOLIATIONS BY CURVES

Geometric Definition 3.1. A *foliation by curves* with singularities in the connected complex manifold M is a non-singular holomorphic foliation by leaves of dimension one in $M-V$, where V has codimension bigger than one.

One may obtain an analytic expression for a foliation by curves as follows. Let $L' \to T(M-V)$ be the subline bundle of the tangent bundle on $M-V$ formed by those vectors tangent to the foliation. We want to show that L' has an extension to M as a holomorphic line bundle and that there is a map $X: L \to TM$ whose image restricted to $M-V$ is L'. This is basically a local problem, so restricting to a ball U in M we obtain a map $(X_1, \ldots, X_n): L' \to TU = \overset{n}{\oplus}\mathbb{C}_U$. Assuming $X_1 \not\equiv 0$, Levi's extension theorem guarantees the existence of a function f in U such that $\{f = 0\} = \overline{\{X_1 = 0\}}$. Hence $(1/f)X_1: L' \to \mathbb{C}_U$ is an isomorphism. This shows that L' may be locally extended, and then the codimension bigger than one hypothesis implies that these local extensions glue to a global one (see [6] for more details). This suggests the following definition:

Analytic Definition 3.2. A *foliation by curves* with singularities in the connected complex manifold M is described by a non-identically zero holomorphic map $X: L \to TM$ from a line bundle L to the tangent bundle of M.

Remark. The second definition is more extensive than the first one, since X could vanish on a set of codimension one. Using Douady's parametrization of quotient sheaves we see that the foliations satisfying the first definition form an analytic open set, and those satisfying the second condition form the closure of this set [6]. The geometric definition also corresponds to Baum-Bott's full sheaves. We will adopt from now on the analytic definition.

Definition 3.3. A foliation by curves F has *isolated singularities* (or *Poincaré type singularities*) if $\mathrm{Sing}\,F$ is discrete (or if around every singular point P of F we may find a holomorphic vector field tangent to the foliation with 0 not in the convex hull of the eigenvalues of the linear part of X at P, respectively).

Using the long exact sequence (2.2) of hypercohomology groups to foliations by curves we obtain the local and semi-local rigidity Theorem [7]:

THEOREM 3.4. *Let F be a foliation by curves defined in the ball B in $\mathbb{C}^n$ with 0 as the only singular point, then the map β: $\mathbb{H}^1_d(B, F) \to \mathbb{H}^1_d(B^*, F)$ associating to an infinitesimal deformation of F the induced transversal deformation in the punctured ball B^* is an isomorphism if $n > 2$ or $n = 2$ and F has a Poincaré type singularity.*

This result is then globalised to

THEOREM 3.5. *Let F be a foliation by curves in the complex manifold M with isolated singularities and assume that $\dim M > 2$ or $\dim M = 2$ and F has only Poincaré type singularities; then there is a long exact sequence*

$$H^1(M, F) \to \mathbb{H}^1_D(M, F) \xrightarrow{\beta} \mathbb{H}^1_d(M', F) \to H^2(M, F), \tag{3.1}$$

where β is as above, $H^q(M, F)$ are the cohomology groups of the invertible sheaf F and $M' = M - \mathrm{Sing}\,F$. ·

If M is the projective space $\mathbb{C}P^n$, then Serre's well known computations of the cohomology of line bundles imply that β in (3.1) is bijective for $n > 2$ and injective for $n = 2$. One may obtain a more general version of this theorem using the Kodaira-Nakano vanishing theorem.

We may interpret β as the derivative of a map between versal spaces for foliations by curves with Poincaré type singularities as follows. By multiplying a vector field with a Poincare type singularity by a suitable constant we obtain that the eigenvalues λ_i have negative real part. If we integrate it with real time we obtain that solutions near the singularity tend to it, and they intersect transversely small spheres. Hence the foliation has a cone like structure near the singularity, with base a foliation by one real dimensional leaves in a small sphere.

If F is a holomorphic foliation by curves with Poincare type singularities, denote by $\bar{M}$ the differentiable manifold with boundary obtained by removing small balls around the singular points of F, and F^{Tr} the transversely holomorphic foliation in $\bar{M}$ induced by F. It follows from the cone like structure of the foliation around the singular points that as long as the balls are sufficiently small we obtain transversely holomorphic equivalent foliations; so F^{Tr} is uniquely defined modulo transversely holomorphic equivalence. This construction may be carried out for families, and it yields ([7]):

THEOREM 3.6. *Let F be a foliation by curves in the compact manifold M with only Poincaré type singularities at $\{P_i\}$ and F^{Tr} the associated transversely holomorphic foliation in $M - \bigcup B_i$. Denote by V^{hol} and V^{Tr} the versal spaces of deformation of F and F^{Tr} respectively, and let $\Phi: V^{hol} \to V^{Tr}$ be the holomorphic map obtained from the versal properties of V^{Tr} applied to the transversely holomorphic family associated to the versal family over V^{hol}; then the derivative of Φ at 0 may be embedded in a long exact sequence of the form*

$$\cdots \longrightarrow H^1(M, F) \longrightarrow T_o V^{hol} \xrightarrow{D\Phi_o} T_o V^{Tr} \longrightarrow H^2(M, F) \to \cdots \tag{3.2}$$

where $H^q(M, F)$ is the cohomology of the invertible sheaf $F \subset \Theta_M$, defining the foliation.

COROLLARY 3.7. *Let F be a foliation as above in $\mathbb{C}P^n$, $n > 2$; then the map Φ is locally an isomorphism of smooth manifolds of dimension*

$$\binom{d+n}{d+1}(d+n+2) - 1 - \dim\left[\frac{PSL(n+1, \mathbb{C})}{G(F)}\right],$$

where $-d$ is the Chern class of F and $G(F)$ is the stabilizer subgroup of F in $PSL(n+1, \mathbb{C})$.

PROPOSITION 3.8. *Let F be a foliation by curves in $\mathbb{C}P^2$ with only Poincaré type singularities then the map Φ above is a germ of an embedding of complex manifold of codimension $\frac{1}{2}(d-1)(d-2)$ for $d > 0$, where $-d$ is the Chern class of F.*

Proof of the Proposition. It follows from the computations of the cohomology groups of line bundle in $\mathbb{C}P^2$ and homological algebra applied to the leaf diagram that for $q > 1$, $\mathbb{H}^q_D(\mathbb{C}P^2, F) = \mathbb{H}^q_d(\mathbb{C}P^2, F) = 0$, for $q > 0$ that $\mathbb{H}^q_d(\mathbb{C}P^2, F) \simeq \mathbb{H}^q_d(\mathbb{C}P^2 - \mathrm{Sing}\, F, F)$ and the short exact sequence

$$0 \to \mathbb{H}^1_D(\mathbb{C}P^2, F) \to \mathbb{H}^1_d(\mathbb{C}P^2, F) \to H^2(\mathbb{C}P^2, F) \to 0$$

It follows then from [5] that V^{Tr} is smooth, that $d\Phi(0)$ is injective and it has codimension $\dim H^2(\mathbb{C}P^2, \mathcal{O}(-d)) = \dim H^0(\mathbb{C}P^2, \mathcal{O}(d-3)) = \frac{1}{2}(d-2)(d-1)$ if $d \geq 3$, and 0 otherwise.

Example 3.9. A Lefschetz pencil in $\mathbb{C}P^2$ is obtained by immersing $\mathbb{C}P^2$ in a projective space by means of homogeneous polynomials and then intersecting with a pencil of hyperplanes. It is described also by a rational function $f: \mathbb{C}P^2 \to \mathbb{C}P^1$ defined outside of the intersection of $\mathbb{C}P^2$ with the axis of the pencil and having only Morse type critical points, all with distinct critical values. It can be shown that all Lefschetz pencils give rise to topologically equivalent foliations and that the transverse structure is determined by the moduli of the critical values in $\mathbb{C}P^1$. One can also see that the set of critical values that appear has codimension $\dim H^2(\mathbb{C}P^2, F)$ in the appropiate symmetric product of $\mathbb{C}P^1$. This exhibits explicit examples of transversely holomorphic foliations that do not arise from holomorphic foliations of $\mathbb{C}P^2$.

Example 3.10. Let F_i be holomorphic foliations by curves with Poincaré type singularities in the complex manifolds M_i, $i=1, \dots, r$, and suppose given a pairing between the singular points of F_i such that corresponding points have equivalent Poincaré-Dulac normal forms [1]. Remove small balls around each singular point and invert locally the orientation of the leaves of one of the pairs and glue both boundaries as in the usual double construction. In this form we obtain a transversely holomorphic foliation from holomorphic foliations with Poincaré type singularities.

References

[1] ARNOLD, V.: *Chapïtres supplementaires de la théorie des équations differentielles ordinaires*, editions MIR, Moscou 1980.

[2] DOUADY, A.: *Le problème des modules pour les variétés analytiques complexes* (d'après M. Kuranishi), Seminaire Bourbaki, Exposé 277, 1964/65.

[3] ——: 'Le problème des modules pour les sous-spaces analytiques compacts d'un espace analytique donné', *Ann. Inst. Fourier* 16 (1966), 1-95.

[4] DUCHAMP, T. and M. KALKA: 'Deformation theory of holomorphic foliations', *J. Diff. Geom.* 14 (1979), 317-337.

[5] GIRBAU, J., A. HAEFLIGER, and D. SUNDARARAMAN: 'On deformations of transversely holomorphic foliations', *Journal für die Reine und Angewandte Mathematik* 345 (1983), 122-147.

[6] GOMEZ-MONT, X.: 'Universal families of foliations by curves', in *Proc. Conf. Dyn. Syst. Dijon 1985*, ed. by D. Cerveau, R. Moussu, to appear.

[7] ——: 'The tranverse dynamics of a holomorphic flow', publicación preliminar, *Inst. Mat. U.N.A.M.* 109 (1986).

[8] KODAIRA, K. and D.C. SPENCER: 'Multifoliate structures', *Ann. Math.* 74 (1961), 52-100.

[9] KURANISHI, M.: 'On the locally complete families of complex analytic structures', *Ann. of Math.* 75 (1962), 536-577.

[10] POURCIN, G.: 'Théorème de Douady au dessus de S. Ann. Scuola Norm. Sup. di Pisa 23 (1969), 451-459.

HURWITZ PAIRS AND OCTONIONS

Shôji Kanemaki
Science University of Tokyo
Department of Mathematics
Shinjuku-ku, Tokyo, Japan

ABSTRACT. It has been exhibited that all Hurwitz pairs (V^n, S^p) in case of $p \leqq n \leqq 8$ are only those for $(n,p) = (1,1)$, $(2,2)$, $(4,3)$, $(4,4)$, $(8,5)$, $(8,6)$, $(8,7)$, and $(8,8)$ ([1], [2]). A new matrix-multiplication law on $M_2(\mathbb{H})$ compatible with the non-associative algebra $\mathbb{O}$ of octonions is introduced here. Our method makes the verification of the Hurwitz pair $(\mathbb{R}^8, \mathbb{R}^8)$ easier and gives us a systematic view in the study of the pairs in case of $p \leqq n \leqq 8$.

0. INTRODUCTION

Recently, the theory of Hurwitz pairs has been studied by J. Ławrynowicz, J. Rembieliński, J. Kalina and O. Suzuki in connection with Clifford algebras, complex analysis and soliton equations ([1], [2], [3]). In these papers they gave an interesting interpretation of the notion of 'Hurwitz pairs' in view of not only number theory but also algebras, analysis and physics.

The author first learnt that the Hurwitz pair $(\mathbb{R}^8, \mathbb{R}^8)$ could be expressed by octonion algebra in terms of real coefficients. Its verification gave rise to a simple but long calculation. In order to avoid such tiresome calculations, he introduced a new matrix-multiplication law, which is a generalization of the usual one. Then this trial has gained its present form.

The author wishes to express his hearty thanks to Dr. Osamu Suzuki for his valuable suggestions and kind encouragement.

1. PRELIMINARIES

1°) Hurwitz Pairs. Let $(V, \|\cdot\|)$ be the n-dimensional Euclidean space with the usual Euclidean norm $\|\cdot\|$ and $(S, \|\cdot\|)$ the p-dimensional Euclidean space. A pair (V,S) of Euclidean spaces is said to satisfy the *Hurwitz condition* if there exists a bilinear mapping $f : V \times S \to V$ such that the equality $\|f(x,y)\| = \|x\| \cdot \|y\|$ holds for any $x \in V$ and $y \in S$.

Suppose that two pairs (V_1, S) and (V_2, S) satisfying the Hurwitz

J. Ławrynowicz (ed.), Deformations of Mathematical Structures, 215–223.

condition are given. The direct sum $V = V_1 \oplus V_2$ forms a pair (V,S) of Euclidean spaces satisfying Hurwitz condition in a natural manner. In this case, the mapping f $(= f_1 \oplus f_2)$ preserves its subspace V_1 and V_2 of V, where f_i $(i = 1,2)$ denotes the bilinear mapping for (V_i, S).

A pair (V,S) satisfying the Hurwitz condition is called *irreducible* if its bilinear mapping f does not preserve any subspace of V, except the trivial $\{0\}$ and V itself. A *Hurwitz pair* is a pair (V,S) of Euclidean spaces which satisfies the Hurwitz condition and is irreducible. Hurwitz pairs in the case of $p \leqq n \leqq 8$ are listed as follows ([1], [2]):

n	1	2	4	4	8	8	8	8
p	1	2	3	4	5	6	7	8

2°) Quaternions. We shall recall the quaternions. The direct sum $\mathbb{H} = \mathbb{C} \oplus \mathbb{C}$ of the complex number field $\mathbb{C}$ may be considered the 4-dimensional vector space over the real field $\mathbb{R}$. The multiplication law on the space $\mathbb{H}$ is defined by

$$(z,u)(v,w) = (zv - \bar{w}u,\ u\bar{v} + wz) \quad \text{for} \quad z,u,v,w \in \mathbb{C},$$

where $\bar{w}$ denotes the complex conjugate of w. The space $\mathbb{H}$ forms an associative division algebra which is non-commutative and its element is called a *quaternion*.

We put $1 = (1,0)$, $\mathbb{i} = (i,0)$, $\mathbb{j} = (0,1)$, $\mathbb{k} = (0,i)$, where i $(= \sqrt{-1})$ denotes the complex pure-imaginary unit. Then it follows that $\mathbb{i}^2 = \mathbb{j}^2 = \mathbb{k}^2 = -1$, $\mathbb{i}\mathbb{j} = \mathbb{k} = -\mathbb{j}\mathbb{i}$, $\mathbb{j}\mathbb{k} = \mathbb{i} = -\mathbb{k}\mathbb{j}$, $\mathbb{k}\mathbb{i} = \mathbb{j} = -\mathbb{i}\mathbb{k}$. Every quaternion p has the form $p = x_0 1 + x_1 \mathbb{i} + x_2 \mathbb{j} + x_3 \mathbb{k}$ for $x_0, x_1, x_2, x_3 \in \mathbb{R}$. We may rewrite p as the sum of the real part and the pure-imaginary part:

$$p = p_0 + \hat{p} \quad \text{for} \quad p_0 \in \mathbb{R}, \quad \hat{p} \in \hat{\mathbb{R}}^3 = \mathbb{R}\mathbb{i} + \mathbb{R}\mathbb{j} + \mathbb{R}\mathbb{k};$$

in this case $p_0 = x_0 1$ and $\hat{p} = x_1 \mathbb{i} + x_2 \mathbb{j} + x_3 \mathbb{k}$. The quaternion conjugate $\bar{p}$ of p is given by $\bar{p} = p_0 - \hat{p}$. The product $\hat{p}\,\hat{q}$ of any two $\hat{p}$ and $\hat{q}$ belonging to the 3-space $\hat{\mathbb{R}}^3$ of the pure-imaginary quaternions is written as

$$\hat{p}\,\hat{q} = -\langle\hat{p}, \hat{q}\rangle\, 1 + \hat{p} \times \hat{q},$$

where $\langle\hat{p}, \hat{q}\rangle$ denotes the Euclidean inner product and $\hat{p} \times \hat{q}$ the vector cross product of $\hat{p}$, $\hat{q}$ in the 3-space $\hat{\mathbb{R}}^3$ ([4]). The quaternion algebra is non-commutative. Furthermore, it has the property:

$$\overline{p\,q} = \bar{q}\,\bar{p} \quad \text{for} \quad p,\ q \in \mathbb{H}.$$

In fact,

$$p\,q = (p_0 + \hat{p})(q_0 + \hat{q}) = p_0\, q_0 - \langle\hat{p}, \hat{q}\rangle + p_0\,\hat{q} + q_0\,\hat{p} + \hat{p} \times \hat{q}.$$

Then

$$\overline{p\,q} = p_0\,q_0 - <\hat{p}, \hat{q}> - p_0\,\hat{q} - q_0\,\hat{p} - \hat{p}\times\hat{q}.$$

On the other hand, $\bar{q}\,\bar{p} = q_0\,p_0 - <\hat{q}, \hat{p}> - p_0\,\hat{q} - q_0\,p + \hat{q}\times\hat{p}$. Thus $\overline{p\,q} = \bar{q}\,\bar{p}$. In particular, we note that

$$p\,\bar{p} = (p_0 + \hat{p})(p_0 - \hat{p}) = p_0^{\,2} + <\hat{p}, \hat{p}> = \bar{p}\,p,$$

which is real and, in addition,

$$p\,\bar{p} = x_0^{\,2} + x_1^{\,2} + x_2^{\,2} + x_3^{\,2} \geqq 0$$

for the form $p = x_0 1 + x_1 \mathbb{i} + x_2 \mathbb{j} + x_3 \mathbb{k}$. Therefore, introducing the norm $|\cdot|$ to $\mathbb{H}$ by $|p|^2 = p\,\bar{p}$, the space $\mathbb{H}$ can be seen as the 4-dimensional Euclidean space.

We apply the following Lemma 1 to prove Theorem 1.

LEMMA 1. *For any* $p, q \in \mathbb{H}$ *the relations hold:*

$$p\,\bar{q} + q\,\bar{p} = \bar{p}\,q + \bar{q}\,p, \qquad p\,q + \overline{p\,q} = q\,p + \overline{q\,p}.$$

P r o o f. According to the same notation mentioned above, we have

$$p\,\bar{q} + q\,\bar{p} = (p_0 + \hat{p})(q_0 - \hat{q}) + (q_0 + \hat{q})(p_0 - \hat{p}) = 2(p_0\,q_0 + <\hat{p}, \hat{q}>),$$

$$\bar{p}\,q + \bar{q}\,p = (p_0 - \hat{p})(q_0 + \hat{q}) + (q_0 - \hat{q})(p_0 + \hat{p}) = 2(p_0\,q_0 + <\hat{p}, \hat{q}>),$$

from which the first desired relation is obtained. Making a substitution of $\bar{q}$ instead of q in this relation, we have the second one.

2. OCTONIONS AND THE MATRIX-REPRESENTATION

We consider the direct sum $\mathbb{O} = \mathbb{H} \oplus \mathbb{H}$ of the quaternion algebras $\mathbb{H}$ and itself. The space $\mathbb{O}$ is the 8-dimensional vector space over $\mathbb{R}$. The multiplication law on the space is defined by

$$(p, q)(r, s) = (pr - \bar{s}q,\ q\bar{r} + sp) \quad \text{for} \quad p,q,r,s \in \mathbb{H},$$

where $\bar{s}$ denotes the quaternion conjugate of s. An element of $\mathbb{O}$ is called an *octonion*, by which we mean a Cayley number, and its space $\mathbb{O}$ the *octonion algebra*. The algebra $\mathbb{O}$ is divisional, but it is not commutative and, further, it is not associative ([4], [5]).

We set $1 = (1,0)$, $e_1 = (\mathbb{i},0)$, $e_2 = (\mathbb{j},0)$, $e_3 = (\mathbb{k},0)$, $e_4 = (0,1)$, $e_5 = (0,\mathbb{i})$, $e_6 = (0,\mathbb{j})$, and $e_7 = (0,\mathbb{k})$. Then it follows that

$$e_\alpha^{\,2} = -1 \quad (\alpha = 1,2,\ldots,7) \quad \text{and} \quad e_\alpha e_\beta + e_\beta e_\alpha = 0 \quad (\alpha \neq \beta).$$

The following figure suggests the rule of the multiplication, $e_\alpha e_\beta$, among $\{e_1,e_2,\ldots,e_7\}$. Each vertex shows e_α and the arrow on the 1-chain the direction of the cycle.

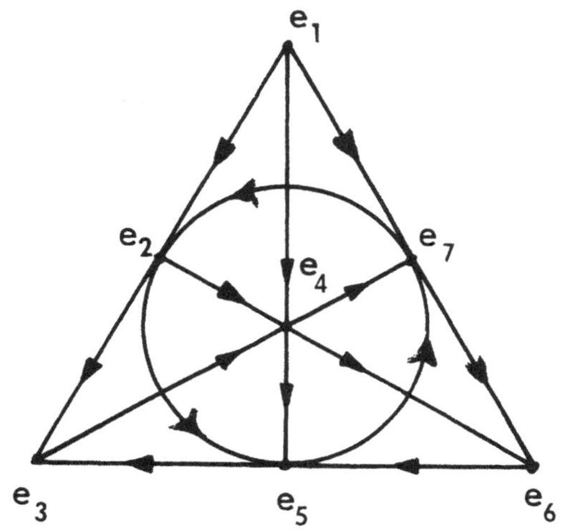

For example, $e_1e_2 = e_3$, $e_2e_3 = e_1$, $e_3e_1 = e_2$; $e_2e_1 = -e_3$, $e_3e_2 = -e_1$, $e_1e_3 = -e_2$. The ordered set $[1, e_1, e_2, e_3]$ forms the quaternion algebra. Hence, there are 7 quaternion algebras $[1, e_1, e_2, e_3]$, $[1, e_1, e_4, e_5]$, ... in the octonion algebra.

An octonion x has the form

$$x = x_0 1 + x_1 e_1 + x_2 e_2 + x_3 e_3 + x_4 e_4 + x_5 e_5 + x_6 e_6 + x_7 e_7$$

in real coefficients. As we can see that $e_5 = e_1e_4$, $e_6 = e_2e_4$, $e_7 = e_3e_4$ by the figure, then

$$x = (x_0 1 + x_1 e_1 + x_2 e_2 + x_3 e_3) + (x_4 1 + x_5 e_1 + x_6 e_2 + x_7 e_3) e_4.$$

Under the fundamental identification of $x_0 1 + x_1 \mathbf{i} + x_2 \mathbf{j} + x_3 \mathbf{k}$ with $x_0 1 + x_1 e_1 + x_2 e_2 + x_3 e_3$, the octonion x is written as $x = p + qe_4$ for $p, q \in \mathbb{H}$, and we understand the octonion $x = (p, q) \in \mathbb{O} = \mathbb{H} \oplus \mathbb{H}$ in this sense.

We denote by $M_2(\mathbb{H})$ the set of all 2×2 matrices with quaternion coefficients. For elements $P = (p_{ij})$ and $Q = (q_{ij})$ of $M_2(\mathbb{H})$ we define the sum $P + Q = (p_{ij} + q_{ij})$ and the scalar products $pP = (pp_{ij})$, $Pp = (p_{ij}p)$ $(p \in \mathbb{H})$. Therefore, $M_2(\mathbb{H})$ forms a left $\mathbb{H}$-module or right $\mathbb{H}$-module according to pP or Pp.

Next, we introduce a new matrix-multiplication law on $M_2(\mathbb{H})$ as follows: Consider the involution $\nu : \mathbb{H} \times \mathbb{H} \to \mathbb{H} \times \mathbb{H}$ given by $\nu(p, q) = (q, p)$ and denote by $\mu : \mathbb{H} \times \mathbb{H} \to \mathbb{H}$ the quaternion-multiplication, $\mu(p, q) = pq$ for $p, q \in \mathbb{H}$. $\lambda^m = \mu \cdot \nu^m$ denotes the composed mapping of μ and ν^m $(\nu^2 = \nu \cdot \nu,\ \nu^3 = \nu \cdot \nu \cdot \nu, \ldots)$ for positive integer m. Define the product PQ of $P = (p_{ij})$ and $Q = (q_{ij})$ on $M_2(\mathbb{H})$ by

$$PQ = [r_{ij}], \qquad r_{ij} = \Sigma_{k=1}^{2} \lambda^{i+j+k-1}(p_{ik}, q_{kj}), \tag{*}$$

namely

$$\begin{bmatrix} p_{11} & p_{12} \\ p_{21} & p_{22} \end{bmatrix} \begin{bmatrix} q_{11} & q_{12} \\ q_{21} & q_{22} \end{bmatrix} = \begin{bmatrix} p_{11}q_{11} + q_{21}p_{12} & q_{12}p_{11} + p_{12}q_{22} \\ q_{11}p_{21} + p_{22}q_{21} & p_{21}q_{12} + q_{22}p_{22} \end{bmatrix}.$$

It follows that $M_2(\mathbb{H})$, considered together with the matrix-multiplication $\tilde{\mu} : M_2(\mathbb{H}) \times M_2(\mathbb{H}) \to M_2(\mathbb{H})$ defined by $\tilde{\mu}(P,Q) = PQ$ $(P,Q \in M_2(\mathbb{H}))$, is a non-associative division algebra. This matrix-multiplication law is identical with the usual one if the coefficient field is restricted to the commutative field $\mathbb{R}$ or $\mathbb{C}$.

We define, in a natural manner, the transpose ${}^tP$ of P, the quaternion conjugate $\overline{P}$ of P by

$${}^t\begin{bmatrix} p_{11} & p_{12} \\ p_{21} & p_{22} \end{bmatrix} = \begin{bmatrix} p_{11} & p_{21} \\ p_{12} & p_{22} \end{bmatrix}, \qquad \overline{\begin{bmatrix} p_{11} & p_{12} \\ p_{21} & p_{22} \end{bmatrix}} = \begin{bmatrix} \overline{p}_{11} & \overline{p}_{12} \\ \overline{p}_{21} & \overline{p}_{22} \end{bmatrix},$$

and $P^* = {}^t\overline{P}$. There are two possible definitions of the determinant of P:

$$\det_{row} P = p_{11}p_{22} - p_{12}p_{21}, \qquad \det_{column} P = p_{11}p_{22} - p_{21}p_{12}.$$

Both definitions coincide if and only if $p_{12}p_{21} = p_{21}p_{12}$, and in this case we write $\det_{row} P = \det_{column} P = \det P$.

<u>THEOREM 1</u>. *Let* j *be a mapping* $j : \mathbb{O} \to M_2(\mathbb{H})$ *defined by*

$$j(x) = \begin{bmatrix} p & -q \\ \overline{q} & \overline{p} \end{bmatrix} \qquad \text{for } x = (p,q) \in \mathbb{O} = \mathbb{H} \oplus \mathbb{H}$$

and put $M_{\mathbb{O}} = \{j(x) \mid x \in \mathbb{O}\}$. *Then* $j : \mathbb{O} \to M_{\mathbb{O}}$ *is an algebra-isomorphism and isometry with respect to the norms*

$$\|x\|^2 = |p|^2 + |q|^2 \text{ on } \mathbb{O}, \qquad \|j(x)\|^2 = (1/2)\operatorname{tr} j(x)^* j(x) \text{ on } M_{\mathbb{O}}.$$

Moreover, the following relations are obtained:

$$\det j(x) = \|j(x)\|^2, \qquad \det j(xy) = \det j(x) \cdot \det j(y)$$

for $x, y \in \mathbb{O}$.

P r o o f. First, we can see that j is an injective $\mathbb{R}$-linear mapping. It satisfies $j(xy) = j(x)\,j(y)$ for $x,y \in \mathbb{O}$. In fact, by putting $y = (r,s)$, we have

$$j(x)j(y) = \begin{bmatrix} p & -q \\ \overline{q} & \overline{p} \end{bmatrix} \begin{bmatrix} r & -s \\ \overline{s} & \overline{r} \end{bmatrix} = \begin{bmatrix} pr - \overline{s}q & -sp - q\overline{r} \\ r\overline{q} + \overline{p}\,\overline{s} & -\overline{q}s + \overline{r}\,\overline{p} \end{bmatrix} =$$

$$= \begin{bmatrix} pr - \bar{s}q & -(sp + q\bar{r}) \\ \overline{sp + q\bar{r}} & \overline{pr - \bar{s}q} \end{bmatrix} = j(xy).$$

Hence, $j : \mathbb{O} \to M_{\mathbb{O}}$ is an algebra-isomorphism. Since

$$j(x)^* j(x) = \begin{bmatrix} |p|^2 + |q|^2 & 0 \\ 0 & |p|^2 + |q|^2 \end{bmatrix},$$

then $\|j(x)\|^2 = |p|^2 + |q|^2 = \|x\|^2$ holds, namely j is an isometry. It follows from $\det j(x) = |p|^2 + |q|^2 = \|j(x)\|^2$ and Lemma 1 that

$$\begin{aligned} \det j(x)j(y) &= (pr - \bar{s}q)(-\bar{q}s + \bar{r}\bar{p}) + (r\bar{q} + \bar{p}\bar{s})(sp + q\bar{r}) \\ &= |p|^2|r|^2 + |q|^2|s|^2 + \overline{|q|^2|r|^2} + \overline{|p|^2|s|^2} \\ &\qquad - (pr\bar{q}s + \overline{pr\bar{q}s}) + (r\bar{q}s)p + \overline{(r\bar{q}s)p} \\ &= \det j(x) \cdot \det j(y). \end{aligned}$$

Q.E.D.

When restricting the above discussion to $M_2(\mathbb{C})$ and $M_2(\mathbb{R})$, we obtain the following corollaries:

COROLLARY 1. *A mapping* $j : \mathbb{H} \to M_{\mathbb{H}} (\subset M_2(\mathbb{C}))$, *defined by*

$$j(x) = \begin{bmatrix} z & -u \\ \bar{u} & \bar{z} \end{bmatrix} \quad \textit{for} \quad x = (z,u) \quad \mathbb{H} = \mathbb{C} \oplus \mathbb{C},$$

is an algebra-isomorphism and isometry with respect to the norms $\|x\|^2 = |z|^2 + |u|^2$ *on* $\mathbb{H}$ *and* $\|j(x)\|^2 = (1/2)\mathrm{tr}\, j(x)^* j(x)$ *on* $M_{\mathbb{H}}$, *and satisfies*

$$\det j(x) = \|j(x)\|^2, \qquad \det j(xy) = \det j(x) \cdot \det j(y).$$

COROLLARY 2. *A mapping* $j : \mathbb{C} \to M_{\mathbb{C}} (\subset M_2(\mathbb{R}))$, *defined by*

$$j(x) = \begin{bmatrix} a & -b \\ b & a \end{bmatrix} \quad \textit{for} \quad x = (a,b) \in \mathbb{C} = \mathbb{R} \oplus \mathbb{R},$$

is an algebra-isomorphism and isometry with respect to the norms $\|x\|^2 = |a|^2 + |b|^2$ *on* $\mathbb{C}$ *and* $\|f(x)\|^2 = (1/2)\mathrm{tr}\, {}^t j(x) j(x)$ *on* $M_{\mathbb{C}}$, *and satisfies*

$$\det j(x) = \|j(x)\|^2, \qquad \det j(xy) = \det j(x) \det j(y).$$

3. HURWITZ PAIRS IN THE CASE OF $p \leq n \leq 8$

In this section we shall apply Theorem 1 to the verification of the pairs $(\mathbb{R}^n, \mathbb{R}^p)$, $p \leq n \leq 8$, to satisfy the Hurwitz condition by an easy calculation. The number-spaces $\mathbb{R}$, $\mathbb{C}$ $(= \mathbb{R}^2)$ and $\mathbb{H}$ $(= \mathbb{R}^4)$ may be treated as the subspaces of the Euclidean 8-space $\mathbb{O}$ $(= \mathbb{R}^8)$. We put $K = \mathbb{R}, \mathbb{C}, \mathbb{H}, \mathbb{O}$, accordingly, $r = 0, 1, 2, 3$, and hence $K = \mathbb{R}^{2^r}$. The mapping $j : K \to M_K$ is an algebra-isomorphism and isometry. The multiplication $\mu : K \times K \to K$ $(\mu(x,y) = xy;\ x,y \in K)$ and the matrix-multiplication $\tilde{\mu} : M_K \times M_K \to M_K$ $(\tilde{\mu}(j(x), j(y)) = j(x)j(y);\ x,y \in K)$ commute the diagram:

$$\begin{array}{ccc} K \times K & \xrightarrow{\ \mu\ } & K \\ \Big\downarrow{\scriptstyle j\times j} & & \Big\downarrow{\scriptstyle j} \\ M_K \times M_K & \xrightarrow{\ \tilde{\mu}\ } & M_K \end{array} ,$$

namely $\tilde{\mu} \circ (j \times j) = j \circ \mu$ holds. In consideration of this situation we have the following assertions:

THEOREM 2. *The pairs* $(\mathbb{R}^{2^r}, \mathbb{R}^{2^r})$, $r = 0,1,2,3$, *are Hurwitz pairs, each of which is given the bilinear mapping* f *by the multiplication* μ: $f = \mu$, *i.e.*

$$f(x,y) = xy \qquad \textit{for} \quad x,y \in \mathbb{R}^{2^r} = K.$$

Proof. Here we describe the proof in the case of $K = \mathbb{O}$. The remaining cases are similar to $\mathbb{O}$. By Theorem 1 we have

$$\|f(x,y)\|^2 = \|xy\|^2 = \|j(xy)\|^2 = \det j(xy) = \|j(x)\|^2 \cdot \|j(y)\|^2$$
$$= \|x\|^2 \cdot \|y\|^2,$$

which shows that $(\mathbb{O}, \mathbb{O})$ satisfies the Hurwitz condition. The relations

$$j(p,0)j(r,s) = j(pr,sp), \qquad j(0,q)j(r,s) = j(-\bar{s}q, q\bar{r})$$

for $p,q,r,s \in \mathbb{H}$ imply that the pair is irreducible. Q.E.D.

Remark. Our method of checking the Hurwitz condition, described in the proof of this theorem, is simpler than the following consideration: We choose arbitrary elements $x = \Sigma_{i=0}^{7} x_i e_i$ and $y = \Sigma_{j=0}^{7} y_j e_j$ of $\mathbb{O}$ with respect to the canonical basis e_i. The $\mathbb{R}$-linear mapping $f(y,\cdot)\colon \mathbb{O} \to \mathbb{O}$ for each y given by $f(y,\cdot)(x) = f(y,\cdot)$ may be considered a matrix which maps a vector x to the vector $f(x,y)$, say $= x'$. It follows that

$$x'_0 = y_0x_0 - y_1x_1 - y_2x_2 - y_3x_3 - y_4x_4 - y_5x_5 - y_6x_6 - y_7x_7,$$

$$x'_1 = y_1x_0 + y_0x_1 - y_3x_2 + y_2x_3 - y_5x_4 + y_4x_5 + y_7x_6 - y_6x_7,$$

$$x'_2 = y_2x_0 + y_3x_1 + y_0x_2 - y_1x_3 - y_6x_4 - y_7x_5 + y_4x_6 + y_5x_7,$$

$$x'_3 = y_3x_0 - y_2x_1 + y_1x_2 + y_0x_3 - y_7x_4 + y_6x_5 - y_5x_6 + y_4x_7,$$

$$x'_4 = y_4x_0 + y_5x_1 + y_6x_2 + y_7x_3 + y_0x_4 - y_1x_5 - y_2x_6 - y_3x_7,$$

$$x'_5 = y_5x_0 - y_4x_1 + y_7x_2 - y_6x_3 + y_1x_4 + y_0x_5 + y_3x_6 - y_2x_7,$$

$$x'_6 = y_6x_0 - y_7x_1 - y_4x_2 + y_5x_3 + y_2x_4 - y_3x_5 + y_0x_6 + y_1x_7,$$

$$x'_7 = y_7x_0 + y_6x_1 - y_5x_2 - y_4x_3 + y_3x_4 + y_2x_5 - y_1x_6 + y_0x_7.$$

It has to be calculated so that $\Sigma_{i=0}^{7} x'^{2}_{i} = (\Sigma_{i=0}^{7} x_i^2)(\Sigma_{j=0}^{7} y_j^2)$. This verification is simple but long.

THEOREM 3. *Any pair* $(\mathbb{R}^{2^r}, \mathbb{R}^p)$ $(1 \leqq p \leqq 2^r,\ 0 \leqq r \leqq 3)$ *satisfies the Hurwitz condition for which the bilinear mapping is given by the restriction of the multiplication* μ. *All Hurwitz pairs in* $(\mathbb{R}^n, \mathbb{R}^p)$ $(p \leqq n \leqq 8)$ *are* $(\mathbb{R}^{2^r}, \mathbb{R}^p)$ *such that*

$$[2^{r-1}] + 1 \leqq p \leqq 2^r, \quad 0 \leqq r \leqq 3.$$

Proof. First we explain the notation $K|_p$. (i) *Case* $(\mathbb{R}^8, \mathbb{R}^p)$: Let $\mathbb{R}^7$ correspond to $\mathbb{O}|_7 = \{(r,s) \in \mathbb{O} = \mathbb{H} \oplus \mathbb{H} \mid s = s_0 1 + s_1 \mathbf{i} + s_2 \mathbf{j}$ for $s_0, s_1, s_2 \in \mathbb{R}\}$. Similarly,

$$\mathbb{R}^6 \text{ to } \mathbb{O}|_6 = \{(r,s) \mid s = s_0 1 + s_1 \mathbf{i}\}, \quad \mathbb{R}^5 \text{ to } \mathbb{O}|_5 = \{(r,s) \mid s \in \mathbb{R}\},$$

$$\mathbb{R}^4 \text{ to } \mathbb{O}|_4 = \{(r,0) \mid r \in \mathbb{H}\}, \quad \mathbb{R}^3 \text{ to } \mathbb{O}|_3 = \{(r,0) \mid r = r_0 1 + r_1 \mathbf{i} + r_2 \mathbf{j}\},$$

$$\mathbb{R}^2 \text{ to } \mathbb{O}|_2 = \{(r,0) \mid r = r_0 1 + r_1 \mathbf{i}\}, \quad \mathbb{R}^1 \text{ to } \mathbb{O}|_1 = \{(r,0) \mid r \in \mathbb{R}\}.$$

Accordingly, (ii) *Case* $(\mathbb{R}^4, \mathbb{R}^p)$: $\mathbb{R}^3$ to $\mathbb{H}|_3 = \{(z,u) \in \mathbb{H} = \mathbb{C} \oplus \mathbb{C} \mid u \in \mathbb{R}\}$, $\mathbb{R}^2$ to $\mathbb{H}|_2 = \{(z,0) \mid z \in \mathbb{C}\}$, $\mathbb{R}^1$ to $\mathbb{H}|_1 = \{(z,0) \mid z \in \mathbb{R}\}$. (iii) *Case* $(\mathbb{R}^2, \mathbb{R}^p)$: $\mathbb{R}^1$ to $\mathbb{C}|_1 = \{(a,0) \mid a \in \mathbb{R}\}$. Consequently, we obtain the subspaces $K|_p$ for $1 \leqq p \leqq 2^r$ in the space K. The bilinear mapping, say - the same letter, $f\colon K \times K|_p \to K$ is the restriction of f on $K \times K$ described in Theorem 2 to $K \times K|_p$. Thus, the relation $\|f(x,y)\| = \|x\|\ \|y\|$ $(x \in K,\ y \in K|_p)$ is valid. Since

$$j(p,0)j(r,0) = j(pr,0), \quad j(0,q)j(r,0) = j(0,q\bar{r}) \quad \text{for } p,q,r \in \mathbb{H},$$

the pairs $(\mathbb{R}^8, \mathbb{R}^p)$ for $1 \leqq p \leqq 4$ are reducible and those for $5 \leqq p \leqq 8$ are irreducible. Q.E.D.

References

[1] ŁAWRYNOWICZ, J. and REMBIELIŃSKI, J.: 'Hurwitz pairs equipped with complex structures', in: Seminar on deformations, Łódź-Warsaw 1982/84, Proceedings (*Lecture Notes in Math.* 1165), Springer, Berlin-Heidelberg-New York-Tokyo 1985, pp. 184-195.

[2] —— and ——: 'Supercomplex vector spaces and spontaneous symmetry breaking', in: *Seminari di Geometria 1984*, Università di Bologna, Bologna 1985, pp. 131-154.

[3] KALINA, J., ŁAWRYNOWICZ, J., and SUZUKI, O.: 'A field equation defined by a Hurwitz pair', *Suppl. Rend. Circ. Mat. Palermo* (2) 9 (1985), 117-128.

[4] GREUB, W. HALPERIN, S., and VANSTONE, R.: *Connections, Curvature, and Cohomology*, vol. I, Academic Press, 1972.

[5] KANEMAKI, S.: 'Cayley systems of Hermitian structures', *Colloquia Mathematica Societatis János Bolyai* 31. *Differential Geometry*, ed. by G.Y. Soós and J. Szenthe, North-Holland, 1982.

HERMITIAN PRE-HURWITZ PAIRS AND THE MINKOWSKI SPACE

Shôji Kanemaki
Department of Mathematics
Science University of Tokyo
Shinjuku-ku, Tokyo
Japan

and

Osamu Suzuki
Department of Mathematics
College of Humanities and Sciences
Nihon University
Setagaya-ku, Tokyo, Japan

ABSTRACT. A. Hurwitz introduced pairs $(\mathbb{R}^n, \mathbb{R}^p)$ of Euclidean spaces, each of which possesses the property $\langle f(x,y), f(x,y)\rangle = \langle x,x\rangle\langle y,y\rangle$ $(x \in \mathbb{R}^n, y \in \mathbb{R}^p)$ for a bilinear mapping $f: \mathbb{R}^n \times \mathbb{R}^p \to \mathbb{R}^n$ for some positive integers n and p. He showed that these pairs determine the spaces of real numbers, complex numbers, quaternions, and octonions if, in particular, $n = p = 1, 2, 4$, and 8, respectively. Here concepts of pre-Hurwitz pairs and hermitian pre-Hurwitz pairs are newly introduced. It is shown that there exists a hermitian pre-Hurwitz pair which determines the Minkowski space, by making use of the Dirac γ-matrices which are familar in quantum field theory. This suggests that hermitian pre--Hurwitz pairs will play an important role in both mathematics and physics.

INTRODUCTION

A. Hurwitz [1] dealt with the problem of finding all bidimensions (n,p) of pairs $(\mathbb{R}^n, \mathbb{R}^p)$ of Euclidean spaces under some condition, called the Hurwitz condition. J. Ławrynowicz and J. Rembieliński [3] reformulated the problem in view of Clifford algebras and discussed it in connection with complex structures. S. Kanemaki (these Proc., pp. 215 - - 223) showed a unified observation of pairs in cases of $p \leq n \leq 8$ by using the extended matrix-representation of octonions, whilst Ławrynowicz and Rembieliński [2] initiated pairs of pseudo-Euclidean spaces and discussed the Hurwitz problem for them.

In this paper, we consider not only pairs of pseudo-Euclidean spaces but also hermitian pre-Hurwitz pairs, each of which consists of a Hermitian space and a pseudo-Euclidean space. In Section 1, we introduce hermitian pre-Hurwitz pairs, $(\mathbb{C}^m(\Lambda), \mathbb{R}^p(\Sigma))$, and state the preliminaries. We can see that for a pre-Hurwitz pair (V^{2m}, S^p) the pseudo-Euclidean space V^{2m} is reducible to a Hermitian space $\mathbb{V}^m$ under some assumption. This gives rise to a reduction of real matrices of degree $2m$, $M_{2m}(\mathbb{R})$, to the half-sized complex matrices, $M_m(\mathbb{C})$. Especially, an equivalent condition for $(\mathbb{C}^m(\Lambda), \mathbb{R}^p(\Sigma))$ to be a hermitian pre-Hurwitz pairs is shown in Proposition 3. In Section 2, the classification of pre-Hurwitz pairs in case of $(n,p) = (4,4)$ is de-

J. Ławrynowicz (ed.), Deformations of Mathematical Structures, 225–232.

scribed in Theorem 1. By this we can see that there does not exist any pre-Hurwitz pair in cases of $p \leq n = 4$ having the Minkowski space as a member of the pair except for $(n,p) = (4,1)$. In the final section, our result is exhibited, which says the following. The pair $(\mathbb{C}^2(I), \mathbb{R}^4(I))$, which is related to quaternions, forms a hermitian pre-Hurwitz pair determining the negative definite metric vector space $\Sigma_{a=1}^{3} \mathbb{R}(i\sigma_a)$, expressed in terms of the Pauli matrices σ_a. In view of this we obtain that the pairs $(\mathbb{C}^4(\Lambda_{2,2}), \mathbb{R}^5(\Sigma_{2,3}))$ and $(\mathbb{C}^4(\Lambda_{2,2}), \mathbb{R}^6(\Sigma_{2,4}))$ form hermitian pre-Hurwitz pairs which determine the Minkowski space $\Sigma_{\mu=0}^{3} \mathbb{R}(i\gamma^\mu)$ and the metric vector space $\Sigma_{\mu=0}^{3} \mathbb{R}(i\gamma^\mu) + \mathbb{R}\gamma_5$ of signature (4.1), respectively, in terms of the Dirac γ-matrices and the matrix γ_5.

The authors would like to express their hearty thanks to Profs. I. Furuoya, H. Hagiwara, and J. Yamashita for their helpful discussions.

1. HERMITIAN PRE-HURWITZ PAIRS

Let $V = \mathbb{R}^n(\Lambda_{r,s})$ be the n-dimensional pseudo-Euclidean space of signature (r,s), $n = r+s$, on which the inner product $<.,.>$ is defined by $<x,x'> = {}^t x \Lambda x'$ for $x, x' \in \mathbb{R}^n$ and $\Lambda_{r,s} (\equiv \Lambda) = \mathrm{diag}(1, \overset{r}{\ldots}, 1, -1, \overset{s}{\ldots}, -1)$ with respect to the canonical basis $\{e_i\}$ for the n-column vector space $\mathbb{R}^n$. We set $S \equiv \mathbb{R}^p(\Sigma_{r',s'})$.

Analogously, let $\mathbb{V} \equiv \mathbb{C}^m(\Lambda_{r'',s''})$ denote the m-dimensional Hermitian space of signature (r'',s''), on which the inner product $<.,.>$ is defined by $<z,z'> = z^* \Lambda z'$ for $z, z' \in \mathbb{V}$, where $z^* = {}^t\bar{z}$, with respect to the canonical basis $\{e_j\}$ for the complex m-column vector space $\mathbb{C}^m$.

A pair (V,S) (resp. $(\mathbb{V}, S)$) is called a *pre-Hurwitz pair* (resp. *hermitian pre-Hurwitz pair*) if it satisfies the *Hurwitz condition*: there exists a bilinear mapping $f: V \times S \to V$ (resp. $F: \mathbb{V} \times S \to \mathbb{V}$) such that

$$<f(x,y), f(x,y)> = <x,x><y,y> \quad \text{for } x \in V,\ y \in S \tag{1.1}$$

$$(\text{resp. } <F(z,y), F(z,y)> = <z,z><y,y> \quad \text{for } z \in \mathbb{V},\ y \in S) \tag{1.2}$$

We note that by a bilinear mapping F we have meant a mapping F such that the mapping $F(\cdot,y): \mathbb{V} \to \mathbb{V}$ for a fixed y is $\mathbb{C}$-linear and that $F(z,y)$ for a fixed z is $\mathbb{R}$-linear. We can see that if two (V_i, S) $(i = 1, 2)$ are pre-Hurwitz pairs, then so is the direct sum $(V_1 \oplus V_2, S)$. A pre-Hurwitz pair (V,S) is called *irreducible* if there does not exist any non-trivial proper subspace V' of V such that (V',S) is a pre-Hurwitz pair with respect to the restriction $f|_{V'}$ of the bilinear mapping f for the pair (V,S). A pre-Hurwitz pair is called a *Hurwitz pair* if it is irreducible. Similarly, a hermitian Hurwitz pair can be defined. It is known that pairs $(\mathbb{R}^n(I), \mathbb{R}^p(I))$ of Euclidean spaces for (n,p) such that $(n,p) = (1,1), (2,2), (4,3), (4,4), (8,5), (8,6), (8,7), (8,8), \ldots$ are Hurwitz pairs (cf. [3] and S. Kanemaki, these Proc., pp. 215-224).

PROPOSITION 1. (1) *If* $(\mathbb{R}^n(I), \mathbb{R}^p(I))$ *is a Hurwitz pair, then so is* $(\mathbb{R}^n(-I), \mathbb{R}^p(I))$, *and the converse is true as well.*
(2) *There does not exist any pre-Hurwitz pair formed with the spaces* $\mathbb{R}^n(\Lambda)$ *and* $\mathbb{R}^p(\Sigma)$ *for* $(\Lambda, \Sigma) = (I, -I)$, $(-I, -I)$, $(I, \textit{indefinite})$, $(-I, \textit{indefinite})$.

This is obtained by comparing signs of both sides of the equality (1.1).

PROPOSITION 2. *If a pair* $(\mathbb{C}^m(\Lambda_{r'',s''}), \mathbb{R}^p(\Sigma_{r',s'}))$ *is a hermitian pre-Hurwitz pair, then the pair* $(\mathbb{R}^{2m}(\Lambda_{2r'',2s''}), \mathbb{R}^p(\Sigma_{r',s'}))$ *is a pre-Hurwitz pair.*

Proof. We define an $\mathbb{R}$-linear mapping $\rho: \mathbb{R}^{2m}(\Lambda) \to \mathbb{C}^m(\Lambda)$ by $\rho(e_{2j-1}) = \mathbf{e}_j$ and $\rho(e_{2j}) = i\mathbf{e}_j$ $(j = 1, 2, \ldots, m)$. The mapping ρ is a bijection and hence it is an isometry, because $\langle x,x\rangle = \langle \tilde{x},\tilde{x}\rangle$ holds for any $x \in \mathbb{R}^{2m}$, where we have put $\rho(x) = \tilde{x}$. Suppose that $(\mathbb{C}^m(\Lambda), \mathbb{R}^p(\Sigma))$ is a hermitian pre-Hurwitz pair. The bilinear mapping F determines the $\mathbb{R}$-linear transformation $f(\cdot,y) = \rho^{-1}\circ F(\cdot,y)\circ\rho$ on the space $\mathbb{R}^{2m}$. It follows that $\langle f(x,y), f(x,y)\rangle = \langle F(x,y), F(x,y)\rangle = \langle\tilde{x},\tilde{x}\rangle\langle y,y\rangle = \langle x,x\rangle\langle y,y\rangle$ holds for $x \in \mathbb{R}^{2m}$ and $y \in \mathbb{R}^p$. Hence $(\mathbb{R}^{2m}(\Lambda), \mathbb{R}^p(\Sigma))$ is a pre-Hurwitz pair.

We shall see the relation $F(\tilde{x},y) = \rho(f(x,y))$ $(x \in \mathbb{R}^{2m},\ y \in \mathbb{R}^p)$ in view of matrices. f and F determine p real matrices $C_\alpha = (C^j_{i\alpha})$ $(\alpha = 1, 2, \ldots, p)$ of degree $2m$, and p complex matrices $C_{(\alpha)} = (C^k_{j\,(\alpha)})$ of degree m, given by

$$f(e_i, \varepsilon_\alpha) = \sum_{j=1}^{2m} C^j_{i\,\alpha}\, e_j, \qquad F(\mathbf{e}_j, \varepsilon_\alpha) = \sum_{k=1}^{m} C^k_{j\,(\alpha)}\, \mathbf{e}_k, \tag{1.3}$$

relative to the basis $\{e_i\}$ for $\mathbb{R}^{2m}$, $\{\varepsilon_\alpha\}$ for $\mathbb{R}^p$, $\{\mathbf{e}_j\}$ for $\mathbb{C}^m$. Operating the bijection ρ on both sides of $f(e_{2j-1}, \varepsilon_\alpha) = \Sigma_{k=1}^m (C_{2j-1}{}^{2k-1}{}_\alpha\, e_{2k-1} + C_{2j-1}{}^{2k}{}_\alpha\, e_{2k})$, we have $F(\mathbf{e}_j, \varepsilon_\alpha) = \Sigma_{k=1}^m (C_{2j-1}{}^{2k-1}{}_\alpha + i\, C_{2j-1}{}^{2k}{}_\alpha)\mathbf{e}_k$. On the other hand, $iF(\mathbf{e}_j, \varepsilon_\alpha) = \Sigma_{k=1}^m (C_{2j}{}^{2k-1}{}_\alpha + i C_{2j}{}^{2k}{}_\alpha)\mathbf{e}_k$, from which it follows that $F(\mathbf{e}_j, \varepsilon_\alpha) = \Sigma_{k=1}^m (C_{2j}{}^{2k}{}_\alpha - i C_{2j}{}^{2k-1}{}_\alpha)\mathbf{e}_k$. Thus the relation between the system of real matrices C_α and that of complex matrices $C_{(\alpha)}$ is expressed as

$$C^k_{j\,(\alpha)} = C_{2j-1}{}^{2k-1}{}_\alpha + iC_{2j-1}{}^{2k}{}_\alpha = C_{2j}{}^{2k}{}_\alpha - iC_{2j}{}^{2k-1}{}_\alpha. \tag{1.4}$$

PROPOSITION 3. *A pair* $(\mathbb{C}^m(\Lambda), \mathbb{R}^p(\Sigma))$ *forms a hermitian pre-Hurwitz pair if and only if there exists a set* $\{S_{(\mu)}\}_{\mu=1}^{p-1}$ *of complex matrices* $S_{(\mu)}$ *of degree* m *such that*

$$S_{(\mu)}S_{(\nu)} + S_{(\nu)}S_{(\mu)} = -2\,\tilde{\eta}_{\mu\nu}\, I, \qquad S_{(\mu)}{}^* = -\Lambda\, S_{(\mu)}\, \Lambda,$$

$(\mu, \nu = 1, 2, \ldots, p-1)$, *where* $\tilde{\eta}_{\mu\nu} = \eta_{\mu\nu}/\eta_{pp}$ *for* $\Sigma = (\eta_{\alpha\beta})$ *and* $S_{(\mu)}{}^* = {}^t\bar{S}_{(\mu)}$, *the Hermitian conjugate of* $S_{(\mu)}$.

<u>Proof</u>. Suppose that the pair forms a hermitian pre-Hurwitz pair. There exists a bilinear mapping F satisfying (1.2). Putting $\tilde{x} = \Sigma_{j=1}^{m} z^j e_j$ with complex coefficients z^j, the complex vector $F(\tilde{x},y)$ reduces to $F(\tilde{x},y) = \Sigma_{j=1}^{m} \Sigma_{\alpha=1}^{p} y^\alpha z^j F(e_j, \varepsilon_\alpha)$. By (1.2) and (1.3) the relation

$$\sum_{\alpha,\beta} y^\alpha y^\beta \{ \langle z^j C^k_{j(\alpha)} e_k,\ z^h C^m_{h(\beta)} e_m \rangle - \eta_{\alpha\beta} \langle \tilde{x}, \tilde{x} \rangle \} = 0$$

holds for any vector y belonging to the space $\mathbb{R}^p$. Hence

$$\sum_{j,h} \bar{z}^j z^h \{ (\bar{C}^k_{j(\alpha)} C^m_{h(\beta)} + \bar{C}^k_{j(\beta)} C^m_{h(\alpha)}) \langle e_k, e_m \rangle - 2\eta_{\alpha\beta} \langle e_j, e_k \rangle = 0$$

holds for any $\tilde{x} = (z^j)$. Therefore, the relation

$$C_{(\alpha)}{}^* \Lambda C_{(\beta)} + C_{(\beta)}{}^* \Lambda C_{(\alpha)} = 2\eta_{\alpha\beta} \Lambda \qquad (\alpha, \beta = 1, 2, \ldots, p) \quad (1.6)$$

holds. Here we put

$$S_{(\mu)} = C_{(p)}{}^{-1} C_{(\mu)} \qquad (\mu = 1, 2, \ldots, p-1). \quad (1.7)$$

Since $C_{(p)}{}^{-1} = \eta_{pp} \Lambda C_{(p)}{}^* \Lambda$, then $S_{(\mu)}{}^* = (\eta_{pp} \Lambda C_{(p)}{}^* \Lambda C_{(\alpha)})^* = -(\eta_{pp} \Lambda C_{(\alpha)}{}^* \Lambda C_{(p)})^* = -\Lambda S_{(\mu)} \Lambda$. Since $C_{(\mu)}{}^*$ reduces to $C_{(\mu)}{}^* = S_{(\mu)}{}^* C_{(p)}{}^* = -\Lambda S_{(\mu)} \Lambda C_{(p)}{}^*$, then the desired relation in (1.5) is obtained by means of (1.6).

Conversely, suppose that there exists a set $\{S_{(\mu)}\}$ of matrices $S_{(\mu)}$ of degree m satisfying (1.5). We choose a matrix $C_{(p)}$ such that $C_{(p)}{}^{-1} = \eta_{pp} \Lambda C_{(p)}{}^* \Lambda$. Actually, when $\eta_{pp} = 1$, we may choose $C_{(p)} = I$, and when $\eta_{pp} = -1$, we have $r'' = s''$, so we may choose

$$C_{(p)} = \begin{bmatrix} 0 & & & & 1 \\ & & & \overset{r''}{\cdot} & \\ & & 1 & & \\ & & -1 & & \\ & \underset{r''}{\cdot} & & & \\ -1 & & & & 0 \end{bmatrix}.$$

Accordingly, we can define $C_{(\mu)}$ by $C_{(\mu)} = C_{(p)} S_{(\mu)}$ in terms of $S_{(\mu)}$. Consequently, the set $\{C_{(\alpha)}\}_{\alpha=1}^{p}$ determines the bilinear mapping F by (1.3) and satisfies the condition (1.6), hence (1.2).

The relation $S_{(\mu)} S_{(\nu)} + S_{(\nu)} S_{(\mu)} = -2\tilde{\eta}_{\mu\nu} I$ mentioned above determines the $(p-1)$-dimensional metric K-vector space (W_K, Q) with a quadratic form Q, given by $Q(\sum_\mu a_\mu S_{(\mu)}) = (\sum_\mu a_\mu S_{(\mu)})^2$ $(= \sum_\mu \tilde{\eta}_{\mu\mu} a_\mu^2 I)$ for $a_\mu \in K$, where $K = \mathbb{R}$ or $\mathbb{C}$. The space (W_K, Q) gives rise to the Clifford algebra $C(W_K, Q)$ over the coefficient field K.

COROLLARY 1 ([2], [3]). *A pair* $(\mathbb{R}^n(\Lambda), \mathbb{R}^p(\Sigma))$ *forms a pre-Hurwitz pair if and only if there exists a set* $\{S_\mu\}_{\mu=1}^{p-1}$ *of real matrices* S_μ *of degree* n *such that*

$$S_\mu S_\nu + S_\nu S_\mu = -2\,\tilde{\eta}_{\mu\nu} I, \quad {}^tS_\mu = -\Lambda S_\mu \Lambda \qquad (\mu,\nu = 1,2,\ \ldots,\ p-1) \tag{1.8}$$

In particular, in the case of $\Lambda = I_n$, $\Sigma = I_p$, *a pair in question is a pre-Hurwitz pair if and only if there is a set* $\{S_\mu\}_{\mu=1}^{p-1}$ *(in* $M_n(\mathbb{R})$*) such that*

$$S_\mu S_\nu + S_\nu S_\mu = -2\,\delta_{\mu\nu} I, \qquad {}^tS_\mu = -S_\mu . \tag{1.9}$$

2. PRE-HURWITZ PAIRS IN CASE OF $(n,p) = (4,4)$

The purpose of this section is to classify pre-Huwwitz pairs in the case of $n = p = 4$ and then to learn whether the Minkowski space appears as a member of some pre-Hurwitz pair under the condition $p \leq n = 4$.

(1°) $(\mathbb{R}(\Lambda), \mathbb{R}(\Sigma))$: This forms a Hurwitz pair if and only if $\Lambda = \pm 1$ and $\Sigma = 1$.

(2°) $(\mathbb{R}^2(\Lambda), \mathbb{R}^2(\Sigma))$: This forms a pre-Hurwitz pair if and only if one of the following conditions holds: (1) $\Lambda = \Sigma = I$, (2) $\Lambda = -\Sigma = -I$, (3) $\Lambda = \Sigma = \mathrm{diag}(1, -1)$. The first two are Hurwitz pairs, but the third one is not a Hurwitz pair. The verification of this is similar to that of the following discussion.

(3°) $(\mathbb{R}^4(\Lambda), \mathbb{R}^4(\Sigma))$: We shall divide the verification into five cases.

Case (i) $\Lambda = \Sigma = I$. We know that $(\mathbb{R}^4(I), \mathbb{R}^4(I))$ is a Hurwitz pair. There exists a triple $\{S_1, S_2, S_3\}$ of matrices S_μ defined by

$$S_1 = \left[\begin{array}{cc|cc} & -1 & & \\ 1 & & & \\ \hline & & & 1 \\ & & -1 & \end{array}\right], \quad S_2 = \left[\begin{array}{cc|cc} & & -1 & \\ & & & -1 \\ \hline 1 & & & \\ & 1 & & \end{array}\right], \quad S_3 = \left[\begin{array}{cc|cc} & & & -1 \\ & & 1 & \\ \hline & -1 & & \\ 1 & & & \end{array}\right],$$

satisfying (1.9) in Corollary 1. The bilinear mapping f is given by $f(x,y) = (y_1S_1 + y_2S_2 + y_3S_3 + y_4 I)\,x$ for any column vector x and $y = (y_\alpha)$.

Case (ii) $\Lambda = -\Sigma = -I$. This is obtained by Proposition 1.

Case (iii) $\Lambda = \Sigma = \mathrm{diag}(1, 1, -1, -1)$. The matrix $\Sigma = (\eta_{\alpha\beta})$ implies $(-\tilde{\eta}_{\mu\nu}) = \mathrm{diag}(-1, 1, 1)$. There exists a triple $\{S_\mu\}$ defined by

$$S_1 = \left[\begin{array}{cc|cc} & -1 & & \\ 1 & & & \\ \hline & & & -1 \\ & & 1 & \end{array}\right], \quad S_2 = \left[\begin{array}{cc|cc} & & 1 & \\ & & & -1 \\ \hline 1 & & & \\ & -1 & & \end{array}\right], \quad S_3 = \left[\begin{array}{cc|cc} & & & 1 \\ & & 1 & \\ \hline & 1 & & \\ 1 & & & \end{array}\right],$$

satisfying (1.8) in Corollary 1. This pair is a pre-Hurwitz pair. But it is not a Hurwitz pair, because the linear transformation $f(\cdot, y)$ for each y has an invariant subspace $\{{}^t(x_1\ x_2\ x_1\ x_2) \mid x_i \in \mathbb{R}\}$ in $\mathbb{R}^4$.

Case (iv) $\Lambda = \mathrm{diag}(1, -1, -1, -1)$. Suppose that there exists a triple $\{S_\mu\}$ of matrices of degree 4 satisfying (1.8). Then S_μ have the form

$$S_\mu = \begin{bmatrix} 0 & a & b & c \\ a & 0 & p & q \\ b & -p & 0 & r \\ c & -q & -r & 0 \end{bmatrix},$$

from which it follows that $S_\mu^2 \neq \pm I$. This is a contradiction.

Case (v) $\Lambda = \mathrm{diag}(1, 1, -1, -1)$, $\Sigma = \mathrm{diag}(1, -1, -1, -1)$. Suppose that $(\mathbb{R}^4(\Lambda), \mathbb{R}^4(\Sigma))$ is a pre-Hurwitz pair. The matrices S_μ $(\mu = 1, 2, 3)$ described in (1.8) are of the form:

$$S_\mu = \begin{bmatrix} a_\mu J & k_\mu X_\mu \\ k_\mu {}^t X_\mu & \varepsilon a_\mu J \end{bmatrix}, \text{ where } X_\mu = \begin{bmatrix} x_\mu & y_\mu \\ -\varepsilon y_\mu & \varepsilon x_\mu \end{bmatrix}, \quad x_\mu^2 + y_\mu^2 = 1,$$

and, in addition, $k_\mu^2 = a_\mu^2 + 1$, $J = \begin{bmatrix} 0 & -1 \\ 1 & 0 \end{bmatrix}$, $\varepsilon = \pm 1$. It follows from the equality of the (1,1)-components of both sides of the first matrix-equation in (1.8) that $a_\mu a_\nu = k_\mu k_\nu (x_\mu x_\nu + y_\mu y_\nu)$ $(\mu \neq \nu)$. Putting $x_\mu = \cos\theta_\mu$, $y_\mu = \sin\theta_\mu$, and $\lambda_\mu = a_\mu / k_\mu$, we have $\lambda_\mu \lambda_\nu = \cos(\theta_\mu - \theta_\nu)$. In particular, $\lambda_2\lambda_3 = \cos\{(\theta_1 - \theta_3) - (\theta_1 - \theta_2)\} = (\lambda_1\lambda_3)(\lambda_1\lambda_2) + \sin(\theta_1 - \theta_3)\sin(\theta_1 - \theta_2)$, from which the relation $(\lambda_1\lambda_2)^2 + (\lambda_2\lambda_3)^2 + (\lambda_3\lambda_1)^2 - 2(\lambda_1\lambda_2\lambda_3)^2 = 1$ is obtained. This equality holds only when $\lambda_1^2 = \lambda_2^2 = \lambda_3^2 = 1$. On the other hand, we have $\lambda_\mu^2 = a_\mu^2/(a_\mu^2 + 1) < 1$. This is a contradiction. Thus, we have

<u>THEOREM 1</u>. *A pair* $(\mathbb{R}^4(\Lambda), \mathbb{R}^4(\Sigma))$ *of pseudo-Euclidean spaces forms a pre-Hurwitz pair if and only if one of the following conditions holds:* (1) $\Lambda = \Sigma = I$, (2) $\Lambda = -\Sigma = -I$, (3) $\Lambda = \Sigma = \mathrm{diag}(1, 1, -1, -1)$. *The first two are Hurwitz pairs, but the third one is not a Hurwitz pair.*

We have the following corollary in consideration of Case (iv) in the verification of Theorem 1.

<u>COROLLARY 2</u>. *There does not exist any pre-Hurwitz pair in cases of* $2 \leq p \leq n = 4$ *possessing the Minkowski space as a member of the pair. The pair* $(\mathbb{R}^4(\Lambda), \mathbb{R}^1(1))$ *of the Minkowski space and the real line is a pre-Hurwitz pair.*

3. THE EXISTENCE OF A HERMITIAN PRE-HURWITZ PAIR DETERMINING THE MINKOWSKI SPACE

In this section, we are to deal with the problem of finding a hermitian pre-Hurwitz pair which determines the Minkowski space by the quartet $\{S_{(\mu)}\}$ of certain complex matrices, described in Proposition 3.

First, we note that the Hurwitz pair $(\mathbb{R}^4(I), \mathbb{R}^4(I))$ related to quaternions ((3°) – Case (i) in Sect. 2) can be interpreted as follows. The pair is identified with $(\mathbb{C}^2(I), \mathbb{R}^4(I))$, in a sense of Proposition 2.

PROPOSITION 4. *The pair* $(\mathbb{C}^2(I), \mathbb{R}^4(I))$ *of the positive definite Hermitian space* $\mathbb{C}^2(I)$ *and the Euclidean space* $\mathbb{R}^4(I)$ *forms a hermitian pre-Hurwitz pair and determines the negative definite metric* $\mathbb{R}$*-vector space* $\Sigma_{a=1}^3 \mathbb{R}(i\sigma_a)$, *where* σ_a *denote the Pauli matrices defined by*

$$\sigma_1 = \begin{bmatrix} 0 & 1 \\ 1 & 0 \end{bmatrix}, \quad \sigma_2 = \begin{bmatrix} 0 & -i \\ i & 0 \end{bmatrix}, \quad \sigma_3 = \begin{bmatrix} 1 & 0 \\ 0 & -1 \end{bmatrix}. \tag{3.1}$$

Proof. We define the triple $\{S_{(a)}\}_{a=1}^3$ by

$$S_{(a)} = i\sigma_a \qquad (a = 1, 2, 3). \tag{3.2}$$

It follows from the properties of σ_a that the conditions $S_{(a)}S_{(b)} + S_{(b)}S_{(a)} = -2\delta_{ab} I$ and $S^*_{(a)} = -S_{(a)}$ hold. Hence, the pair is a hermitian pre-Hurwitz pair by Proposition 3. Its bilinear mapping F has the form $F(x,y) = (\Sigma_{a=1}^3 y_a(i\sigma_a) + y_4 I)\tilde{x}$ for any column vector $\tilde{x} \in \mathbb{C}^2$ and $y = (y_\alpha) \in \mathbb{R}^4$.

The Dirac γ-matrices γ^μ $(\mu = 0, 1, 2, 3)$ are characterized by the relation $\gamma^\mu\gamma^\nu + \gamma^\nu\gamma^\mu = 2\eta^{\mu\nu} I$, $(\eta^{\mu\nu}) = \mathrm{diag}(1, -1, -1, -1)$, in the space $M_4(\mathbb{C})$ of complex matrices of degree 4. The Pauli-Dirac represention of γ^μ is given by

$$\gamma^0 = \begin{bmatrix} I & 0 \\ 0 & -I \end{bmatrix}, \quad \gamma^a = \begin{bmatrix} 0 & \sigma_a \\ -\sigma_a & 0 \end{bmatrix} \qquad (a = 1, 2, 3). \tag{3.3}$$

It follows that $\gamma^{0*} = \gamma^0$, $\gamma^{a*} = -\gamma^a$, and that $\{\gamma^\mu\}$ determines the Minkowski space $M_{1,3} = \Sigma_{\mu=0}^3 \mathbb{R}\gamma^\mu$ of signature (1.3). Taking account of these Dirac matrices, we see that $(\mathbb{C}^m(\Lambda), \mathbb{R}^p(\Sigma))$ should be of $(m,p) = (4,5)$.

Now define the quartet $\{S_{(\mu)}\}$ by

$$S_{(\mu)} = i\gamma^\mu \qquad (\mu = 0, 1, 2, 3), \tag{3.4}$$

and $\Lambda = \Lambda_{2,2}$, $\Sigma = \Sigma_{2,3}$. The matrix $\Sigma_{2,3}$ implies $(-\tilde{\eta}_{\mu\nu}) =$

diag(-1, 1, 1, 1), hence, (1.5) is satisfied. Then, by Proposition 3, the pair $(\mathbb{C}^4(\Lambda_{2,2}), \mathbb{R}^5(\Sigma_{2,3}))$ is a hermitian pre-Hurwitz pair determining the Minkowski space $M_{3,1} = \Sigma_{\mu=0}^{3} \mathbb{R} S_{(\mu)}$ of signature (3.1). In this case, the bilinear mapping F can be written as $F(\tilde{x},y) = (\Sigma_{\mu=0}^{3} y_\mu i\gamma^\mu + y_4 I)\tilde{x}$ for $\tilde{x} \in \mathbb{C}^4$ and $y = (y_\alpha) \in \mathbb{R}^5$. Thus, we have

THEOREM 2. *The pair* $(\mathbb{C}^4(\Lambda_{2,2}), \mathbb{R}^5(\Sigma_{2,3}))$ *of the twistor space* $\mathbb{C}^4(\Lambda_{2,2})$ *and the pseudo-Euclidean space* $\mathbb{R}^5(\Sigma_{2,3})$ *of signature* (2.3) *forms a pre-Hurwitz pair and determines the Minkowski space* $M_{3,1} = \Sigma_{\mu=0}^{3} \mathbb{R}(i\gamma^\mu)$.

The matrix γ_5 is defined by $\gamma_5 = i\gamma^0\gamma^1\gamma^2\gamma^3$, which is customarily used in physics. According to the Pauli-Dirac representation of γ^μ, it has the form

$$\gamma_5 = \begin{bmatrix} 0 & I \\ I & 0 \end{bmatrix}. \tag{3.5}$$

It follows that $\gamma_5 = I$, $\gamma_5\gamma^\mu + \gamma^\mu\gamma_5 = 0$, and $\gamma_5^* = \gamma_5$. Now define the quintet $\{S_{(\mu)}\}$ by

$$S_{(\mu)} = i\gamma^\mu \quad (\mu = 0, 1, 2, 3), \qquad S_{(4)} = \gamma_5, \tag{3.6}$$

and $\Lambda = \Lambda_{2,2}$, $\Sigma = \Sigma_{2,4}$. The matrix $\Sigma_{2,4}$ implies $(-\tilde{\eta}_{\mu\nu}) =$ diag(-1, 1, 1, 1, 1); hence, the quintet satisfies (1.5). By Proposition 3 the pair $(\mathbb{C}^4(\Lambda_{2,2}), \mathbb{R}^4(\Sigma_{2,4}))$ forms a hermitian pre-Hurwitz pair determining the metric $\mathbb{R}$-vector space $M_{4,1} = \Sigma_{\mu=0}^{4} \mathbb{R} S_{(\mu)}$ of signature (4.1). Its bilinear mapping F has the form $F(\tilde{x},y) = (\Sigma_{\mu=0}^{3} y_\mu i\gamma^\mu + y_4 I + y_5\gamma_5)\tilde{x}$ for $\tilde{x} \in \mathbb{C}^4$ and $y = (y_\alpha) \in \mathbb{R}^6$. Consequently, we obtain

THEOREM 3. *The pair* $(\mathbb{C}^4(\Lambda_{2,2}), \mathbb{R}^6(\Sigma_{2,4}))$ *forms a hermitian pre-Hurwitz pair and determines the metric* $\mathbb{R}$*-vector space* $M_{4,1} = \Sigma_{\mu=0}^{3} \mathbb{R}(i\gamma^\mu) + \mathbb{R}\gamma_5$.

References

[1] HURWITZ, A.: 'Über die Komposition der quadratischen Formen', *Math. Ann.* 88 (1923), 1-25.

[2] ŁAWRYNOWICZ, J. and J. REMBIELIŃSKI: 'Pseudo-Euclidean Hurwitz pairs and generalized Fueter equations', in *Clifford algebras and their applications in mathematical physics, Proceedings, Canterbury 1985*, D. Reidel Publ. Co., Dordrecht-Boston-Lancaster 1986, pp.39.

[3] ——— and ———: 'Supercomplex vector spaces and spontaneous symmetry' in: *Seminari di Geometria 1984*, Università di Bologna, Bologna 1985, pp. 131-154.

Part III

Real Analytic Geometry

edited by

Stancho Dimiev

Institute of Mathematics
Bulgarian Academy of Sciences
Sofia, Bulgaria

Julian Ławrynowicz

Institute of Mathematics
Polish Academy of Sciences
Łódź, Poland

and

Karlheinz Spallek

Fakultät für Mathematik
Ruhr-Universität Bochum
Bochum, BRD

MORPHISMS OF KLEIN SURFACES AND STOILOW'S TOPOLOGICAL THEORY OF ANALYTIC FUNCTIONS

Cabiria Andreian Cazacu
University of Bucharest
Faculty of Mathematics
Bucureşti 1, Romania

ABSTRACT. It is proved (Theorem 1) that for every surface X (orientable or non-orientable, with border or without) there exists an interior transformation $T : X \to D$, where D denotes the closed disc. The Klein covering (X, T, D) is shown to be complete and it can present folds on ∂D. This generalizes the Stoilow theorem that for every orientable surface X without border there exists an interior transformation $T : X \to S$. The generalized theorem is then applied to prove the existence of a dianalytic structure on every surface (Theorem 2).

1. INTRODUCTION

In the Seminar he held at the Institute of Mathematics of the Romanian Academy in Bucharest as well as in various lectures he delivered at the University of Kiev in 1957, at the International Colloquium on the Theory of Functions, Helsinki 1957 [13] and at the 5th Congress of the Austrian Mathematicians, Innsbruck 1960 [14], Simion Stoilow initiated the study of the topological function theory for non-orientable Riemann surfaces, yet only the monograph 'Foundations of the theory of Klein surfaces' [2] by N.L. Alling and N. Greanleaf did initiate the systematic and general research in this direction.

The methods of Stoilow's topological theory of analytic functions and Riemann surfaces can usefully be applied in the study of the morphisms of Klein surfaces since, as we proved in [6], the non-constant morphisms of Klein surfaces are topologically equivalent to interior transformations in the sense of Stoilow (i.e. to continuous, open, and 0-dimensional (light) maps); this result generalizes Stoilow's topological characterization of non-constant analytic functions ([12], Ch. V, III, 5).

A Klein surface after Alling and Greenleaf is a pair consisting of an orientable or non-orientable two-manifold X, with border (boundary) or without border, and a dianalytic structure $\mathcal{X}$. It will be denoted by $\mathcal{X}$ and its border by ∂X ([2], Ch. 1, Sect. 2).

Klein surfaces are two-manifolds with countable basis and in what follows we shall call such manifolds *surfaces*. Further, for the sake of

J. Ławrynowicz (ed.), Deformations of Mathematical Structures, 235–246.

simplicity we shall only consider connected surfaces.

After Stoilow's definition of the Riemann covering ([12], Ch. II, I,2 and Ch. V, III, 4) we call a triple (X,T,Y), where $T: X \to Y$ is an interior transformation between the surfaces X and Y, a *Klein covering*.

In previous papers we studied the ramification of such a covering [4] as well as other properties related to the composition of morphisms [5] and to the lifting of dianalytic structures ([3], Sect. 2). Now, we concentrate on Stoilow's characterization of the Riemann surfaces ([10], [11], and [12], Ch. V, III, 4): A two-manifold X without border is homeomorphic to a Riemann surface or, equivalently, a two-manifold X without border is triangulable and orientable, iff there exists an interior transformation of X into the sphere S.

Namely, in Section 2 we shall generalize the following part of this characterization which is called in the present paper

STOILOW'S THEOREM. *For every orientable surface* X *without border there exists an interior transformation* $T: X \to S$.

Our main result is Theorem 1 from below. In Section 3 we apply this theorem in order to prove the existence of a dianalytic structure on every surface (Theorem 2).

Throughout this paper arc and curve mean Jordan arc and Jordan curve, respectively. It will result from the context whether they are considered sets of points or paths.

2. GENERALIZED STOILOW'S THEOREM

Stoilow's Theorem has soon received other proofs, among which that by M. Heins [7] and that by L.V. Ahlfors presented in R. Nevanlinna's monograph ([9], Ch. II, Sect. 4, 2.90) are not only very simple but also construct a surjective interior transformation T such that the Riemann covering (X,T,S) is *complete* in Ahlfors-Sario's sense ([1], Ch. I, 21.A).

In 1953 an extension of Stoilow'sTheorem to the non-orientable case was given by

R.J. WILLE'S THEOREM [16]. *If* X *is a surface without border, orientable or non-orientable, then there exists an interior transformation* $T: X \to P$, *where* P *is the real projective plane.*

Wille's transformation T is again surjective and the covering (X,T,P) complete.

Our aim is to extend these results to arbitrary bordered or unbordered surfaces and we shall prove

THEOREM 1. *For every surface* X *(orientable or non-orientable, with border or without) there exists an interior transformation* $T: X \to D$, *where* D *denotes the closed disc. The Klein covering* (X,T,D) *is complete and can present folds on* ∂D.

P r o o f. Observe that the theorem is known or follows immediately from *Alling-Greenleaf's Theorem* (Th. 1.7.2 in [2]) for compact surfaces and for surfaces without border: *each compact surface can be represented as a cover of the disc.* In fact Theorem 1.7.2 indicates several cases, the covered surface being sometimes the sphere, but the sphere S is the double Schottky cover of the disc and P is represented by this theorem as a double cover of D with a ramification of order 1 in the interior of D and a fold over ∂D. Thus Theorem 1.7.2 and the theorems by Stoilow and Wille prove Theorem 1 in the case mentioned above.

Therefore it remains to consider only the *non-compact bordered case* and we shall adapt Wille's method based on a Kerékjártó-Stoilow exhaustion of the non-compact surface X by a sequence of polyhedrons $\{X_n\}_{n\in\mathbb{N}}$ ([8], V, Sect. 1 and [12], Ch. IV, II, 2). For the orientable bordered case such an exhaustion was built by Ecaterina Visinescu-Gionea in 1967 in her doctoral dissertation [15] and her method can also be applied to the non-orientable case. By means of ramifications in the interior and identifications on the border, as in the proof of Theorem 1.7.2, we shall construct, successively with copies of disc D, a sequence of compact surfaces $\{Z_n\}_{n\in\mathbb{N}}$ such that for every $n\in\mathbb{N}$, Z_n and X_n will be homeomorphic and the homeomorphism $h_n : X_n \to Z_n$ together with the natural projection $p_n : Z_n \to D$ will define the interior transformation $T_n = p_n \circ h_n : X_n \to D$. The construction will be continued so that

$$Z_{n+1} \supset Z_n \quad \text{and} \quad h_{n+1}|X_n = h_n .$$

Thus a non-compact surface $Z = \bigcup_n Z_n$ will be obtained, being homeomorphic with X by a homeomorphism $h : X \to Z$, $h|X_n = h_n$ and having a natural projection $p : Z \to D$, $p|Z_n = p_n$. The map $T = p \circ h : X \to D$ will give the interior transformation we are looking for.

Now, we present the successive steps of this proof which can apply to all the non-compact surfaces bordered or not.

2.1. The exhaustion $\{X_n\}$

As in [15], we consider a triangulation of the surface X such that ∂X consists of sides of triangles. The triangulation will be replaced by subdivision whenever it is necessary.

A polyhedron Π on X forms a compact bordered surface with $\partial\Pi$ consisting of μ mutually disjoint curves, but if Π is considered a part of X these curves may be classified as follows:

1) curves B_j, $j = \overline{1,l}$, which are closed components of ∂X,

2) curves Γ_j, $j = \overline{1,m}$, which lie in the interior of X and consist of sides of triangles, each side belonging to only one triangle of Π,

3) curves C_j, $j = \overline{1,k}$, which are formed by pairs of arcs B_{ji} from ∂X and arcs Γ_{ji} from int X (except for the end points which lie on ∂X);

$l + m + k = \mu$. The set $\partial\Pi \cap \partial X$ will be denoted by $B\Pi$.

Proceeding as in [15] we construct an exhaustion sequence of polyhedrons $\{X_n\}_{n \in \mathbb{N}}$ of X with the properties:

1) For every n, $X_n \subset \operatorname{int} X_{n+1} \cup \operatorname{int} BX_{n+1}$, where $\operatorname{int} BX_{n+1}$ means the interior of the topological subspace BX_{n+1} of X.

2) $\bigcup_n X_n = X$.

3) There is no arc on $X \setminus \operatorname{int} X_n$ to join two distinct curves in ∂X_n, for any $n \in \mathbb{N}$.

4) When a point of the curve B of ∂X belongs to X_n, the whole curve B belongs to X_n.

The curves of type B, Γ, C and the arcs of type B, Γ in $\partial\Pi$ will be denoted by a supplementary index n for $\Pi = X_n$.

2.2. The covering of D by a compact surface

In order to construct the surfaces Z_n we need a uniform procedure to represent compact surfaces as coverings of D, while Alling and Greenleaf's Theorem 1.7.2 provides several cases, some boundary curves of the considered compact surface being mapped topologically on the whole ∂D and others covering twice arcs of ∂D. However, by passing from Z_n to Z_{n+1} we have to identify border points with the same projection on ∂D (see 2.3 below) and it is possible that points to be identified belong, according to Theorem 1.7.2, to curves of different types. Thus we cannot apply directly this theorem and we have to give new constructions even for the covering of the disc by compact surfaces. In the following we provide two such constructions: all the boundary curves of the compact surface are homeomorphically projected on the whole ∂D in Lemma 1 while all of them cover twice mutually disjoint arcs of ∂D in Lemma 2. Both lemmas will be used in 2.3 in order to obtain in two different ways the construction of Z for a non-compact surface X.

N o t a t i o n s. For an orientable or non-orientable compact surface X we denote the number of the boundary curves by $\mu = \mu(X) \geqq 0$ and the genus by $g = g(X) \geqq 0$ if X is orientable and by $\mathfrak{g} = \mathfrak{g}(X)$ otherwise; in the first case we set $\mathfrak{g} = \mathfrak{g}(X) = 2g$. If (X,T,D) is a complete Klein covering, we denote by n the number of the sheets and by r the total order of ramification.

LEMMA 1. *For every compact surface* X *there exists a complete Klein covering* (X,T,D), *each boundary curve of* X *being topologically projected onto* ∂D. *The covering may present folds.*

P r o o f. By the classical homeomorphy theorem of the compact surfaces it is sufficient to construct a Klein covering (Z,p,D), where Z consists of copies of D, which are glued together by ramification points interior to D and by identification of boundary arcs or curves, such that Z is orientable or non-orientable as X and has the same genus and the same number of boundary curves as X. Here p denotes again the natural projection $Z \to D$.

R e m a r k 1. Let (Z_o, p_o, D) be a complete Klein covering as in Lemma 1, where Z_o consists of n_o copies of D, p_o is the natural projection $Z_o \to D$, r_o the total ramification order, $\mu_o = \mu(Z_o)$ and $g_o = g(Z_o)$. If we glue to Z_o a new copy of D by means of two ramification points of the first order lying on the same sheet of the covering (Z_o, p_o, D), different from the ramification points of this covering and projected by p_o on two different interior points of D, and if we repeat this procedure l times, we obtain a compact surface Z and a complete Klein covering (Z,p,D), again as in Lemma 1, with $n = n_o + l$ sheets, $r = r_o + 2l$, $\mu = \mu(Z) = \mu_o + l$ and, by the Hurwitz formula, with $g = g(Z) = g_o$.

1) In order to construct an orientable surface Z with an arbitrary genus $g \geq 0$ and an arbitrary $\mu \geq 2$ we take for Z_o a surface consisting of two copies of D which are glued together by $r_o = 2g + 2$ ramification points of the first order. Evidently, $n_o = 2$, $\mu_o = 2$ and, again by the Hurwitz formula, $g_o = g$. If we take $l = \mu - 2 \geq 0$ we obtain the requested covering (Z,p,D) with $n = \mu$.

For an orientable surface Z with an arbitrary genus $g \geq 0$ and $\mu = 1$ or $\mu = 0$ we identify with Z_o one or two copies of D along the whole ∂D. The covering (Z,p,D) will have $n = 3$ or 4 and one or two folds over ∂D, respectively. (Obviously, in the particular cases $\mu = 1$, $g = 0$ and $\mu = 0$, $g = 0$ we also have simpler unramified coverings with $n = 1$ and $n = 2$, respectively.)

2) In the non-orientable case we begin with a double cover (Z_o, p_o, D) ramified in $g \geq 1$ interior points of D and with the two sheets identified along the whole ∂D. The surface Z_o will have $g_o = g \geq 1$ and $\mu_o = 0$. Remark 1 will give (Z,p,D) with $n = 2 + l$, the non-orientable surface Z having $\mu = l \geq 0$ boundary curves and the genus $g \geq 1$.

LEMMA 2. *For every compact surface* X *there exists a complete Klein covering* (X,T,D), *the boundary curves of* X *being projected as two-sheeted covers of disjoint arcs of* ∂D. *The covering* (X,T,D) *presents folds and the number of sheets is* 2 *or* 4 *in the orientable case (as the genus* $g = 0$ *or* $g \geq 1$*) and* 2 *in the non-orientable case.*

P r o o f. 1) In the orientable case consider again the double cover (Z_o, p_o, D) from Lemma 1.1), denote by D_1 and D_2 the two copies of D which form Z_o, and take two other copies D_3 and D_4. Identify ∂D_1 with ∂D_3 and ∂D_2 with ∂D_4, if $\mu = 0$, or μ mutually disjoint arcs of ∂D_2 with the same arcs of ∂D_4, if $\mu \geq 1$. Thus we obtain an orientable surface Z of an arbitrary genus $g \geq 0$ and an arbitrary number $\mu \geq 0$ of boundary curves.

In the case $g = 0$ the construction is simpler. We take for Z a surface consisting of two copies of D identified along the whole ∂D, if $\mu = 0$, or along μ mutually disjoint arcs of ∂D, if $\mu \geq 1$.

The fact that Z has the genus g follows from the Hurwitz formula for total coverings with folds, each end point of a fold being counted as a ramification of order 1/2 [4], or from the formula of addition of the characteristics.

The covering (Z,p,D) has the properties from Lemma 2.

2) In the non-orientable case the representation given by Alling and Greenleaf in Theorem 1.7.2 has these properties, so we shall use it, namely: the non-orientable surface Z with arbitrary $g \geqq 1$ and arbitrary $\mu \geqq 0$ is obtained from two copies of D glued together by g ramification points in $\operatorname{int} D$, and by the identification of the whole boundaries, if $\mu = 0$, or of μ mutually disjoint arcs on ∂D, if $\mu \geqq 1$.

Remark 2. In Lemma 2 the projections of the identified arcs on ∂D can arbitrarily be chosen if $\mu \geqq 1$.

Remark 3. In both lemmas the homeomorphism $h : X \to Z$ has a great degree of arbitrariness. Further we shall use the following modifications of h:

1) Let C be an arbitrary boundary curve of X and V an annular neighbourhood of C, separated from $X \setminus V$ by a curve C'. The homeomorphism h can be replaced by a homeomorphism $\tilde{h} : X \to Z$ such that h and $\tilde{h}$ coincide on $X \setminus V$, $\tilde{h}(C) = h(C)$, but $\tilde{h} : C \to h(C)$ is arbitrarily chosen, the curves $\tilde{h}(C)$ and $h(C)$ having, obviously, the same orientation if C is oriented.

2) Let $\Gamma = A_1 A_2$ be an arc of a boundary curve C of X. In a simply connected neighbourhood V of Γ let $A_1' A_2'$ and $A_1'' A_2''$ be arcs on C with the same orientation as $A_1 A_2$, such that $A_1' A_2'$ lies in the $\operatorname{int} A_1'' A_2''$ and $A_1 A_2$ in the $\operatorname{int} A_1' A_2'$. Further, let

$$X' = A_1' A_1 \; A_2 A_2' \quad \text{and} \quad X'' = A_1'' A_1' \; A_2' A_2''$$

be quadrilaterals in V, X' being separeted from $X \setminus X'$ by an arc Γ' with the end points A_1' and A_2' and X'' being separated from $X \setminus (X' \cup X'')$ by an arc Γ'' with the end points A_1'' and A_2''. We obtain from h a homeomorphism $\tilde{h} : \tilde{X} = X \setminus X' \to Z$ such that h and $\tilde{h}$ coincide on $X \setminus (X' \cup X'')$, $\tilde{h}(X'') = h(X' \cup X'')$, $\tilde{h}(\Gamma') = h(\Gamma)$ and $\tilde{h} : \Gamma' \to h(\Gamma)$ is arbitrarily chosen with $\tilde{h}(A_i') = h(A_i)$, $i = 1,2$.

3) We can replace h by a homeomorphism $\tilde{h} : X \to \tilde{Z}$, $\tilde{h} = \sigma \circ h$, where $\sigma : Z \to \tilde{Z}$ is a homeomorphism which reverses the orientation of all the boundary curves. For instance, σ may be a reflection in a fixed diameter of D for each sheet of Z in Lemma 1 and a reflection which interchanges the sheets D_1 and D_2, D_3 and D_4, if $n = 4$, or D_1 and D_2, if $n = 2$, in Lemma 2.

After any of these modifications we take once again the ancient notation, i.e. write h, X and Z for $\tilde{h}$, $\tilde{X}$ and $\tilde{Z}$, respectively.

2.3. The covering of D by a non-compact surface X

2.3.1. Construction of Z by means of Lemma 1. We begin with an exhaustion sequence $\{X_n\}_{n \in \mathbb{N}}$ of X as in 2.1. and, applying Lemma 1 to X_o, we obtain Z_o, h_o and $T_o = p_o \circ h_o$.

By induction, suppose that we have realized the step $n \geq 0$, constructing Z_n, h_n, and T_n. In order to pass to the step $(n+1)$ observe that $X_{n+1} \setminus X_n$ decomposes into a finite number of polyhedrons, which we denote by X_{nt}. There are two possibilities:

Case 1. X_{nt} is separated from X_n by a curve Γ_n of ∂X_n, which we denote by Γ_{nt}, since there is a bijective correspondence between the curves Γ_n of ∂X_n and the polyhedrons X_{nt} adjacent to them, $\Gamma_{nt} = \partial X_n \cap \partial X_{nt}$.

Case 2. X_{nt} is separated from X_n by an arc or a set of arcs Γ_{ntj} of a curve C_n of ∂X_n. The curve C_n can be adjacent to several polyhedrons X_{nt}, but for each of them $\partial X_n \cap \partial X_{nt} \subset C_n$. At the same time $\partial X_n \cap \partial X_{nt}$ is included in one or several curves $C^i_{nt} \subset \partial X_{nt}$. Each arc Γ_{ntj} belongs to a curve C^i_{nt} and it will also be denoted by Γ^i_{ntl}, $l = l(j)$, when it is regarded as an arc of C^i_{nt}. We abbreviate the notation by writing l instead of $l(j)$.

By means of Lemma 1, we construct for each X_{nt} the covering (Z_{nt}, p_{nt}, D) and denote by h_{nt} the corresponding homeomorphism $X_{nt} \to Z_n$. Our aim is to glue successively the polyhedrons Z_{nt} to Z_n and so to obtain Z_{n+1} and the homeomorphism $h_{n+1} : X_{n+1} \to Z_{n+1}$.

Case 1. We fix an orientation on Γ_{nt} and modify, if necessary, h_{nt} following Remark 3.3) and 3.1) so that for the new homeomorphism h_{nt} we have

$$p_{nt} \circ h_{nt} = p_n \circ h_n \quad \text{on } \Gamma_{nt}. \tag{1}$$

Then we identify the points $h_{nt}(P)$ and $h_n(P)$, for each point $P \in \Gamma_{nt}$.

Case 2. We fix an orientation on the curve C_n containing $\partial X_n \cap \partial X_{nt}$ and consider successively the different curves C^i_{nt} of ∂X_{nt}. For such a curve C^i_{nt}, let ν be the number of the arcs Γ_{ntj} of $C_n \cap C^i_{nt}$. We renumber these arcs in the order determined by the orientation on C_n and denote them again by Γ_{ntj}, $j = \overline{1,\nu}$. The arc Γ_{nt1} determines an orientation on C^i_{nt}. We write $\Gamma_{nt1} = \Gamma^i_{nt1}$ and renumber Γ^i_{ntl} in the order given by this orientation on C^i_{nt}, beginning with Γ^i_{nt1}, $l = \overline{1,\nu}$, so that now we have $\Gamma_{ntj} = \pm\Gamma^i_{ntl}$.

If for all the indices $j, l = j$, $\Gamma_{ntj} = \Gamma^i_{ntl}$ and the orientations of the projections of $\gamma_{ntj} = h_n(\Gamma_{ntj})$ and $\gamma^i_{ntl} = h_{nt}(\Gamma^i_{ntl})$ coincide on ∂D, then, modifying if necessary h_{nt} along C^i_{nt} by Remark 3.1), we obtain (1) on $C_n \cap C^i_{nt}$ and glue Z_{nt} to Z_n along the images of this intersection under h_{nt} and h_n.

In the opposite case, it is again possible to apply this procedure for some of the arcs Γ_{ntj}, but not for all of them, so that we shall present a general method of gluing Z_{nt} to Z_n with respect to Γ_{ntj} by means of Remark 3.2), in one of the following two ways:

(i) If $\Gamma_{ntj} = \Gamma^i_{ntl}$ and the projections of γ_{ntj} and γ^i_{ntl} have the same orientation on ∂D or if $\Gamma_{ntj} = -\Gamma^i_{ntl}$ and these projections have opposite orientations on ∂D, then we modify X_{nt} in the neighbourhood of Γ^i_{ntl} taking away two quadrilaterals

$$X'_{ntj} = A'_1 A_1 A_2 A'_2 \quad \text{and} \quad X''_{ntj} = A'_2 A'_1 A''_1 A''_2.$$

Here $A_1 A_2 = \Gamma_{ntj}$, $A_1 A'_1 A''_1$ and $A_2 A'_2 A''_2$ are arcs on BX_{nt}, while $A'_1 A'_2$ and $A''_1 A''_2$ are arcs in $\operatorname{int} X_{nt}$, except for their end points. After this modification the new X_{nt} will be delimited from X''_{ntj} by the new arc $\Gamma^i_{ntl} = A''_1 A''_2$ or $A''_2 A''_1$, as we have had $\Gamma^i_{ntl} = \Gamma_{ntj}$ or $-\Gamma_{ntj}$ at the beginning. We represent X'_{ntj} and X''_{ntj} by h'_{ntj} and h''_{ntj} topologically onto a copy of D denoted by Z'_{ntj} and Z''_{ntj}, respectively, so that we have

$$\begin{aligned} p'_{ntj} \circ h'_{ntj} &= p_n \circ h_n && \text{on } A_1 A_2, \\ p''_{ntj} \circ h''_{ntj} &= p'_{ntj} \circ h'_{ntj} && \text{on } A'_1 A'_2, \\ p_{nt} \circ h_{nt} &= p''_{ntj} \circ h''_{ntj} && \text{on } A''_1 A''_2, \end{aligned} \tag{2}$$

with the evident notations for the projections. Thus it is possible to glue Z'_{ntj}, Z''_{ntj} and Z_{nt} to Z_n, Z'_{ntj} and Z''_{ntj}, respectively.

R e m a r k 4. In order to obtain conditions (2) it is necessary that the projections of γ_{ntj} and γ^i_{ntl} do not cover the whole ∂D. This can be performed, for instance, by a modification as in Remark 3.1) applied to h_o and to all h_{nt}, such that every arc Γ on a curve C is mapped onto an arc γ with the length of its projection $< L/2$, where L is the length of ∂D.

(ii) If $\Gamma_{ntj} = -\Gamma^i_{ntl}$ and the projections of γ_{ntj} and of γ^i_{ntl} have the same orientation on ∂D or if $\Gamma_{ntj} = \Gamma^i_{ntl}$ and these projections have opposite orientations on ∂D, we suppose that

$$p_n(\gamma_{ntj}) \cap p_{nt}(\gamma^i_{ntl}) = \emptyset. \tag{3}$$

Then we take out of X_{nt} only one quadrilateral $X_{ntj} = A'_1 A_1 A_2 A'_2$, the new arc Γ^i_{ntl} being now $A'_2 A'_1$ or $A'_1 A'_2$, as we have had $\Gamma^i_{ntl} = -\Gamma_{ntj}$ or Γ_{ntj}. Indeed, we can choose the homeomorphism

$$h_{ntj} : X_{ntj} \to Z_{ntj},$$

where Z_{ntj} is a copy of D, such that

$$\begin{aligned} p_{ntj} \circ h_{ntj} &= p_n \circ h_n && \text{on } A_1 A_2, \\ p_{nt} \circ h_{nt} &= p_{ntj} \circ h_{ntj} && \text{on } A'_1 A'_2, \end{aligned} \tag{4}$$

which permits to glue Z_{ntj} and Z_{nt} to Z_n and Z_{ntj}, respectively. In Remark 5 below we prove that the hypothesis (3) can always be fulfilled, but even without (3) the gluing can be performed, for instance by taking out of X_{nt} three quadrilaterals and proceeding as in the point (i).

We repeat the procedure for all the indices j, i, t and, further, for all the curves C_n. In this way we obtain Z_{n+1} from Z_n and from all the surfaces Z_{nt}, and define h_{n+1} by the relations: $h_{n+1} = h_n$ on X_n and $h_{n+1} = h_{nt}$ on every X_{nt}.

The boundary of X_{n+1} consists of curves Γ_{n+1}, each of them passing from a boundary curve Γ of a polyhedron X_{nt}; curves B_{n+1} from BX_n or from BX_{nt} for some index t; and curves C_{n+1} formed by arcs Γ of one or several X_{nt} and by arcs B from BX_{nt}, but also from BX_n. While $p_{n+1} \circ h_{n+1}$ represents each curve Γ_{n+1} topologically onto ∂D, a curve C_{n+1} has this property only if it coincides with a curve C of X_{nt}. Otherwise, C_{n+1} is mapped by h_{n+1} on the border of several sheets of Z_{n+1}, each arc Γ_{n+1} of C_{n+1} together with the adjacent arcs, say B_n' and B_n'', from the corresponding BX_{nt} being mapped topologically onto an arc of the boundary of the same sheet.

Remark 5. It is useful to modify h_{n+1} on C_{n+1} by means of Remark 3.1) so that the projections of the images of all the arcs Γ_{n+1} of C_{n+1} are mutually disjoint. The modification of h_{n+1} is to be performed only on the arcs $B_n' \Gamma_{n+1} B_n''$ by keeping h_{n+1} fixed on the complementary arcs, so that it should not have an influence on X_n and on h_n. Let us number the arcs $B_n' \Gamma_{n+1} B_n''$ on C_{n+1}, denoting them by τ_k and setting $\Lambda_k = h_{n+1}(\tau_k)$ and $\lambda_k = p_{n+1}(\Lambda_k)$. We form disjoint maximal systems

$$S_m = \{\lambda_{k_1}, \ldots, \lambda_{k_\nu}\}, \quad \nu = \nu(m),$$

of arcs λ_k with nonvoid intersection

$$\lambda^m = \bigcap_{t=1}^{\nu} \lambda_{k_t},$$

divide each λ^m into $2\nu + 1$ equal arcs and denote the even subarcs obtained by λ_j^m, $j = \overline{1,\nu}$.

Let k be one of the indices k_t. We modify h_{n+1} on τ_k so that the new h_{n+1} maps Γ_{n+1} of τ_k onto

$$\Lambda_k^m = (p_{n+1}|\Lambda_k)^{-1}(\lambda_k^m)$$

and the arcs B_n', B_n'' of τ_k onto the two arcs of $\Lambda_k \setminus \text{int}\, \Lambda_k^m$, respectively. If we make this operation at each step n of our construction, then it is easy to modify h_{nt} on every C_{nt}^i, for all the indices l at once, in order to fulfil (3).

Summing up, we have constructed a complete Klein covering (X,T,D), each compact border of X being homeomorphic under T with ∂D, since it forms the border of a sheet, and each other border of X having points in infinitely many sheets of the covering. Such a non-compact border is decomposed into infinitely many arcs. Each of these arcs is homeomorphic under T with an arc of ∂D and its end points are ramifications of order 1/2 [4], as the end points of the folds produced in the gluing process.

2.3.2. Construction of Z by means of Lemma 2. With the same notations and device as in 2.3.1., we represent X_o and X_{nt} after Lemma 2, implying differences in the gluing method.

Case 1. The curve $h_n(\Gamma_{nt})$ on ∂Z_n covers doubly its projection arc a_1a_2 on ∂D. Denote by A_1 and A_2, respectively, the points of Γ_{nt} which correspond to a_1 and a_2 under the map $T_n = p_n \circ h_n$. They divide Γ_{nt} into two arcs, each of which is homeomorphic to a_1a_2 under the map T_n. We can choose $h_{nt}(\Gamma_{nt})$ on ∂Z_{nt} such that it also covers a_1a_2 doubly, and then, if necessary, modify h_{nt} by Remark 3.1) in order to fulfil (1). Thus the identification of Z_{nt} and Z_n becomes possible. The points A_1 and A_2 determine ramifications of the first order for the obtained covering $X_n \cup X_{nt} \to D$.

Case 2. According to Remarks 2 and 3.1) it is easy to verify by induction that we can choose h_o and h_{nt} such that for every n all the arcs Γ_{ntj} of all the curves C_n of ∂X_n are mapped by T_n onto mutually disjoint arcs of ∂D. Indeed, if this is true for a step $n \geq 0$, we choose h_{nt} such that all the curves of ∂X_{nt} are doubly projected under $p_{nt} \circ h_{nt}$ onto mutually disjoint arcs of the complement of $\bigcup_{t,j} p_n(\gamma_{ntj})$ with respect to ∂D and that $p_{nt} \circ h_{nt}$ is a homeomorphism for each arc Γ of a curve C of ∂X_{nt} and map all the arcs Γ of C onto mutually disjoint arcs of $p_{nt} \circ h_{nt}(C)$.

The gluing of Z_n and Z_{nt} along γ_{ntj} and γ^i_{ntl} can be realized now by Remark 3.2) with the use of a single quadrilateral $X_{ntj} = A_1' A_1 A_2 A_2'$, whose image Z_{ntj} will consist, by Lemma 2, of two copies of the disc D identified along a convenient subarc of ∂D. Here the same two cases as in 2.3.1. appear:

(i) $\Gamma_{ntj} = A_1 A_2$, $\Gamma^i_{ntl} = A_1' A_2'$ or $A_2' A_1'$

as Γ_{ntj} has been Γ^i_{ntl} or $-\Gamma^i_{ntl}$ at the beginning of the modification,

$$p_n(\gamma_{ntj}) = a_1 a_2 \quad \text{and} \quad p_{nt}(\gamma^i_{ntl}) = a_1' a_2' \quad \text{or} \quad a_2' a_1',$$

respectively, and the arcs $a_1 a_2$ and $a_1' a_2'$ have the same orientation on ∂D. We choose Z_{ntj} so that its border Λ is doubly projected by p_{ntj} onto an arc λ of ∂D which contains in its interior the arcs $a_1 a_2$ and $a_1' a_2'$. Further, we choose on Λ the arcs $\alpha_1\alpha_2$ and $\alpha_1'\alpha_2'$, which are homeomorphic under p_{ntj} to a_1a_2 and $a_1'a_2'$; one on a branch of Λ with respect to the covering $(\Lambda, p_{ntj}, \lambda)$ and the other on the other branch, since $\alpha_1, \alpha_2, \alpha_2', \alpha_1'$ have to determine a cyclic order on Λ. Then it is again possible by Remark 3.1) to choose h_{ntj} such that (4) is verified and to glue successively Z_{ntj} and Z_{nt} to Z_n and Z_{ntj}, respectively.

Evidently, there are special cases when the gluing is possible directly without the use of X_{ntj}.

(ii) With the previous notations the arcs a_1a_2 and $a_2'a_1'$ will have the same orientation this time. By construction we take them disjoint. Hence the procedure from (i) can be applied again, with the difference that $\alpha_1\alpha_2$ and $\alpha_2'\alpha_1'$ have to be taken on the same branch of Λ with respect to the covering $(\Lambda, p_{ntj}, \lambda)$.

In this way, we have obtained for every non-compact surface X a complete covering (X,T,D) with folds such that every compact border of X is projected as a double cover of an arc of ∂D, each other border having common arcs with infinitely many sheets of the covering.

3. DIANALYTIC STRUCTURES

One of the most important applications of Theorem 1 consists in the introduction of a dianalytic structure on any surface X, which completes Theorem 1.7.1 in [2].

Let us consider the disc D with its natural structure of Klein surface. The existence of an interior transformation $T : X \to D$, given by Theorem 1, together with the local behaviour of such a transformation, which is that of a local morphism of Klein surfaces [6], assure by means of Theorem 1.5.2 in [2] the existence of a dianalytic structure $\mathcal{X}$ on X for which T is a morphism $\mathcal{X} \to D$. Thus we have proved

THEOREM 2. *There exists on each surface* X *a dianalytic structure* $\mathcal{X}$, *which organizes* X *as a Klein surface*.

References

[1] AHLFORS, L.V. and SARIO, L.: *Riemann surfaces*, Princeton University Press, Princeton, New Jersey 1960.

[2] ALLING, N.L. and GREENLEAF, N.: 'Foundations of the theory of Klein surfaces', *Lecture Notes in Math.* 219, Springer-Verlag, Berlin-Heidelberg-New York 1971.

[3] ANDREIAN CAZACU, C.: 'Betrachtungen über rumänische Beiträge zur Theorie der nicht orientierbaren Riemannschen Flächen', *Analele Univ. Bucuresti* 31 (1982), 3-13.

[4] ——: 'Ramification of Klein coverings', *Ann. Acad. Scient. Fenn., A.I. Math.* 10 (1985), 47-56.

[5] ——: 'On the morphisms of Klein surfaces', *Rev. Roum. de Math. Pures et Appl.* 31 (1986), 461-470.

[6] ——: 'Interior transformations between Klein surfaces', *ibid.* 33 (1988), 21-26.

[7] HEINS, M.: 'Interior mappings of an orientable surface into S^2', *Proc. Amer. Math. Soc.* 3 (1951), 951-952.

[8] KERÉKJÁRTÓ, B.: *Vorlesungen über Topologie I*, Springer-Verlag, Berlin 1923.

[9] NEVANLINNA, R.: *Uniformisierung*, Springer-Verlag, Berlin-Göttingen-Heidelberg 1953.

[10] STOILOW, S.: 'Sur la caractérisation topologique des surfaces de Riemann', *C. R. Acad. Sci. Paris* 200 (1935), 189-190.

[11] ——: 'Sur les transformations intérieures et la caractérisation topologique des surfaces de Riemann', *Compositio Math.* 3 (1936), 435-440.

[12] ——: *Leçons sur les principes topologiques de la théorie des fonctions analytiques*, Gauthier-Villars, Paris 1938, II éd. 1956, éd. russe 1964.

[13] ——: 'Sur la théorie topologique des recouvrements riemanniens', *Ann. Acad. Sci. Fenn.* 250,35 (1958), 1-7.

[14] ——: 'Einiges über topologische Funktionentheorie auf nicht orientierbaren Flächen', *Revue Roum. de Math. Pures et Appl.* 19, 4 (1974), 503-506.

[15] VIŞINESCU, E.: 'Topological properties of the bordered Riemann surfaces', *ibid.* 15,8 (1970), 1281-1297.

[16] WILLE, R.J.: 'Sur la transformation intérieure d'une surface non orientable dans le plan projectif, *Indagationes Math.* 56 (1953), 63-65.

GENERALIZED GRADIENTS AND ASYMPTOTICS OF THE FUNCTIONAL TRACE

Thomas P. Branson and Bent Ørsted
Department of Mathematics
The University of Iowa
Iowa City, IA 52242 USA

Department of Mathematics
Odense University
DK-5230 Odense M, Denmark
and
Department of Mathematics
The University of Iowa
Iowa City, IA 52242 USA

ABSTRACT. Let G be a generalized gradient on a compact, Riemannian n-manifold M, carrying an $O(n)$-irreducible tensor bundle F to another such bundle E, and suppose that G^*G is elliptic but not conformally covariant. Let Tr_0 denote the trace of the compression to the Hodge sector $\mathcal{R}(G)$ in the decomposition $L^2(E) = \mathcal{R}(G) \oplus \mathcal{N}(G^*)$. Then if $\omega \in C^\infty(M)$, $\mathrm{Tr}_0\, \omega \exp(-t\, G G^*)$ has a small-time asymptotic expansion in powers of t, plus a $\log t$ term in the case of even n. All coefficients except that of t^0 are integrals of local expressions. For $G = d: C^\infty(M) \to C^\infty(\Lambda^1(M))$ and $n = 4$, the coefficient of $\log t$ is nonzero for some ω whenever the scalar curvature is not constant.

1. INTRODUCTION. ASYMPTOTICS OF THE FUNCTIONAL TRACE

Let (M,g) be a smooth, compact, Riemannian manifold of dimension $n \geq 2$. Suppose that F is a Riemannian vector bundle over M, and that $D: C^\infty(F) \to C^\infty(F)$ is a formally self-adjoint differential operator with positive definite leading symbol. In particular, D is elliptic and has even order 2ℓ; we shall assume that $\ell > 0$. Suppose further that $B: C^\infty(F) \to C^\infty(F)$ is an auxiliary differential operator of order $b \geq 0$. Then (see [G], Theorem 1.7.7):

<u>THEOREM 1.1</u>. *Let* (M,g), n, F, D, ℓ, B, *and* b *be as above. Then* $B\exp(-tD)$ *is an infinitely smoothing operator on* $L^2(F)$, *with a kernel function* $H(t,x,y) = H[B\exp(-tD)](t,x,y) \in F_x \otimes F^*_y$ *which is smooth on* $(t,x,y) \in (0,\infty) \times M \times M$. *On the diagonal* $\{y = x\}$, *the fiberwise trace of* H *has an asymptotic expansion*

---

*TB supported by N.S.F. (U.S.) Grant DMS-8696098.

Both authors supported by NATO Collaborative Research Grant 720/84.

J. Ławrynowicz (ed.), Deformations of Mathematical Structures, 247–262.

$$\operatorname{tr}_{F_x} H(t,x,x) \sim \sum_{i=-[b/2]}^{\infty} t^{(2i-n)/2\ell} U_i(x), \qquad t \downarrow 0, \tag{1.1}$$

where the $U_i(x) = U_i[B \exp(-tD)](x)$ *are universal polynomial expressions in the jets of* D *and* B *satisfying certain homogeneity conditions.*

The unexplained "homogeneity conditions" will be spelled out (below, in Sec. 2) only for "geometric" operators D and B. The prototypical case of (1.1) is that in which D is the Laplacian on functions and $B = 1$; this is the classical *Minakshisundaram-Pleijel expansion* [MP, BGM]. (1.1) can be integrated:

THEOREM 1.2 ([G], Sec. 1.6 and Theorem 1.7.7). *In the setting of Theorem 1.1,* $B \exp(-tD)$ *is trace class on* $L^2(F)$ *for all* $t > 0$, *and its* L^2 *trace has the asymptotic expansion*

$$\begin{aligned}\operatorname{Tr}_{L^2(F)} B \exp(-tD) &= \int_M \operatorname{tr}_{F_x} H(t,x,x)\,dx \\ &\sim \sum_{i=-[b/2]}^{\infty} t^{(2i-n)/2\ell} \int_M U_i(x)\,dx, \qquad t \downarrow 0,\end{aligned} \tag{1.2}$$

where dx *is the Riemannian measure.*

It is natural to ask whether small-time asymptotic expansions like (1.2) exist when the ellipticity/positivity condition on D is relaxed. For example, D might be the "Laplacian" of some sub-Riemannian geometry [Str] on a compact manifold; the case of CR geometry has been the subject of intensive recent work [BGS, FS, JS, Sa, Sta]. In his recent thesis, Xu [X] answers this question in the case in which B is the identity and D is a member of a large class which includes operators that are locally "sums of squares of vector fields" in the sense of Hörmander [Hö]. Xu's expansions are of the form

$$\operatorname{Tr}_{L^2} \exp(-tD) \sim \sum_\alpha c_\alpha t^{m_\alpha} (\log t)^{n_\alpha} + \sum_\alpha d_\alpha t^{p_\alpha}, \qquad t \downarrow 0,$$

where $\{m_\alpha\}$ and $\{p_\alpha\}$ are sequences of rational numbers tending to $+\infty$, and the n_α are nonnegative integers. It is interesting that Hadamard's asymptotic expansion of the fundamental solution of the wave equation [Ha] enters in an essential way (see also [M]), since it was Hadamard's "transport equation" approach that originally inspired Minakshisundaram and Pleijel.

We shall go in a different direction and treat a class of non-elliptic operators already present in Riemannian geometry. A special case of our result concerns the familiar operators

$$d\colon C^\infty(M) \to C^\infty(\Lambda^1(M)), \qquad \delta = d^*\colon C^\infty(\Lambda^1(M)) \to C^\infty(M).$$

Suppose that $n > 2$, and let U_i be the coefficients in the Minakshisundaram-Pleijel expansion (1.1) for the heat semigroup of the Laplacian $\Delta = \delta d$ on functions:

$$H[\exp(-t\Delta)](t,x,x) \sim \sum_{i=0}^{\infty} t^{(2i-n)/2} U_i(x), \qquad t \downarrow 0.$$

Let $\omega \in C^{\infty}(M)$, and let Tr_0 denote the trace on the direct summand $\mathcal{R}(d)$ in the Hodge decomposition

$$L^2(\Lambda^1(M)) = \mathcal{R}(d) \oplus \mathcal{N}(\delta).$$

Then we have an asymptotic expansion involving the heat semigroup of the nonelliptic operator $d\delta$: $C^{\infty}(\Lambda^1(M)) \to C^{\infty}(\Lambda^1(M))$:

$$\mathrm{Tr}_0\, \omega \exp(-td\delta) \overset{t \downarrow 0}{\sim}$$

$$C(\omega) + \sum_{i=0}^{\infty} t^{(2i-n)/2} \{ \int \omega U_i - \frac{1}{n-2i} \int (\Delta\omega) U_{i-1} \}, \qquad n \text{ odd},$$

$$C(\omega) + \sum_{i \neq n/2} t^{(2i-n)/2} \{ \int \omega U_i - \frac{1}{n-2i} \int (\Delta\omega) U_{i-1} \}$$

$$+ \tfrac{1}{2}(\log t) \int (\Delta\omega) U_{\frac{n}{2}-1}, \qquad n \text{ even}.$$

Here $C(\omega)$ is a constant, not necessarily gotten by integrating a local expression. The $\log t$ term is not always zero; for example, on a compact Riemannian 4-manifold of nonconstant scalar curvature K, we get something nonzero when we put $\omega = K$, since $U_1 = K/6$.

We would like to thank Antoni Pierzchalski, Walter Seaman, and Tuong Ton-That for helpful discussions.

2. GEOMETRIC OPERATORS

In this section, we give a precise description of the "homogeneity conditions" mentioned in the statement of Theorem 1.1, in the case of operators which occur naturally in Riemannian geometry. We keep the assumptions of Sec. 1, and suppose further that F is a subbundle of the tensor bundle of some contravariant/covariant degree $\begin{bmatrix} p \\ q \end{bmatrix}$, defined by imposing symmetry and trace conditions.

Definition 2.1. A *local scalar invariant* S is a universal scalar-valued expression on Riemannian manifolds (M,g) built polynomially from g, its inverse $g^{\#} = (g^{\alpha\beta})$, the Riemann curvature tensor R, the Riemannian covariant derivative ∇, tensor product, contractions, and permutation of arguments (indices) in tensor expressions. S has *level* m if it obeys the dilation law

$$\bar{g} = A^2 g, \quad 0 < A \in \mathbb{R} \Rightarrow \bar{S} = A^{-m} S. \tag{2.1}$$

A differential operator $D: C^\infty(F) \to C^\infty(E)$, where E is a bundle of $\begin{bmatrix} r \\ s \end{bmatrix}$ tensors, is *geometric* if it is built polynomially from the ingredients above. D has *level* m if

$$\bar{g} = A^2 g, \quad 0 < A \in \mathbb{R} \Rightarrow \bar{D} = A^{-m+s-r-q+p} D. \qquad (2.2)$$

R e m a r k 2.2. Level m local scalar invariants can alternately be described as universal linear combinations of expressions

$$\mathrm{trace}(R_{ijk\ell|\alpha} \cdots R_{stuv|\beta}, \quad (2+|\alpha|)+\ldots+(2+|\beta|)=m \qquad (2.3)$$

where $\alpha, \ldots, \beta$ are multi-indices, indices after the bar denote covariant derivatives, and "trace" represents some partitioning of all the indices into pairs, the raising of one index in each pair, and contraction to a scalar, (See, e.g., [BFG], Sec. 5.7) m is necessarily even here. Similarly, each term in a level m geometric dif erential operator has "derivative count" m, where we view R as already containing two derivatives (of g).

It is an easy exercise to get from the statement of [G], Theorem 1.7.7 to:

THEOREM 2.3. *Suppose that the hypotheses of Theorem* 1.1 *are satisfied, that* D *and* B *are geometric operators, that the level (as well as the order) of* D *is* 2ℓ, *and that* B *has a consistent level* m. *Then the* U_i *appearing in* (1.1) *have level* $2i+m$.

R e m a r k 2.4. By the definitions, we must have $\mathrm{ord}\, B = b \leq m \in 2\mathbb{Z}$ so $2i+m$ is automatically even and nonnegative.

3. GENERALIZED GRADIENTS

Let F be a natural tensor bundle, as above, which is also irreducible under the action of the orthogonal group; that is, each fiber F_x carries an irreducible representation of the orthogonal group $O(g_x)$. The covariant derivative ∇ carries sections of F to sections of $T^*M \otimes F$, which in turn splits as a direct sum of natural tensor bundles irreducible under the orthogonal group. If E is one of these irreducible pieces and $P: T^*M \otimes F \to E$ is the orthogonal projection (a bundle map) we call $G = P\nabla : C^\infty(F) \to C^\infty(E)$ a *generalized gradient*. (See also [SW, F].) The simplest example involve objects very familiar in geometry: let $F = T^*M$. Then $T^*M \otimes F$, the bundle of covariant two-tensors, breaks up as a direct sum of three irreducible pieces: the two-forms $\Lambda^2(M)$, the "pure traces" $\mathbb{R}g$, and the trace-free symmetric two-tensors $(TFS)_2$. The corresponding generalized gradients are, respectively, $d/2$, $-(\mathrm{atr}\,\delta)/n$, and the *Ahlfors operator* S [A, P]. Here the *antitrace* atr carries a scalar c to cg and

$$S: \eta \to \tfrac{1}{2}(\nabla_\alpha \eta_\beta + \nabla_\beta \eta_\alpha + \tfrac{2}{n}(\delta\eta) g_{\alpha\beta}).$$

After identifying covectors with vectors with vectors through g, we recognize S as the operator whose kernel consists of the conformal vector fields: if $X^\alpha = g^{\alpha\beta}\eta_\beta$,

$$S\eta = \tfrac{1}{2} L_X g - \tfrac{1}{n}(\operatorname{div} X)\, g.$$

More generally, if $1 \le k \le n-1$, $T^*M \otimes \Lambda^k$ splits into irreducibles as

$$\Lambda^{k+1} \oplus \operatorname{atr} \Lambda^{k-1} \oplus (TFB)_{k+1},$$

where $\operatorname{atr}: \Lambda^{k-1} \to T^*M \otimes \Lambda^k$ is a bundle injection acting by

$$(\operatorname{atr}\eta)_{\alpha_1 \ldots \alpha_{k+1}} = \sum_{s=2}^{k+1} (-1)^s \eta_{\alpha_2 \ldots \hat{\alpha}_s \ldots \alpha_{k+1}} g_{\alpha_1 \alpha_s},$$

and $(TFB)_{k+1}$ is the bundle of covariant, totally trace-free $(k+1)$-tensors satisfying the generalized Bianchi identity

$$\sum_{s=1}^{k+1} (-1)^{k(s-1)} \theta_{\alpha_s \ldots \alpha_{k+1} \alpha_1 \ldots \alpha_{s-1}} = 0.$$

(See [F], Sec. 4 for a description of $(TFB)_{k+1}$ in terms of the representation theory of $O(n)$, or [B3] for a more tensorial view.) The corresponding generalized gradients are, respectively, $d/(k+1)$, $-(\operatorname{atr}\delta)/(n-k+1)$, and an operator which we shall call S_k.

Since orthogonal-invariant tensor bundles are defined by symmetry and trace conditions, a generalized gradient G, its formal adjoint G^*, and G^*G, GG^* are all geometric differential operators. G^*G (resp. GG^*) is formally self-adjoint, and is elliptic with positive definite leading symbol if and only if the leading symbol of G is injective (resp. surjective). In our example of $F = \Lambda^k(M)$, the second order operators produced are δd, $d\delta$, $S_k^* S_k$, and $S_k S_k^*$. Of these, the only elliptic operators are δd on 0-forms, $d\delta$ on n-forms, and the

$$S_k^* S_k = \frac{k}{k+1}\,\delta d + \frac{n-k}{n-k+1}\, d\delta - W_k.$$

Here W_k is the Lichnerowicz-Weitzenböck operator $\delta d + d\delta - \nabla^*\nabla$ on k-forms. (see [B3].) $S_k S_k^*$ has no chance of being elliptic, as

$$(\text{fiber dimension } (TFB)_{k+1}) - (\text{fiber dimension } \Lambda^k)$$

$$= \frac{n!}{(k+1)\ (n-k+1)}\,(n+1)(kn-(k^2+1)) > 0.$$

4. CONFORMAL VARIATION

In this section, we review results of [B2, BØ1, BØ2] on conformal variation of local invariants, geometric operators, and heat kernel traces. Again let (M,g) be a compact, Riemannian n-manifold, and consider a conformal curve of metrics $g(u) = e^{2u\omega} g$, where u is a real parameter and $\omega \in C^\infty(M)$. Our first variation operator will be denoted by a dot:

$$\bullet = \frac{d}{du}\bigg|_{u=0}.$$

Definition 4.1. Let ω be an indeterminate element of $C^\infty(M)$. We define *ω-augmented* local scalar invariants and geometric differential operators $B(\omega)$ by adding $\nabla\omega$ to the list of ingredients in Definition 2.1, and requiring that $B(c\omega) = cB(\omega)$, $c \in \mathbb{R}$. We extend the definitions (2.1), (2.2) of the level to ω-augmented objects.

Remark 4.2. The basic level m, ω-augmented local scalar invariant has a form analogous to (2.3), viz.

$$\operatorname{trace}(R_{ijk\ell|\alpha} \cdots R_{stuv|\beta}\, \omega_{|\gamma}), \qquad |\gamma| \geq 1,$$

$$(2 + |\alpha|) + \ldots + (2 + |\beta|) + |\gamma| = m.$$

Repeated integration by parts shows that a level m, ω-augmented local scalar invariant $Q(\omega)$ is equal, modulo exact divergences, to ω times an ordinary level m local scalar invariant L; for the above basic expression, this L is

$$(-1)^{|\gamma|} \operatorname{trace}((R_{ikj\ell|\alpha} \cdots R_{stuv|\beta})_{|\gamma}).$$

In particular, $\int_M Q(\omega) = \int_M \omega L$.

THEOREM 4.3. a) ([B2, Sec. 1.) *The conformal variation, under* $\dot{g} = 2\omega g$, *of a level* m *local scalar invariant* L *is*

$$\dot{L} = -m\omega L + L'(\omega),$$

where $L'(\omega)$ *is a level* m, *ω-augmented local scalar invariant. If* $L'(\omega)$ *is identically zero for all* (M,g,ω), *then* L *is a relative conformal invariant, i.e.,*

$$\bar{g} = \Omega^2 g, \qquad 0 < \Omega \in C^\infty(M) \Rightarrow \bar{L} = \Omega^{-m} L.$$

b) ([B2], Sec. 1.) *Let* D *be a level* m *geometric differential operator carrying a bundle* F *of* $\left[\begin{smallmatrix} p \\ q \end{smallmatrix}\right]$ *tensors to a bundle* E *of* $\left[\begin{smallmatrix} r \\ s \end{smallmatrix}\right]$ *tensors. Given* $a \in \mathbb{R}$, *the conformal variation of* D *may be written*

$$\dot{D} = (-m + s - r - q + p)\omega D + a[D, \omega] + D'^{(a)}(\omega),$$

where $D^{\cdot(a)}(\omega)$ *is a level* m, ω*-augmented geometric differential operator, and the* ω *in the commutator term represents multiplication by* ω. *If* $D^{\cdot(a)}(\omega)$ *is identically zero for all* (M,g,ω), *then* D *is a conformal covariant of bidegree* $(a, a+m-s+r+q-p)$, *i.e.,*

$$\bar{g} = \Omega^2 g, \qquad 0 < \Omega \in C^\infty(M) \Rightarrow \bar{D} = \Omega^{-(a+m-s+r+q-p)} D\Omega^a$$

(*where the* Ω^a *to the right of* D *is a multiplication operator*).

Some examples of conformally covariant operators are the exterior derivative d (bidegree (0,0)), the conformal Laplacian $\Delta + (n-2)K/4(n-1)$ on functions, $K =$ scalar curvature (bidegree $((n-2)/2, (n+2)/2)$, the Maxwell operator δd on $(n-2)/2$-forms for n even (bidegree (0,2)), and, extending the definitions above to cover tensor-spinor bundles as in [BØ2], the Dirac operator P (bidegree $((n-1)/2, (n+1)/2)$. There are also series of order 2, level 2 operators $D_{2,k}$ on k-forms introduced in [B1] for $n \neq 1, 2$, and order 4, level 4 operators $D_{4,k}$ on k-forms introduced in [B2] for $n \neq 1, 2, 4$. The Ahlfors operator $S = S_1$ is conformally covariant of bidegree (-2,-2) (see, e.g., [ØP, B3]). In fact, Fegan's main result in [F] is that *all* generalized gradients are conformally covariant; he is able to write their conformal bidegrees in terms of the root structure of O(n) and the weights of its representations. In particular, he calculates the conformal bidegree of the S_k introduced in the last section to be $(-(k+1), -(k+1))$ ([F], Sec. 4.1).

All of Fegan's results are stated in terms of SO(n)-irreducible bundles, but a careful check of the weight arguments shows that they are really O(n) results. When an O(n)-irreducible representation splits under SO(n), it splits into a contragredient pair of representations; for example, the self-dual and antiself-dual 2ℓ-forms $\Lambda_+^{2\ell}$, $\Lambda_-^{2\ell}$ when $n = 4\ell$; and the conformal weights for the contragredient representations agree. In the self-dual/anti-self-dual example, from the bundle point of view, the upshot is (taking an oriented, 4ℓ-dimensional M) that $d: C^\infty(\Lambda_+^{2\ell} \oplus \Lambda_-^{2\ell}) \to C^\infty(\Lambda^{2\ell+1})$, and not just $d: C^\infty(\Lambda_\pm^{2\ell}) \to C^\infty(\Lambda^{2\ell+1})$, is conformally covariant.

If D is a conformal covariant of bidegree (a,b), it is easy to see that the formal adjoint D* is also a conformal covariant of bidegree

$$(2(r-s)+n-b,\ 2(p-q)+n-a). \tag{4.1}$$

(Just compute with the conformal variation of the Riemannian measure, $(d\mu)^\bullet = n\omega d\mu$, or see [B3], sec. 2.) Thus the conformal bidegree of a formally self-adjoint geometric operator, if it has one, is fixed: for example, knowing only that $D_{2,k}$ and $D_{4,k}$ are conformally covariant, we can conclude that they have conformal bidegrees $((n-2k-2)/2), (n-2k+2)/2)$ and $((n-2k-4)/2, (n-2k+4)/2)$, respectively. The formal adjoint G* of a generalized gradient G is conformally covariant; in particular $\delta: C^\infty(\Lambda^k) \to C^\infty(\Lambda^{k-1})$ has bidegree $(n-2k, n-2k+2)$, and $S_k^*: C^\infty((TFB)_{k+1}) \to C^\infty(\Lambda^k)$ has bidegree $(n-k-1, n-k+1)$.

Remark 4.4. Remark 4.2 and $(d\mu)^{\bullet} = n\omega d\mu$ imply that the conformal variation of the integral of a local invariant L is

$$\left(\int_M L\, d\mu\right)^{\bullet} = \int_M \omega U\, d\mu,$$

where U is a local invariant of the same level as L. Since ω is arbitrary in $C^{\infty}(M)$, U is uniquely determined. If $U = cL$, $c \in \mathbb{R}$, the special case $\omega = 1$ shows that c must be $n - (\text{level } L)$.

THEOREM 4.5 ([BØ1], Sec. 3). a) *Let* $D: C^{\infty}(F) \to C^{\infty}(F)$ *be a geometric differential operator on a tensor bundle* F, *formally self-adjoint with positive definite leading symbol. Suppose that* 2ℓ *is the level and order of* D *for some* $\ell \in \mathbb{Z}^{+}$. *Then the* L^2 *trace of* $\exp(-tD)$ *is conformally differentiable, and its asymptotic expansion is term-by-term conformally differentiable*:

$$(\mathrm{Tr}_{L^2(F)} \exp(-tD))^{\bullet} \overset{t\downarrow 0}{\sim} \sum_{i=0}^{\infty} t^{\frac{2i-n}{2\ell}} \left(\int U_i\right)^{\bullet},$$

where

$$(\mathrm{Tr}_{L^2(F)} \exp(-tD) \overset{t\downarrow 0}{\sim} \sum_{i=0}^{\infty} t^{\frac{2i-n}{2\ell}} \int U_i.$$

b) *If* D *is conformally covariant*, $(\int U_i)^{\bullet} = (n-2i)\int \omega U_i$ *and* (n *even*) $\int U_{n/2}$ *is a conformal invariant.*

By Theorem 2.3, level $U_i = 2i$, above, so $\int U_i$ has the simplest possible conformal variation law for D a conformal covariant.

An outline of the proof of Theorem 4.5 (b) will illuminate the thinking behind the proof of our main theorem in the next section. First, a generalization of a variational formula due to Ray and Singer [RS] in the case of the Laplacian gives

$$(\mathrm{Tr}_{L^2(F)} \exp(-tD))^{\bullet} = -t\, \mathrm{Tr}_{L^2(F)} \dot{D} \exp(-tD). \tag{4.2}$$

Thus, given Theorem 4.5 (a), our problem is to evaluate the right side of (4.2). But for a conformal covariant D,

$$\dot{D} = -(b-a)\omega D + a[D,\omega],$$

where (a,b) is the bidegree of D. Now the commutator term contributes zero to (4.2), as can be seen in two completely different ways: $-t\,\mathrm{Tr}_{L^2(F)} [D,\omega]\exp(-tD)$ is the first variation of the L^2 trace for the isospectral family

$$\exp(-t e^{-u\omega} D e^{u\omega}), \qquad u \in \mathbb{R}.$$

Alternately, the infinitely smoothing character of $\exp(-tD)$ allows us to cyclically permute operators under the trace:

$$\mathrm{Tr}_{L^2(F)}\, D\omega \exp(-tD) = \mathrm{Tr}_{L^2(F)}\, \omega \exp(-tD) D$$

$$= \mathrm{Tr}_{L^2(F)}\, \omega D \exp(-tD).$$

Thus we are reduced to

$$(\mathrm{Tr}_{L^2(F)} \exp(-tD))^{\bullet} = (b-a)t\, \mathrm{Tr}_{L^2(F)}\, \omega D \exp(-tD).$$

But $t\omega D\exp(-tD)$ has kernel $-t\omega(x)(d/dt)H(t,x,y)$, where $H = H[\exp(-tD)]$, the fiberwise trace of which expands asymptotically on the diagonal as

$$-\sum_{i=0}^{\infty} \frac{2i-n}{2\ell}\, t^{\frac{2i-n}{2\ell}} \int_M \omega U_i$$

as $t \downarrow 0$. But the fact that $\mathrm{level}\, D = 2\ell$ forces $b - a = 2\ell$.

5. THE MAIN THEOREM

In this section, we let $G\colon C^{\infty}(F) \to C^{\infty}(E)$ be a generalized gradient on a compact Riemannian n-manifold (M,g), and we assume that the leading symbol of G is injective (so that the formally self-adjoint operator G^*G has positive definite leading symbol). Though GG^* may not be elliptic, we can say something interesting about asymptotics of "partially smoothing" operators built from GG^*. Theorem 5.2 below was first proved in [ØP] in the case where $G = S$, the Ahlfors operator.

LEMMA 5.1. *Suppose* G *is as directly above, and that* F *is a bundle of covariant* q-*tensors. Then* G *has a conformal bidegree*

$$(a,a), \qquad a \in \mathbb{R},$$

and G^* *has conformal bidegree*

$$(A, A+2), \qquad A = -2q + n - a - 2.$$

Proof. That G has a conformal bidegree follows from [F] (recall the discussion following Theorem 4.3); that the components of this bidegree are equal follows form the uniform dilation law (2.2), the fact that E must be a bundle of covariant $(q+1)$-tensors and the fact that $\mathrm{level}\, G = 1$. The conformal bidegree of G^* is then given by (4.1).

THEOREM 5.2. *Suppose* G, a, *and* A *are as above, and let* Tr_0 *denote the trace of the compression of an operator to the Hodge sector* $\mathcal{R}(G)$ *in the decomposition*

$$L^2(E) = \mathcal{R}(G) \oplus \mathcal{N}(G^*).$$

Let $U_i(x)$ *be the coefficients in the expansion*

$$H[\exp(-tG^*G)](t,x,x) \sim \sum_{i=0}^{\infty} t^{\frac{2i-n}{2}} U_i(x), \qquad t \downarrow 0.$$

Then if $\omega \in C^\infty(M)$,

$$(A-a)\mathrm{Tr}_0\, \omega \exp(-tGG^*) \overset{t\downarrow 0}{\sim} \tag{5.1}$$

$$\begin{cases} C(\omega) + \sum_{i\neq n/2} t^{\frac{2i-n}{2}} \{\frac{2}{2i-n}(\int U_i)^{\bullet} + (A+2-a)\int \omega U_i\} \\ \qquad\qquad + (\log t)(\int U_{n/2})^{\bullet}, \quad n \text{ even}, \\ C(\omega) + \sum_{i=0}^{\infty} t^{\frac{2i-n}{2}} \{\frac{2}{2i-n}(\int U_i)^{\bullet} + (A+2-a)\int \omega U_i\}, \quad n \text{ odd}. \end{cases}$$

where $C(\omega)$ *is a constant, and as usual the dot denotes conformal variation via* $\dot{g} = 2\omega g$.

For the proof, it is convenient to state the following lemma.

LEMMA 5.3. *Suppose* $\varphi(t)$ *and* $\psi(t)$ *are continuous real-valued function on* $(0,1]$, *and that* $\varphi(t) - \psi(t) = O(t^k)$, $k > -1$, *as* $t \downarrow 0$. *Let* Φ *and* Ψ *be primitives of* φ *and* ψ *respectively on* $(0,1]$. *Then*

$$\Phi - \Psi = C + O(t^{k+1}) \tag{5.2}$$

for some constant C. *If* $\tilde{\psi}(t)$ *is any function on* $(0,1]$ *with*

$$\lim_{t\downarrow 0} \tilde{\Psi}(t) - \Psi(t) = 0$$

and

$$\Phi - \tilde{\Psi} = \tilde{C} + O(t^{k+1}), \tag{5.3}$$

$\tilde{C} \in \mathbb{R}$, *then* $\tilde{C} = C$.

Proof. Integrating $\varphi(t) - \psi(t) = O(t^k)$, we get

$$O(t^{k+1}) = \int_0^t [\varphi(s) - \psi(s)]ds = -\int_t^1 \varphi(s)ds + \int_t^1 \psi(s)ds + \text{const}$$

$$= \Phi(t) - \Psi(t) + \text{const}.$$

Subtracting (5.3) from (5.2) and taking the limit as $t \downarrow 0$, we get $C = C$.

<u>Proof of Theorem 5.2</u>. Recall first that as a formally self-adjoint, positive semidefinite second-order differential operator with positive definite leading symbol, G^*G has discrete real spectrum $0 \leq \lambda_0 \leq \lambda_1 \leq \dots \nearrow +\infty$, and in fact,

$$\lambda_j \sim \text{const} \cdot j^{2/n}. \tag{5.4}$$

(See, e.g., [G], Sec. 1.) $L^2(F)$ has an orthonormal basis $\{\varphi_j\}$ of eigensections: $G^*G\,\varphi_j = \lambda_j \varphi_j$. Thus the Hodge sector $\mathcal{R}(G)$ admits the orthonormal basis

$$\{\psi_j = G\,\varphi_j/\lambda_j\}_{j \geq j_0}, \tag{5.5}$$

where λ_{j_0} is the first nonzero eigenvalue of G^*G, and $GG^*\psi_j = \lambda\psi_j$.

By the Ray-Singer formula (4.2) and Theorem 4.5,

$$\begin{aligned} -t\,\mathrm{Tr}_{L^2(F)}\,(G^*G)^{\bullet}\exp(-t\,G^*G) \\ = (\mathrm{Tr}_{L^2(F)}\exp(-t\,G^*G))^{\bullet} \overset{t\downarrow 0}{\sim} \Sigma\, t^{\frac{2i-n}{2}} (\textstyle\int U_i)^{\bullet}. \end{aligned} \tag{5.6}$$

Now by Lemma 5.1,

$$\dot{G} = a[G, \omega], \qquad (G^*)^{\bullet} = -2\omega G^* + A[G^*, \omega],$$

so that

$$(G^*G)^{\bullet} = (a - A - 2)\,\omega\, G^*G + a[G^*G, \omega] + (A - a)G^*\omega\, G.$$

As in the proof of Theorem 4.5, the $[G^*G, \omega]$ term contributes zero to the left side of (5.6), and the $\omega\, G^*G$ term contributes something asymptotic to

$$\frac{A+2-a}{2}\,\Sigma\,(n-2i)\,t^{\frac{2i-n}{2}} \textstyle\int \omega U_i.$$

Thus

$$\begin{aligned} (a - A)\,t\,\mathrm{Tr}_{L^2(F)}\,G^*\omega\, G\exp(-t\,G^*G) \overset{t\downarrow 0}{} \\ \sum_{i=0}^{\infty} t^{\frac{2i-n}{2}} \{(\textstyle\int U_i)^{\bullet} - \frac{A+2-a}{2}(n-2i)\textstyle\int \omega\, U_i\}. \end{aligned} \tag{5.7}$$

But working with the left side of (5.7), the estimate (5.4) justifies the manipulations

$$\operatorname{Tr}_{L^2(F)} G^* \omega G \exp(-t G^*G) = \operatorname{Tr}_{L^2(E)} \omega G \exp(-t G^*G) G^*$$
$$= \operatorname{Tr}_{L^2(E)} \omega GG^* \exp(-t GG^*).$$

Since $\omega GG^* \exp(-t GG^*)$ annihilates $N(G^*)$, the trace over $L^2(E)$ on the last line can be replaced by Tr_0.

Now if ω_{jk} are the matrix entries of the compression of multiplication by ω to $R(G)$ in the basis (5.5),

$$\operatorname{Tr}_0 \omega GG^* \exp(-t GG^*) = \sum_{j\geq j_0} \omega_{jj} \lambda_j e^{-t\lambda_j}$$
$$= -\frac{d}{dt} \sum_{j\geq j_0} \omega_{jj} e^{-t\lambda_j} = -\frac{d}{dt} \operatorname{Tr}_0 \omega \exp(-t GG^*),$$

again using (5.4). Thus (5.7) reads

$$(A-a) \frac{d}{dt} \operatorname{Tr}_0 \omega \exp(-t GG^*) \overset{t\downarrow 0}{\sim} \sum_{i=0}^{\infty} t^{\frac{2i-n-2}{2}} \{(\int U_i)^{\bullet} - \frac{A+2-a}{2} (n-2i) \int \omega U_i\}.$$

Applying Lemma 5.3, we get (5.1).

COROLLARY 5.4. *Under the assumptions of Theorem* 5.2, *the metric* g *is conformal critical for the functional* $\int U_{n/2}$ (n *even*) *if and only if* $\operatorname{Tr}_0 \omega \exp(-t GG^*)$ *has an asymptotic expansion in powers of* t (*no* $\log t$ *term*) *for all* $\omega \in C^\infty(M)$.

Remark 5.5. a) If $A=a$, (5.1) says that

$$C(\omega) = 0,$$
$$(\int U_i)^{\bullet} = (n-2i) \int \omega U_i.$$

This checks nicely with Theorem 4.5: since the "final" conformal degree a of G matches up with the "initial" conformal defree A of G^*, the operator G^*G is conformally covariant. An example of this situation is the case $G=d\colon C^\infty(M) \to C^\infty(\Lambda^1(M))$ in dimension two: $A=a=0$, and $\delta d = \Delta$ is conformally covariant (see also Example 5.6 below). For our other examples $G=S_k$, we always have $A-a=n$ (recall the discussion following Theorem 4.3).

b) Setting $\omega \equiv 1$ in (5.1) yields

$$(A-a)(\operatorname{Tr}_{L^2(F)} \exp(-t G^*G) - \dim N(G)) \overset{t\downarrow 0}{\sim} C(1) + (A-a) \sum_{i\neq n/2} t^{\frac{2i-n}{2}} \int \omega U_i.$$

(Note that $N(G) = N(G^*G)$ is finite-dimensional.) In particular,

$$C(1) = \begin{cases} (A-a)(\int U_{n/2} - \dim N(G)), & n \text{ even}, \\ -(A-a)\dim N(G), & n \text{ odd}. \end{cases}$$

c) If GG^* is also elliptic, Theorem 1.2 implies that $C(\omega)$ is the integral of a local expression for n even, and is zero for n odd. In addition, the $\log t$ term must vanish: $\int U_{n/2}$ is conformally invariant for n even. This is exactly what happens on spin manifolds for the Dirac operator $P = P^*$: though P^2 is not conformally covariant, $\int U_{n/2}[\exp(-tP^2)]$ is conformally invariant ([BØ1], Sec. 4.a). In fact, an easy adaptation of the proofs of Theorems 4.5 and 5.2 shows that if D is a conformal covariant with both D^*D and DD^* elliptic, then $\int U_{n/2}[\exp(-tD^*D)]$ is a conformal invariant for even n.

E x a m p l e 5.6. In the case $G = d\colon C^\infty(M) \to C^\infty(\Lambda^1(M))$, $U_i = U_i[\exp(-t\Delta)]$, $\Delta = \delta d$, and

$$(\int U_i)^\bullet = (n-2i)\int \omega U_i + \frac{n-2}{2}\int \omega \Delta U_{i-1}$$

([BØ1], equation (4.1)). Here $a = 0$ and $A = n-2$, so

$$(n-2)\operatorname{Tr}_{\mathcal{R}(d)}\omega \exp(-td\delta) \overset{t\downarrow 0}{\sim} \tag{5.8}$$

$$\begin{cases} C(\omega) + (n-2)\displaystyle\sum_{i\neq n/2} t^{\frac{2i-n}{2}}\{\int\omega U_i + \frac{1}{2i-n}\int\omega\Delta U_{i-1}\} \\ \qquad + \dfrac{n-2}{2}(\log t)\int\omega\Delta U_{(n-2)/2}, & n \text{ even}, \\ C(\omega) + (n-2)\displaystyle\sum_{i=0}^{\infty} t^{\frac{2i-n}{2}}\{\int\omega U_i + \frac{1}{2i-n}\int\omega\Delta U_{i-1}\}, & n \text{ odd}. \end{cases}$$

Thus if $n > 2$ is even, the $\log t$ term is nonzero for some ω if and only if $U_{(n-2)/2}$ is constant. In particular, for $n = 4$, $U_1 = K/6$, so $\log t$ appears whenever the scalar curvature is nonconstant.

A weak check on (5.8) can be made by taking M to be the standard sphere S^n and letting ω be a first-order spherical harmonic. All local invariants on S^n are constant, and odd-order spherical harmonics integrate to zero, so we must have

$$\Sigma\,\omega_{jj}\,e^{-t\lambda_j} \sim C(\omega), \quad t\downarrow 0.$$

But by [B4], Lemma 3.9 (third equation), all ω_{jj} are zero. In fact, can be taken to be any odd function on S^n, and all the same reasoning applies.

6. ARITHMETIC MEANING OF THE TRACE ASYMPTOTICS.

Theorem 5.2 may be interpreted as saying that $\{\omega_{jj}\}$, the sequence of diagonal matrix entries of multiplication by $\omega \in C^\infty(M)$ on $\mathcal{R}(G)$, is a very special sequence of numbers, in the following sense. Suppose n is even, and write (5.1) as

$$\sum_{j \geq j_0} \omega_{jj} e^{-t\lambda_j} = \mathrm{Tr}_0\, \omega \exp(-t\, GG^*) \overset{t \downarrow 0}{\sim} \sum_{i=0}^{\infty} \alpha_i t^{\frac{2i-n}{2}} + \beta \log t. \tag{6.1}$$

Define a zeta function for large $\operatorname{Re} s$ by

$$\zeta(s) = \zeta(s; G, \omega) = \sum_{j \geq j_0} \omega_{jj} \lambda_j^{-s}.$$

(This is possible by (5.4) and the fact that multiplication by ω is a bounded operator.) Since

$$\Gamma(s) = \int_0^\infty e^{-t} t^{s-1}\, dt = \int_0^\infty e^{-\lambda t} (\lambda t)^{s-1}\, d(\lambda t) = \lambda^s \int_0^\infty e^{-\lambda t} t^{s-1}\, dt,$$

we have

$$\zeta(s) = \sum_{j \geq j_0} \omega_{jj} \frac{1}{\Gamma(s)} \int_0^\infty e^{-\lambda_j t} t^{s-1}\, dt$$

$$= \frac{1}{\Gamma(s)} \int_0^\infty (\mathrm{Tr}_0\, \omega\, e^{-t\, GG^*})\, t^{s-1}\, dt.$$

Now the $\int_1^\infty$ part is analytic for all $s \in \mathbb{C}$, and the $\int_0^1$ part becomes, by (6.1),

$$\frac{1}{\Gamma(s)} \int_0^1 \{ \sum_{i=0}^{N} \alpha_i t^{\frac{2i-n}{2}} + \beta \log t + O(t^{N+1}) \} t^{s-1}\, dt$$

$$= \frac{1}{\Gamma(s)} \{ \sum_{i=0}^{N} \frac{2\alpha_i}{2i - n + 2s} - \frac{\beta}{s^2} + \int_0^1 O(t^{n+1})\, t^{s-1}\, dt \}.$$

This analytically continues to a meromorphic function on all of $\mathbb{C}$, with (possible) simple poles at $0, 1, \ldots, n/2$. Thus the behavior of ζ is similar to, but slightly different than that of the analytic continuation of the zeta function of G^*G, which has the possibility of simple poles only at $1, 2, \ldots, n/2$.

Recalling Remark 5.5(c), one might say that $\zeta(s; G, \omega)$ can "feel" the nonellipticity of GG^*, and exhibits this by developing a simple pole at 0.

References

[A] AHLFORS, L.: 'Conditions for quasiconformal deformation in several variables', in: *Contributions to Analysis. A Collection of Papers Dedicated to L. Bers*, Academic Press, New York 1974, 19-25.

[BFG] BEALS, M., C. FEFFERMAN, and R. GROSSMAN: 'Strictly pseudoconvex domains in C^n', *Bull. Amer. Math. Soc.* 8 (1983), 125-322.

[BGS] ——, P. GREINER, and N. STANTON: 'The heat equation on a CR manifold', *J. Diff. Geom.* 20 (1984), 343-387.

[BGM] BERGER, M., P. GAUDUCHON, and E. MAZET: *Le Spectre d'une variete Riemannienne*, Springer-Verlag, Berlin 1971.

[B1] BRANSON, T.: 'Conformally covariant equations on differential forms', *Comm. Partial Differential Equations* 7 (1982), 393-431.

[B2] ——: 'Differential operators canonically associated to a conformal structure', *Math. Scand.* 57 (1985), 293-345.

[B3] ——: 'Geometry of the Ahlfors operator', preprint, *University of Iowa*, 1987.

[B4] ——: 'Group representations arising from Lorentz conformal geometry', *J. Funct. Anal.*, to appear.

[BØ1] —— and B. Ørsted: 'Conformal indices of Riemannian manifolds', *Compositio Math.* 60 (1986), 261-293.

[BØ2] —— and ——: 'Conformal deformation and the heat operator', *Indiana U. Math. J.*, to appear.

[FS] FEFFERMAN, C. and A. SANCHEZ-CALLE: 'Fundamental solutions for second order subelliptic operators', *Ann. Math.* 124 (1986), 247-272.

[F] FEGAN, H.: 'Conformally invariant first order differential operators', *Quart. J. Math. Oxford* 27 (1976), 371-378.

[G] GILKEY, P.: *Invariance theory, the heat equation, and the Atiyah -Singer index theorem*, Publish or Perish, Wilmington, Delaware, 1984.

[Ha] HADAMARD, J.: *Le probleme de Cauchy et les equations aux derivees partielles lineaires hyperboliques*, Hermann et Cie, Paris 1932.

[Hö] HÖRMANDER, L.: 'Hypoelliptic second order differential equations', *Acta Math.* 119 (1967), 147-171.

[JS] JERISON, D. and A. SANCHEZ-CALLE: 'Estimates for the heat kernel for a sum of squares of vector fields', *Indiana U. Math. J.* 35 (1986), 835-854.

[M] MELROSE, R.: 'The trace of the wave group', *Cont. Math.* 27 (1984), 127-169.

[MP] MINAKSHISUNDARAM, S. and Å. PLEIJEL: 'Some properties of the eigenfunctions of the Laplace operator on Riemannian manifolds', *Canad. J. Math.* 1 (1949), 242-256.

[ØP] ØRSTED, B. and A. PIERZCHALSKI: 'The Ahlfors Laplacian on a Riemannian manifold', preprint, *Sonderforschungsbereich Göttingen* 1987.

[P] PIERZCHALSKI, A.: 'On quasiconformal deformations of manifolds and hypersurfaces', *Ber. Univ. Jyväskylä Math. Inst.* 28 (1984), 79-94.

[RS] RAY, D. and I. SINGER: 'R-torsion and the Laplacian on Riemannian manifolds', *Advances in Math.* 7 (1971), 145-210.

[Sa] SANCHEZ-CALLE, A.: 'Fundamental solutions and geometry of sums of squares of vector fields', *Invent. Math.* 78 (1984), 143-160.

[Sta] STANTON, N.: 'The heat equation for the $\bar{\partial}_b$-Laplacian', *Comm. Partial Differential Equations* 9 (1984), 597-686.

[SW] STEIN, E. and G. WEISS: 'Generalization of the Cauchy-Riemann equations and representations of the rotation group', *Amer. J. Math.* 90 (1968), 163-196.

[Str] STRICHARTZ, R.: 'Sub-Riemannian geometry', *J. Diff. Geom.* 24 (1986), 221-263.

[X] XU, C.: 'On the asymptotic expansion of the trace of the heat kernel for a subelliptic operator', *Ph. D. Thesis, M.I.T.*, 1987.

HOLOMORPHIC QUASICONFORMAL MAPPINGS IN INFINITE-DIMENSIONAL SPACES

Petru Caraman
Institute of Mathematics
University of Iaşi
Iaşi, Romania

ABSTRACT. We proved in the previous paper [1] that if a quasiconformal mapping (qc) $f: D \rightleftarrows D_1$ (D, D_1 - domains in the Euclidean $2n$-space R^{2n}) is holomorphic as a map of n complex variables between the domains D_c and D_{1c} (obtained from D and D_1 by complexification), then the infinitesimal ellipsoids transformed by f into infinitesimal spheres have the semi-axes equal by twos. In particular, if $D_c = C^n$, f comes to an affine transformation.

In this paper, we establish successively that this result still holds in l_2, in separable Hilbert spaces, in abstract Wiener spaces, in separable Banach spaces, and for maps between HC^1-surfaces or Riemann-Wiener manifolds.

1. Everywhere in this paper, D and D_1 denote domains in the corresponding spaces. We recall [2] that an ellipsoid in l_2 is the image of a sphere by a non-degenerate affine transformation and the principal characteristic parameter p_1 is the ratio between the supremum a_1 and the infimum a_0 of the semi-axes of the ellipsoid. A homeomorphism $f: l_2 \rightleftarrows l_2$ is said to transform an infinitesimal ellipsoid $E(x)$ of centre $x \in l_2$ into an infinitesimal sphere if it transforms an ellipsoid $E_h(x)$ with centre x and $a_0(x) = h$ into a surface $f(E_h)$ contained between two spheres $S[f(x), R]$ and $S[f(x), r]$ with centre $f(x)$ and radii R and r, respectively, so that

$$\lim_{h\to 0} \frac{R(h)}{r(h)} = 1$$

as E_h shrinks homothetically to x.

A homeomorphism $f: D \rightleftarrows D_1$, $D, D_1 \subset l_2$, is called K-*qc* ($1 \leq K < \infty$) *with a principal characteristic parameter* p_1 if it maps infinitesimal ellipsoids into infinitesimal spheres and $p_1(x) \leq K$.

A family of closed surfaces $\{\sigma_\alpha\}$, $0 < \alpha < 1$, (i.e. homeomorphic images of the unit sphere S) is said to be *regular of parameter* $k = k(x_0)$ *relatively to a point* x_0 in a normed space X if the surfaces σ_α are the images of the spheres $||t|| = \alpha$ in X by a homeo-

J. Ławrynowicz (ed.), Deformations of Mathematical Structures, 263–270.

morphism ψ of the unit ball $\|t\| < 1$ onto a neighbourhood of the point $x_0 = \psi(0)$ and

$$\overline{\lim_{\alpha\to 0}} \frac{\max\limits_{x\in\sigma_\alpha} \|x - x_0\|}{\min\limits_{x\in\sigma_\alpha} \|x - x_0\|} = k < \infty.$$

A homeomorphism is called *regular at* x_0 if it maps a regular family of parameter k relatively to x_0 into a regular family of parameter $k' = k'(x_0)$ $(1 \leq k' < \infty)$ relatively to $f(x_0)$.

A homeomorphism $f : D \rightleftarrows D_1$ (D and D_1 - domains in a measure Banach space B) is *K-qc in Markushevich-Pesin sense* $(1 \leq K < \infty)$ if it is regular in D, the characteristic $q(x) = \inf k(x)\, k'(x)$, where the infimum is taken over all regular families relatively to x, is bounded in D and $q(x) \leq K$ almost everywhere in D.

According to Theorem 1 of our paper [2], the class of K-qc with a principal characteristic parameter and the class of K-qc in Markushevich-Pesin sense coincide in the hypothesis of differentiability with bijective Fréchet derivative.

PROPOSITION. *A sequence* $\{\varepsilon_n\} \subset B_c$ *is a basis of a complex Banach space* B_c *iff the sequence* $\{e_m\} \subset B$ *defined by the relations* $e_{2n-1} = \varepsilon_n$, $e_{2n} = i\,\varepsilon_n$, $n = 1,2,\ldots$, *is a basis of the corresponding real Banach space* B (cf. [8], Chap. 1, & 1, Proposition 1.1, p. 3).

We can observe that B and B_c represent the same set of points, only that B is considered a Banach space over the real field R, while B_c the one over the complex field C. Then we have

$$z = \sum_n z_n \varepsilon_n = \sum_n (x_{2n-1} + i x_{2n})\varepsilon_n = \sum_n (x_{2n-1} e_{2n-1} + x_{2n} e_{2n}) = \sum_n x_n e_n .$$

It is easy to see that a Fréchet differentiable map f transforms infinitesimal ellipsoids of the form

$$\alpha_{mn}(x)\, dx^m\, dx^n = dr^2 \tag{1}$$

into infinitesimal spheres. With this notation and arguing as in Theorem 1 and its corollary of our paper [1], we obtain

THEOREM 1. *The infinitesimal ellipsoid* (1) *of a holomorphic map* $f : D_c \to D_{1c}$ (D_c *and* D_{1c} *are the domains corresponding to* D *and* D_1, *respectively, in the complexification* l_{2c} *of* l_2) *satisfies the relations*

$$\alpha_{2m-1,2n-1} = \alpha_{2m,2n},\ \alpha_{2m-1,2n} = \alpha_{2m,2n-1} \quad (m,n = 1,2,\ldots)$$

and its semi-axes are equal by twos, i.e. $a_{2n-1} = a_{2n}$. $(n = 1,2,\ldots)$.

COROLLARY 1. *A holomorphic* K-*qc in Markushevich-Pesin sense* $f : D_c \rightleftarrows D_{1c}$ *maps certain infinitesimal ellipsoids with the semi-axes equal by twos into infinitesimal spheres.*

COROLLARY 2. *A diffeomorphism* $f : D \rightleftarrows D_1$ *is holomorphic iff the semi-axes of the infinitesimal ellipsoids transformed into infinitesimal spheres by* f *and* f^{-1} *are equal by twos and their association in pairs of equal semi-axes is independent of the point (of* D_c *or* D_{1c}*).*

Remarks. 1. It is easy to see that there do not exist inclusion relations between qc and holomorphic mappings because the semi-axes of an infinitesimal ellipsoid (1) of a qc are, in general, not equal by twos, while $p_1(z)$ in the case of a holomorphic mapping may tend to ∞ as $z \to \partial D_c$.

2. Since a separable Hilbert space is isomorphic to l_2, it follows that Theorem 1 and its corollaries still hold in the case of a separable Hilbert space. We can also observe that since a set $E \subset l_2$ is an ellipsoid iff any non-empty bidimensional section is an ellipse ([2], Lemma 2), it follows that in the case of a separable Hilbert space we may define ellipsoids by this property.

2. Now, let us consider the case of an abstract Wiener space (H,B), i.e. a Banach space B obtained by the completion of a Hilbert space H with respect to a measurable norm. We observe that a complex abstract Wiener space (H_c, B_c) is obtained in the same way by completion of a complex Hilbert space H_c with respect to a measurable norm. Corollary 1 of Theorem 1 may be generalized in the case of an abstract Wiener space as follows:

THEOREM 2. *If a K-qc in Markushevich-Pesin sense* $f : D \rightleftarrows D_1$, $D, D_1 \subset (H,B)$, *is a holomorphic map* $f : D_c \rightleftarrows D_{1c}$, *where* D_c, D_{1c} *are the complexifications of* D, D_1 *(see the proposition from above), then* $\forall\ x \in D$ *the infinitesimal ellipsoids* (1) *contained in* $(D-x) \cap H$ *and mapped by* $\phi(h) = f(x+h)$ *into infinitesimal spheres will have the semi-axes equal by twos.*

Proof. Since f is holomorphic,

$$\lim_{\Delta z \to 0} \frac{\|f(z+\Delta z) - f(z) - A(\Delta z)\|}{\|\Delta z\|} = 0,$$

where $\|\cdot\|$ is the B-norm; it follows that

$$\lim_{h \to 0} \frac{\|\phi(h) - \phi(0) - A(h)\|}{\|h\|} = 0,$$

which is equivalent to

$$\lim_{h \to 0} \frac{|\phi(h) - \phi(0) - A(h)|}{\|h\|} = 0,$$

where $|\cdot|$ is the Hilbert norm. Since there exists a constant $C > 0$ such that $\|\cdot\| \leq C|\cdot|$ (H.H. Kuo [7]), it follows that

$$\lim_{h\to 0} \frac{|\phi(h) - \phi(0) - A(h)|}{|h|} \leq \lim_{h\to 0} \frac{|\phi(h) - \phi(0) - A(h)|}{\frac{1}{C}\,||h||}$$

$$= C \lim_{h\to 0} \frac{|\phi(h) - \phi(0) - A(h)|}{||h||} = 0,$$

that is, ϕ is (complex) Fréchet differentiable at 0. In a similar way we show that ϕ is (complex) Frechet differentiable at any point $h_0 \in (D_c - z) \cap H_c$ implying the holomorphy of ϕ at 0 and then, on account of the isomorphism between H and H_c, its real Fréchet differentiability in a neighbourhood U_0 of 0 in $(D - x) \cap H$.

Next, from the quasiconformality of f in Markushevich-Pesin sense we deduce that

$$\overline{\lim_{\alpha\to 0}} \frac{\sup\limits_{h \in (\sigma_\alpha - x_0) \cap H} ||\phi(h)-\phi(0)||}{\inf\limits_{h \in (\sigma_\alpha - x_0) \cap H} ||\phi(h)-\phi(0)||} \leq \overline{\lim_{\alpha\to 0}} \frac{\sup\limits_{h\in(\sigma_\alpha - x_0)\cap H} ||f(x_0+h)-f(x_0)||}{\inf\limits_{h\in(\sigma_\alpha - x_0)\cap H} ||f(x+h) - f(x)||}$$

$$\leq \overline{\lim_{\alpha\to 0}} \frac{\sup\limits_{x\in\sigma_\alpha} ||f(x) - f(x_0)||}{\inf\limits_{x\in\sigma_\alpha} ||f(x) - f(x_0)||} = k'(x_0) < \infty.$$

Yet, this condition still holds in a whole neighbourhood of 0 since $\forall\ h_1 \in U_1 \subset (D - x_0) \cap H$, such that $x_0 + h + h_1 \in D$,

$$\overline{\lim_{\alpha\to 0}} \frac{\sup\limits_{h \in (\sigma_\alpha^{h_1} - x_0) \cap H} ||\phi(h + h_1) - \phi(h_1)||}{\inf\limits_{h \in (\sigma_\alpha^{h_1} - x_0) \cap H} ||\phi(h + h_1) - \phi(h_1)||} \leq$$

$$\overline{\lim_{\alpha\to 0}} \frac{\sup\limits_{h\in(\sigma_\alpha^{h_1}-x_0)\cap H} ||f(h+h_1+x_0)-f(h_1+x_0)||}{\inf\limits_{h\in(\sigma_\alpha^{h_1}-x_0)\cap H} ||f(h+h_1+x_0)-f(h_1+x_0)||} \leq \overline{\lim_{\alpha\to 0}} \frac{\sup\limits_{x\in\sigma_\alpha^{h_1}} ||f(x)-f(h_1+x_0)||}{\inf\limits_{x\in\sigma_\alpha^{h_1}} ||f(x)-f(h_1+x_0)||}$$

$$= k'(h_1 + x_0) < \infty,$$

where $\sigma_\alpha^{h_1}$ are the images of the spheres $||y|| = \alpha$ by the homeomorphism ψ of the ball $||y|| < 1$ onto a neighbourhood of $x_0 + h_1 = \psi(0)$, hence ϕ will be K-qc in Markushevich-Pesin sense in U_1. Yet, since ϕ is also (real) Fréchet differentiable in U_0, on account of the equivalence of the two preceding definitions of the K-qc (see [2], Theorem 12), it follows that ϕ is a K-qc with a principal characteristic parameter in a neighbourhood $U \subset U_0 \cap U_1$. Then, taking into account the holomorphy of ϕ in the complexified U_0 and the preceding remark, it follows that the infinitesimal ellipsoid (1) has the semi-axes equal by twos.

Let us define the map $f : D_c \rightleftarrows D_{1c}$, where D_c and D_{1c} are obtained by complexification from $D, D_1 \subset (H,B)$, as H_c-*holomorphic* if $\phi(h) = f(x+h)$ is holomorphic in $(D_c - z) \cap H_c$ $\forall$ $z \in D_c$. This generalization of holomorphic mappings allows us to establish

THEOREM 3. *If a K-qc in Markushevich-Pesin sense* $f : D \rightleftarrows D_1$ *is* H_c-*holomorphic, then the ellipsoids* (1) *corresponding to* ϕ *has the semi-axes equal by twos.*

3. Now, we recall that a mapping $f : B \to Y$, where B is an abstract Wiener space and Y a Banach space (H,B), is said to be H-*continuous* (H-*differentiable*) at a point $x \in B$ if $\phi(h) = f(x+h)$ is continuous (differentiable) at 0 with respect to the topology of H. A function $u : U \subset B \to R$ (B - an abstract Wiener space) is called a HC^1-*function* if u is continuous and H-differentiable in U, its derivative $Du : U \to H^*$ is continuous and $Du(x) \in B^*$ $\forall$ $x \in U$, where H^* and B^* are topological duals of H and B, respectively.

A set $\sigma \subset (H,B)$ is (following V. Goodman [4]) a HC^1-*surface* if $\forall$ $\xi \in \sigma$ there exists an open neighbourhood U of ξ in B and HC^1-function $u : U \to R$ such that $Du(x) \neq 0$ and $\sigma \cap U = \{x \in U;\ u(x) = 0\}$.

Let us consider $\forall$ $\xi \in \sigma$ a normal projection N_ξ ([4]), then there exist in ξ a normal versor $h \in B^*$ such that $N_\xi h = h$ and a coordinate neighbourhood $W \subset \sigma$ of ξ such that $|N_w h| > 0$ $\forall$ $w \in W$ and $J_\xi = I - N_\xi$ is a homeomorphism of W into the kernel $\ker N_\xi$. By means of the *local normal surface measure*

$$\rho(W,E) = \frac{1}{\sqrt{2\pi}} \int_{J_\xi(E)} \frac{e^{-\frac{|N_\xi \xi|^2}{2}}}{|N_\xi h|} \, dp_1(\eta),$$

where $E \subset W$ is a Borel set, $\xi = J^{-1}(\eta)$, and p_1 is the abstract Wiener measure in $\ker N_\xi$, V. Goodman [4] defines also a (global) *normal surface measure* m_σ.

Now, we recall (see [3]) that an HC^1-surface σ is named *admissible* if $\forall$ $\xi \in \sigma$ and $\forall$ $\varepsilon > 0$ there exist a neighbourhood $U_\xi \subset (H,B)$ of ξ with $U_\xi \cap \sigma$ contained in a coordinate neighbourhood W and a quasi-isometry i_ξ defined in U_ξ such that the restriction of i_ξ to $U_\xi \cap \sigma$ is the homeomorphism $J_\xi : U_\xi \cap \sigma \rightleftarrows \Delta_\xi \subset \ker N_\xi$ and the maximal distortion $C(i_\xi)$ of i_ξ in U_ξ satisfy the inequalities

$$\sup_{\xi\in\sigma} C(i_\xi) < \infty, \qquad \operatorname{ess\,sup}_{\xi\in\sigma} C(i_\xi) \leq 1+\varepsilon, \tag{2}$$

where ess sup is considered with respect to the normal surface measure m_σ.

A homeomorphism $f:\sigma \rightleftarrows \sigma_1$ (σ, σ_1 - admissible HC^1-surfaces) is said to be a K - *qc in Markushevich-Pesin sense* if $f: U_\xi \cap \sigma \rightleftarrows U_{\xi_1} \cap \sigma_1$, where U_ξ, U_{ξ_1} are the neighbourhoods involoved in the definition of an admissible HC^1-surface, and $\forall\ \varepsilon > 0$ and $\forall\ \xi \in \sigma$ the mapping

$$g_\xi = i_{\xi_1} \circ f \circ i_\xi^{-1} : \Delta_\xi \rightleftarrows \Delta_{\xi_1},$$

where $\Delta_\xi = i_\xi(U_\xi \cap \sigma)$ and $\Delta_{\xi_1} = i_\xi\ (U_{\xi_1} \cap \sigma_1)$ are domains in (H,B), is a K-qc in Markushevich-Pesin sense and its maximal dilatation $K(g_\xi)$ satisfies the following inequalities:

$$\sup_{\xi\in\sigma} K(g_\xi) < \infty, \qquad \operatorname{ess\,sup}_{\xi\in\sigma} K(g_\xi) \leq K(1+\varepsilon). \tag{3}$$

Let (H_c, B_c) be a complex separable abstract Wiener space. A map $F: E \to Y$ of a set $E \subset B_c$ into a complex Banach space Y is called H_c-*continuous at* $z \in E$ if the transformation $\phi(h) = F(x+h)$ defined in $(E-z) \cap H_c$ is continuous at 0 with respect to the topology of H_c. A mapping $F: E \to Y$ is named H_c-*holomorphic at* $z \in E$ if ϕ is holomorphic at $0 \in (E-z) \cap H_c$. If σ and $\sigma_1 \subset (H,B)$ are two admissible HC^1-surfaces, a qc $f:\sigma \rightleftarrows \sigma_1$ is *holomorphic* (H_c-*holomorphic*) if $\forall\ \xi \in \sigma$ $g_\xi : \Delta_{\xi c} \rightleftarrows \Delta_{\xi_1 c}$, where $\Delta_{\xi c}$ and $\Delta_{\xi_1 c}$ are the complexified Δ_ξ and Δ_{ξ_1}, respectively, is holomorphic (H_c-holomorphic).

<u>THEOREM 4</u>. *If a* K-*qc in Markushevich-Pesin sense* $f:\sigma \rightleftarrows \sigma_1$ *between two admissible* HC^1-*surfaces is holomorphic* (H_c-*holomorphic*), *then* $\forall\ \xi \in \sigma$ *and* $\forall\ x \in \Delta_\xi$ *the infinitesimal ellipsoids* (1) *corresponding to the transformation* $\phi_\xi(h) = g_\xi(x+h)$ *will have the semi-axes equal by twos.*

This is a direct consequence of the preceding theorem since g_ξ is a K-qc between two domains Δ_ξ, $\Delta_{\xi_1} \subset (H,B)$, and the holomorphy ($H_c$-holomorphy) of f implies the holomorphy (H_c-holomorphy) of $g_\xi : \Delta_{\xi c} \rightleftarrows \Delta_{\xi_1 c}$ so that we have the hypotheses of the preceding theorem.

<u>4</u>. Finally, following H.H. Kuo [5], let us recall that a Riemann-Wiener manifold σ is a triple (W,τ,g), where W is a C^k-differentiable $(k \geq 3)$ manifold modelled on a separable Banach space B having a Wiener structure, given by the continuous assignement of a norm $\tau(x)$ to each tangent space $T_x(W)$ $\forall\ x \in W$, and a Hilbert structure on the subspace $R_x \subset T_x(W)$, given by the inner product $\langle .,. \rangle_x = g(x)$.

A Riemann-Wiener manifold $\sigma = (W,\tau,g)$ is said to be *admissible* if $\forall\ \xi \in \sigma$ there exists a quasi-isometry i_ξ (defined by means of the norm τ) such that $\forall\ \varepsilon > 0$ there exists a neighbourhood $V_\xi \subset \sigma$ of ξ, which is mapped by i_ξ onto a domain Δ_ξ of a separable Banach space B, and the maximal distortion $C(i_\xi)$ of i_ξ in V_ξ satisfies (2), where, this time, ess sup is considered with respect to the surface measure μ_σ on σ, introduced by H.H. Kuo [6].

Let σ and σ_1 be two admissible Riemann-Wiener manifolds, $f : \sigma \rightleftarrows \sigma_1$ a homeomorphism, f and f^{-1} - AC with respect to μ_σ and μ_{σ_1}, respectively, and $\forall\ \xi \in \sigma$, $\xi_1 = f(\xi)$ let i_ξ, i_{ξ_1} be the quasi-isometries involved in the definition of an admissible Riemann-Wiener manifold. Then f is said to be *K-qc in Markushevich-Pesin sense* if $\forall\ \varepsilon > 0$ there exist two neighbourhoods $V_\xi \subset \sigma$, $V_{\xi_1} = f(V_\xi) \subset \sigma_1$ such that

$$\Delta_\xi = i_\xi(V_\xi), \qquad \Delta_{\xi_1} = i_{\xi_1}(V_{\xi_1})$$

and the homeomorphism

$$g_\xi = i_{\xi_1} \circ f \circ i_\xi^{-1} : \Delta_\xi \rightleftarrows \Delta_{\xi_1}$$

is K-qc in Markushevich-Pesin sense and verifies (3), where, this time, ess sup is relative to μ_σ.

A qc $f : \sigma \rightleftarrows \sigma_1$, where σ, σ_1 are Riemann-Wiener manifolds, is *holomorphic* (*H_c-holomorphic*) if the corresponding homeomorphism $g_\xi : \Delta_{\xi c} \rightleftarrows \Delta_{\xi_1 c}$ is holomorphic (H_c-holomorphic).

THEOREM 5. *If a K-qc in Markushevich-Pesin sense $f : \sigma \rightleftarrows \sigma_1$ between two Riemann-Wiener manifolds is holomorphic (H_c-holomorphic), then $\forall\ \xi \in \sigma$ and $\forall\ \eta \in \Delta_\xi$ the infinitesimal ellipsoids* (1) *corresponding to $\phi_\xi(h) = g_\xi(\eta + h)$ defined in $(\Delta_\xi - \eta) \cap H$ will have the semi-axes equal by twos.*

References

[1] CARAMAN, P.: 'About the connection between the 2n-dimensional quasi-conformal homeomorphisms and the n-dimensional pseudo-conformal transformations', *Rev. Roumaine Math. Pures Appl.* 13 (1968), 1255-1271.

[2] ——: 'Quasiconformal mappings in real normed spaces', *ibid.* 24 (1979), 33-78.

[3] ——: 'Quasiconformal mappings between infinite-dimensional manifolds in abstract Wiener spaces', in Série: Recherches sur les déformations, *Bull. Soc. Sci. Lettr. Łódź* 36, no. 24 (1986), 13 pp.

[4] GOODMAN, V.: 'A divergence theorem for Hilbert spaces', *Trans. AMS* 164 (1972), 411-426.

[5] KUO, H.-H.: 'Integration theory on infinite-dimensional manifolds', *ibid.* 159 (1971), 57-78.

[6] KUO, H.-H.: 'Diffusion and Brownian motion on infinite-dimensional manifolds', *Trans. AMS* **169** (1972), 439-459.

[7] ——: 'Gaussian measures in Banach spaces', *Lect. Notes in Math.* **463**, Springer-Verlag, Berlin-Heidelberg-New York 1975, p. 224.

[8] SINGER, I.: *Bases in Banach spaces I*, Springer-Verlag, Berlin-Heidelberg-New York 1970, 668 pp.

PRODUCT SINGULARITIES AND QUOTIENTS OF LINEAR GROUPS

Klaus Reichard
Fakultät für Mathematik
Ruhr-Universität Bochum
D-4630 Bochum, BRD

and Karlheinz Spallek
Fakultät für Mathematik
Ruhr-Universität Bochum
D-4630 Bochum, BRD

ABSTRACT. In [14] a unique-product factorisation theorem is proved for quite arbitrary space germs (as for Whitney stratified germs). This we are going to apply here for factorisations of quotient singularities according to the announcement in [14].

INTRODUCTION, STATEMENT OF RESULTS

With [14] we first have:

1) To "any"*) germ ${}^{N}A$ of a reduced N-differentiable space with N $\{\infty, \omega$(real analytic), ω^*(complex analytic)$\}$ there exists -up to numbering and C^N-diffeomorphisms of the factors- a unique p-irreducible factorisation into germs ${}^{N}A_j$:

$$ {}^{N}A \simeq {}^{N}A_1 \times \ldots \times {}^{N}A_r , $$

where: ${}^{N}A_1 \simeq {}^{N}(K^{\ell})$ with $\ell \geq 0$; $K = \mathbb{R}_o$, $N \in \{\infty,\omega\}$ or $K = \mathbb{C}_o$, $N = \omega^*$; and the other germs ${}^{N}A_j$ are singular and p-irreducible ($\overset{\circ}{:}$ admit no further product factorisation).

1') Any other factorisation ${}^{N}A \simeq {}^{N}B_1 \times \ldots \times {}^{N}B_s$, where ${}^{N}B_1 \simeq {}^{N}(K^m)$ and no other ${}^{N}B_j$ splits up some ${}^{1}N_K$, gives: $\ell = m$ and each ${}^{N}B_i$, $i \neq 1$, is a product of some ${}^{N}A_j$'s with $j \neq 1$.

In general "${}^{N}A$ p-irreducible" does not imply "${}^{N'}A$ p-irreducible" for $N' < N$. However this holds in the following cases.

2) A is semianalytic, $N = \omega$, $N' = \infty$. Moreover then: Given p-irreducible factorisations ${}^{\infty}A \simeq {}^{\infty}A_1 \times \ldots \times {}^{\infty}A_r \simeq {}^{\infty}B_1 \times \ldots \times {}^{\infty}B_r \curvearrowright A_i, B_i$ are up to C^{∞}-diffeomorphisms semianalytic (: [14] for analytic cases) and we even have modulo numbering ${}^{\omega}A_i \simeq {}^{\omega}B_i$ (: 1) and [8]).

3) A is complex analytic and for example (algebraically) irreducible, $N = \omega^*$, $N' \in \{\infty,\omega\}$. Moreover then: Given p-irreducible factorisations

*) "any": locally compact, kurvenreich ([14]), for ex. semianalytic

J. Ławrynowicz (ed.), Deformations of Mathematical Structures, 271–282.

${}^{\infty}A \simeq {}^{\infty}A_1 \times \ldots \times {}^{\infty}A_r \simeq {}^{\infty}B_1 \times \ldots \times {}^{\infty}B_r \curvearrowright A_i, B_i$ are up to C^{∞}-diffeo's complex analytic and modulo numbering even in pairs holomorphic or antiholomorphic equivalent (: 1), [2] or more generally [14]).

As here, properties of factors A_i in general carry over to the product $\times A_i$ and vice versa. In the following we study such phenomena for quotient singularities:

Any Lie group G operating properly on an N-differentiable space X leads to a quotient-N-differentiable space X/G for $N \in \{\infty, \omega, \omega^*\}$ ([7]). Especially if $G \subset GL(n,K)$ and G is a compact (Lie-) group, the quotient space ${}^{N}(K^n)/G$ exists ([1], [5], [10]). Note that we consider here K as germ $\mathbb{R}_o$ or $\mathbb{C}_o$ at zero; then also ${}^{N}(K^n)/G$ is a germ of a space.

In general, the ω-differentiable space induced by ${}^{\omega^*}(\mathbb{C}^n_o)/G$ is different from ${}^{\omega}(\mathbb{C}^n_o)/G$. One has only a map ${}^{\omega}(\mathbb{C}^n_o)/G \to {}^{\omega^*}(\mathbb{C}^n_o)/G$.

To any $G \subset GL(m,K)$ the largest group $\max^N G$ is associated, that leaves the same C^N-functions as G invariant. We have:

4) Any compact group $G \subset GL(n,\mathbb{C})$ is finite (: classical).
5) If $G \subset GL(n,K)$ is finite, then $G = \max^{\infty} G = \max^{\omega} G$, and if $K = \mathbb{C}_o$ also $G = \max^{\omega^*} G$.
6) If $G \subset GL(n,K)$ is compact, so is $\max^N G$, $N \in \{\infty, \omega, \omega^*\}$.

From now on let $G \subset GL(n,K)$ be a compact (Lie-) group, ${}^{N}A = {}^{N}(K^n)/G$ the quotient germ with $q: K^n \to {}^{N}A$ as some quotient map; and assume always:

if $K = \mathbb{R}$, then $N \in \{\infty, \omega\}$ and G is maximal, i.e.: $G = \max^N G$,
if $K = \mathbb{C}$, then $N = \omega^*$ and G has no reflections.

In case $K = \mathbb{C}$ *define* ${}^{\infty}A$ to be the reduced C^{∞}-space associated to ${}^{\omega^*}A$. Note, that in case $K = \mathbb{R}$ the germ ${}^{\omega}(\mathbb{R}^n)/G$ *is* the reduced C^{∞}-space associated to ${}^{\omega}(\mathbb{R}^n)/G$ (see note, p. 5).

The above assumptions on G are necessary to obtain

7) THEOREM. *The following are equivalent*
a) ${}^{\infty}A \simeq {}^{\infty}A_1 \times \ldots \times {}^{\infty}A_r$ *(not necessarily* p-*irreducible),*
b) $\exists n_i \in \mathbb{N}$, $G_i \subset GL(n_i,K)$ *compact, maximal if* $K = \mathbb{R}$, *without reflections if* $K = \mathbb{C}$, *such that:* $n = \Sigma^r n_i$

$$G \simeq G_1 \times \ldots \times G_r \quad \textit{(up to conjugation in } GL(n,K))$$
$${}^{\infty}A_i \simeq {}^{\infty}(K^{n_i})/G_i \quad \textit{for each } i.$$

In the equivalent situations of 7) we have moreover:

8) THEOREM a) ${}^{\infty}A_i$ *is singular iff* $G_i \neq \mathrm{id}$
b) ${}^{\infty}A \simeq {}^{\infty}A_1 \times \ldots \times {}^{\infty}A_r$ *is* p-*irreducible iff*

$$G \simeq G_1 \times \ldots \times G_r \quad \textit{is irreducible (: has no refined factorisation).}$$

Next assume that $G \subset GL(n,K)$, $\tilde{G} \subset GL(m,K)$ are finite groups with irreducible factorisations

$$G = G_1 \times \ldots \times G_r, \quad \tilde{G} = \tilde{G}_1 \times \ldots \times \tilde{G}_s,$$

quotients ${}^N A$ for G as above, similarly ${}^N\tilde{A}$ for $\tilde{G}$.

9) THEOREM. ${}^\infty A \simeq {}^\infty\tilde{A}$ *iff* $n = m$, $r = s$ *and*
a) $\forall$ i: $G_i = \tilde{G}_i$ *if* $K = \mathbb{R}$ (*up to numbering and conjugation*)
b) $\forall$ i: $G_i = \tilde{G}_i$ *or* $\bar{G}_i = \tilde{G}_i$ *if* $K = \mathbb{C}$ (")
Moreover: ${}^{\omega *}A \simeq {}^{\omega *}\tilde{A}$ *in case* $K = \mathbb{C}$ *iff* $n = m$, $r = s$ *and*
c) $\forall$ i: $G_i = \tilde{G}_i$ (*up to numbering and conjugation*)

Here $\bar{G}_i := \{\bar{g} := \bar{a}\, z \mid g = a\, z \in G_i\}$.

Especially we have now: The C^∞- and C^ω-p-irreducible (in case $K = \mathbb{C}$ also the $C^{\omega *}$-p-irreducible) factorisations of K^n/G are all the "same" and correspond exactly to the irreducible factorisations $G \simeq G_1 \times \ldots \times G_s$. Here the G_i's are
i) up to conjugation if $K = \mathbb{R}$
ii) up to holomorphic or antiholomorphic conj. if $K = \mathbb{C}$
uniquely determined by the C^∞-p-factors of K^n/G.

The proofs require different methods for the case $K = \mathbb{R}$ and for the case $K = \mathbb{C}$. In the more involved case $K = \mathbb{R}$ essentially they are applications of results (for ex. factorisation-theorems) and methods (locally integrable vector fields) of [13], [14] and of [8], [9], described first by the first-named author. The case $K = \mathbb{C}$ uses the "Riemannscher Hebbarkeitssatz".

1. LOCALLY INTEGRABLE VECTOR FIELDS AND TRIVIAL FACTORS

From a quotient singularity we first factor aut trivial factors ${}^N(K^n)$. For this and further use we need some results from [13], [14].

A (tangent) vector field V on an N-differentiable space X is called locally integrable, if through each point $p \in X$ passes an integral curve of V on X. Let $T_p^i X$ denote the *set* of those tangent vectors $v \in T_p X$ of X at p, for which there exists a locally integrable field V in a neighbourhood of p on X with $V(p) = v$. We have ([13], [14]):

10) a) $T_p^i X$ is a vector space. b) A tangent vector field V on X is locally integrable iff $V(p) \in T_p^i X \ \ \forall\, p \in X$.
c) $T_p^i X = T_{p_1}^i X_1 \times T_{p_2}^i X_2$ for $p = (p_1, p_2) \in X := X_1 \times X_2$, if the set of manifold-points of X is dense in X. In this situation we also have:
d) A field $V = (V_1, V_2)$ on $X = X_1 \times X_2$ is locally integrable iff

$V_1 | X_1 \times \{q_2\}$, $V_2 | \{q_1\} \times X_2$ are locally integrable on X_1 resp. X_2 for each $(q_1, q_2) \in X_1 \times X_2$.

e) $r := \dim T_p^i X$ is the largest number s such that $X \simeq Y \times K^s$ (locally near p).

Note, that on real or complex analytic sets *each* differentiable or analytic vectorfield is locally integrable ([13]). However this is *not* true in general on more general spaces, especially not in general on semianalytic sets or even on quotient singularities.

For any germ $A \subset K^n$ let ${}^N A$ denote the associated reduced C^N-space(-germ). For a compact Lie group $G \subset GL(n,K)$ let $q: {}^N(K^n) \to {}^N(K^n)/G$ be a quotient map (which therefore factors any G-invariant mapping ${}^N(K^n) \to X$). The following known fact gives "b)$\curvearrowright$a)" of 7).

11) If $G = G_1 \times G_2$, $G_i \subset GL(n_i, K)$, $n_1 + n_2 = n$, then

$$ {}^N(K^n)/G \simeq {}^N(K^{n_1})/G_1 \times {}^N(K^{n_2})/G_2 $$

($N \in \{\infty, \omega\}$ if $K = \mathbb{R}$; $N = \omega^*$ if $K = \mathbb{C}$).

Because. As quotient map $q = (q_1, \ldots, q_\ell): K^n \to K^\ell$ any finite sequence of G-invariant polynomials q_i can be taken, which generates (as algebra) the set of all G-invariant polynomials on K^n ([1], [5], [10]). Moreover, by taking averages (summing over G) any polynomial q can be turned into a G-invariant polynomial q^*. We have $q = q^*$ iff q is G-invariant. This gives for $q_i = \Sigma\, a_{i\sigma\tau} x^\sigma \cdot y^\tau$, x on K^{n_1}, y on K^{n_2}, $G = G_1 \times G_2$:

$$ q_i^* = \Sigma\, a_{i\sigma\tau}\, x^{\sigma *} \cdot y^{\tau *} $$

Here $x^{\sigma *}$ (resp. $y^{\tau *}$) is G_1- (resp. G_2-)invariant. As generating sequence above we therefore may assume $q = (q_1, \ldots, q_\ell)$ to be of the following type: $q^1 := (q_1, \ldots, q_r)$ (resp. $q^2 := (q_{r+1}, \ldots, q_\ell)$) are G_1- (resp. G_2-) invariant polynomials on K^{n_1} (resp. K^{n_2}) generating the quotient mappings

$$ q^i : {}^N(K^{n_i}) \to {}^N(K^{n_i})/G_i, \quad q : {}^N(K^n) \to {}^N(K^n)/G. $$

This leads to the required result.

Note. The proof shows: If $K = \mathbb{R}$, then ${}^\infty(\mathbb{R}^n)/G$ *is* the reduced C^∞-differentiable space associated to the reduced C^ω-differentiable space ${}^\omega(\mathbb{R}^n)/G$.

12) Remark. If G operates properly on a space X, then any G-equivariant vector field $V: X \to TX$ pushes down to a vector field $V^*: X/G \to T(X/G)$. If V is locally integrable, so is V^*.

B e c a u s e. We have the following commutative diagrams for any $g \in G$:

$$\begin{array}{ccc} X & \xrightarrow{g} & X \\ \downarrow q & & \downarrow q \\ X/G & \xrightarrow{id} & X/G \end{array} \qquad \begin{array}{ccc} TX & \xrightarrow{Tg} & TX \\ \downarrow Tq & & \downarrow Tq \\ T(X/G) & \xrightarrow{id} & T(X/G) \end{array} \qquad \begin{array}{ccc} X & \xrightarrow{V} & TX \\ \downarrow q & & \downarrow Tq \\ X/G & \dashrightarrow^{V^*} & T(X/G) \end{array}$$

Because V is equivariant $(V \circ g = Tg \circ V)$, $Tq \circ V$ is G-invariant and factors therefore over V. The rest is obvious.

Let $G \subset GL(n,K)$ be compact with quotient map $q: {}^N(K^n) \to {}^N(K^n)/G$ and G without reflections in case $K = C_o$.

13) THEOREM. *The following are equivalent for* $K = \mathbb{R}_o$:
a) $Tq(o) \neq o$. b) *There exists a constant* G-*equivariant vector field* $V \neq o$ *on* ${}^N(K^n)$. c) $T_o^i({}^N(K^n)/G) \neq o$. d) ${}^N(K^n)/G \simeq {}^NA \times {}^NK$. e) *There exists a regular curve* k *on* ${}^N(K^n)/G$ *through the origin.* f) $G \subset GL(n-1, K)$ *(up to conjugation in* $GL(n,K)$*)*.

P r o o f. Without restriction: $q(o) = o$. By assumption: G is compact; any metric on $\mathbb{R}^n$ can be changed into a G-invariant metric (taking average over G). So without restriction: G is an orthogonal group.

"a) $\curvearrowright$ b)": $q = (q_1, \ldots, q_\ell)$, and without restriction: $Tq_1(o) \neq o$. q is G-invariant, so q_1, so its linear part $Tq_1(o)$; i.e. $Tq_1(o)$ considered as a row-vector, $g \in G$ as a matrix gives: $Tq_1(o)^t = (Tq_1(o) \cdot g)^t = g^t \cdot Tq_1(o)^t = g^{-1} \cdot Tq_1(o)^t$ for the transposed "t".

"b) $\curvearrowright$ c)": The field V under b) pushes down under Tq to a locally integrable field V^* on the quotient with $V^*(o) \neq o$.

"c) $\leftrightarrow$ d)": [14].

"d) $\curvearrowright$ e)": obvious.

"e) $\curvearrowright$ a)": x^2 on $\mathbb{R}^n$ is G-invariant and factors therefore: $x^2 = H \circ q$. Assume: $Tq(o) = 0$, then $|q(x)| \leq c\,x^2$ for some $c > 0$, all x near 0. Therefore: $|q(x)| \leq c \cdot x^2 = c \cdot H(q(x))$, $|y| \leq c \cdot H(y)$ for all y on the quotient close to 0. For any differentiable curve k on the quotient with $k(o) = 0$ we obtain $H \circ k(t) \geq 0$, therefore $(H \circ k)'(o) = 0$, $H \circ k(t) \leq d \cdot t^2$, $|k(t)| \leq c \cdot H \circ k(t) \leq c \cdot dt^2$, therefore $k'(o) = 0$. Finally, the equivalence "b) $\leftrightarrow$ f)" holds, because G is assumed to be orthogonal. With this 13) is proved.

13') THEOREM. *The following are equivalent for* $K = C_o$:
a + b) *There exist*: $L: K^n \to K^\ell$ *linear*, G-*invariant and* $V: K^n \to K^n$ *constant*, G-*equivariant vector field, with* $L(V) \neq 0$. c) $T_o^i({}^N(K^n)/G) \neq 0$. d) ${}^N(K^n)/G \simeq {}^NA \times {}^NK$. f) $G \subset GL(n-1, K)$ *(up to conjugation)*.

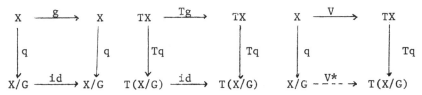

P r o o f. "c) $\curvearrowright$ d)" holds as above. "f) $\curvearrowright$ d)" follows from 11). "c) $\curvearrowright$ a+b)": G has no reflections; the set S of those points in K^n, where q has rank $< n$, therefore is of codimension ≥ 2. Any vector field V^* on the quotient, $V^*(o) \neq 0$, can locally be lifted to a field on K^n outside of S. Because of $\operatorname{codim} S \geq 2$ this lifting can uniquely be extended over $K^n \setminus S$ and then extended to a field $\tilde{V}$ on K^n (Hebbarkeitssatz of Riemann). $\tilde{V}$ is G-equivariant with V^* as push-down. Therefore $Tq(o)(\tilde{V}(o)) = V^*(o) \neq 0$. Now $\tilde{V}(o)$ can be considered as a constant G-equivariant vector field on K^n. With $L := Tq(o)$ statement a+b) follows. Finally, "a+b) $\curvearrowright$ f)" can be seen directly and is left to the reader.

2. FACTORING IN THE REAL CASE $K = \mathbb{R}_o$

Due to the general factorisation theorems of our introduction as well as to 10) and 13) we may first factor our trivial factors $^{\infty}(K^n)$ and restrict then to the following situation ($G \subset GL(n, \mathbb{R})$ compact, $G = \max {}^{\infty}G$, without restriction: G orthogonal):

14) Let $^{\infty}(\mathbb{R}^n_o)/G \simeq {}^{\infty}A_1 \times {}^{\infty}A_2$, $Tq(o) = 0$ for the associated quotient map q. Then there exist $G_i \subset GL(n_i, \mathbb{R})$ compact maximal Lie groups with $n = n_1 + n_2$, $G = G_1 \times G_2$ up to conjugation and $^{\infty}A_i \simeq {}^{\infty}(\mathbb{R}^n_o)/G_i$ for $i = 1, 2$.

P r o o f. For $q = (q_1, q_2)$, $q_i\colon \mathbb{R}^n_o \to A_i$, $A_1^* := q_2^{-1}(o)$; $A_2^* := q_1^{-1}(o)$ we obtain step by step a) - e):

a) Each A_i^* is G-invariant; the tangent spaces $T_o A_i^* \subset \mathbb{R}^n$ are G-invariant vector spaces; $^{\infty}A_i^*/G \simeq {}^{\infty}A_i$.

B e c a u s e. For example any G-invariant function on A_i^* can be extended over K^n, then can be made G-invariant over K^n and finally can be pushed down to $A_1 \times A_2$ and restricted onto $A_i \times \{0\}$. This gives the last statement.

Consider now q as a map $\mathbb{R}^n \to \mathbb{R}^s \supset ({}^{\infty}\mathbb{R}^n)/G$ and its differential as a map $Tq\colon \mathbb{R}^n \ni x \to \operatorname{Hom}(\mathbb{R}^n, \mathbb{R}^s) \simeq \mathbb{R}^{n \cdot s}$. We obtain next:

b) $Tq_1 \mid A_2^* = 0$, $Tq_2 \mid A_1^* = 0$,

$T(Tq_1)(o) \mid T_o A_2^* = 0$, $T(Tq_2)(o) \mid T_o A_1^* = 0$ with $A_i^* := {}^{\infty}A_i^*$.

B e c a u s e. If for example $x \in A_2^*$ gives $Tq_1(x) \neq 0$, take a curve h in $\mathbb{R}^n$, with $h(o) = x$ and $q_1 \circ h$ regular. But $q_1 \circ h(o) = 0$, so $k := (q_1 \circ h, o)$ would be a regular curve on $^{\infty}A_1 \times {}^{\infty}A_2$ through the origin, contradicting $Tq(o) = 0$ and 13). So $Tq_1 \mid A_2^* = 0$, similarly $Tq_2 \mid A_1^* = 0$. The rest then follows.

c) $A_i^* \subset T_o A_i^*$ for $i = 1, 2$.

B e c a u s e. The identical vector field $V^*\colon x \to x$ in $\mathbb{R}^n$ is G-equivariant (G is assumed to be linear) and therefore pushes down to a locally integrable vector field $V = (V_1, V_2)$ on $^{\infty}A_1 \times {}^{\infty}A_2$. Then each $V_i \mid A_i \times \mathrm{const}$ is locally integrable on A_i (10), d) and 12)). Connect a given $x \in A_1^*$ to 0 by the integral curve $h(t) = e^{t-1}x$ of V^*. Then $q \circ h =: (k_1, k_2)$ is an integral curve of V. We have: $k_2(1) = 0$; $V_2 \mid A_1 \times 0 = 0$ (by b)); $V_1 \mid A_1 \times 0$ locally integrable (10, d)). Then also $(k_1, 0)$ is an integral curve of V. By uniqueness of integral curves we now obtain $k_2 = 0$; hence $k = (k_1, 0)$, and k lies on $A_1 \times 0$, h on A_1^*, i.e. $s \cdot x$ A_1^* $\forall$ $0 < s \leq 1$. This gives c) for $i = 1$ and similarly for $i = 2$.

d) $T_o A_1^* \cap T_o A_2^* = 0$.

<u>Because</u>. x^2 is G-invariant and factors therefore by $x^2 = H \circ q$. Considering differential-maps as above and using $q(o) = 0$, $Tq(o) = 0$ we obtain by Taylor:

$$2x = T(H \circ q)(x) = TH(o)(T(Tq)(o)(x))$$

Therefore: $0 = T(Tq)(o)(x) = (T(Tq_1)(o)(x),\ T(Tq_2)(o)(x))$ iff $x = 0$. This gives d).

e) $T_o A_1^* + T_o A_2^* = \mathbb{R}^n$.

<u>Because</u>. G is orthogonal, there exists therefore a G-invariant vector space $V_3 \subset \mathbb{R}^n$ with $\mathbb{R}^n = T_o A_1^* \oplus T_o A_2^* \oplus V_3$. Write $V_i := T_o A_i^*$, $\bar{q}_i := q_i \mid V_i$ for $i = 1, 2$ and look at these as germs at 0. Then $\bar{q}_i : V_i \to A_i$ is surjectiv ($A_i^* \subset V_i$!) and G-invariant. Especially $\bar{q} := (\bar{q}_1, \bar{q}_2)$ factors like follows

$$\begin{array}{ccc} V_1 \times V_2 \times V_3 & \xrightarrow{\text{projection}} & V_1 \times V_2 \\ \Big\downarrow q & & \Big\downarrow \bar{q} \\ {}^{\infty}A_1 \times {}^{\infty}A_2 & \overset{f}{\dashrightarrow} & {}^{\infty}A_1 \times {}^{\infty}A_2 \end{array}$$

$\bar{q}$ is onto, hence also f. But $f \mid A_1^{\infty} \times \{0\} = \mathrm{id}$, $f \mid \{0\} \times A_2^{\infty} = \mathrm{id}$, hence $Tf(o) = \mathrm{id}$ and f is a diffeo. This means, that on $V_1 \times V_2$ and $V_1 \times V_2 \times V_3$ we have the same G-invariant functions. If z are coordinates on V_3, z^2 is a G-invariant function on V_3, so on $V_1 \times V_2 \times V_3$; but z^2 is zero on $V_1 \times V_2$. So $z = 0$, so $\dim V_3 = 0$, and e) is proved.

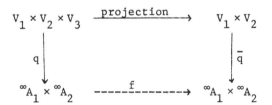

Changing linerly coordinates if necessary, we may assume now $\mathbb{R}^n = \mathbb{R}^{n_1} \times \mathbb{R}^{n_2}$, $V_i = \mathbb{R}^{n_i}$. Let $G_i := G \mid \mathbb{R}^{n_i}$, then $G \subset G_1 \times G_2$. Any G_i-invariant function g on V_i can be considered as G-invariant function on $V_1 \times V_2 = \mathbb{R}^n$, therefore can be pushed down by q to ${}^{\infty}A_1 \times {}^{\infty}A_2$. So g can be pushed down by $\bar{q}_i$ on ${}^{\infty}A_i$, therefore $\bar{q}_i : V_i \to {}^{\infty}A_i$ is a quotient map to G_i, and by 11 (and its proof) also $\bar{q} = (\bar{q}_1, \bar{q}_2)$ is a quotient map to $G_1 \times G_2$. So by the above diagram (with $V_3 = 0$) G and $G_1 \times G_2$ have the same invariant functions. By $G \subset G_1 \times G_2$ and the maximality of G we obtain: $G = G_1 \times G_2$, and moreover the G_i are maximal themselves. This proves 14).

The theorems 7), 8) now follow for $K = \mathbb{R}_0$:
For "7), b) $\rightsquigarrow$ a)" see 11). For "7), a) $\rightsquigarrow$ b)" start with a p-irreducible factorisation ${}^\infty A = {}^\infty B_1 \times \ldots \times {}^\infty B_s$ according to 1), especially ${}^\infty B_1 \simeq {}^\infty(\mathbb{R}^\ell)_o$. Using 10), we may apply 13) inductively to obtain $G \subset GL(n-\ell, \mathbb{R})$; i.e.: $G = \mathrm{id} \times \tilde{G}$ with $\tilde{G} \subset GL(n-\ell, \mathbb{R})$, $q = (q^1, q^2)$, $q^1: \mathbb{R}^n \to \mathbb{R}^\ell$ projection, $q^2: \mathbb{R}^{n-\ell} \to \mathbb{R}^s$ quotient map for $\tilde{G}$ with $Tq^2(0) = 0$. Then apply 14) inductively to obtain 7), b) for a p-irreducible factorisation. Now factoring out ${}^\infty A_j = {}^\infty(\mathbb{R}_o^{\ell_j}) \times {}^\infty A_j^*$ with largest possible ℓ_j, 10) gives $\Sigma\, \ell_j = \ell$. By 1') each ${}^\infty\tilde{A}_j$ is a product of ${}^\infty(\mathbb{R}_0^{\ell_j})$ with some of the ${}^\infty B_i$'s, $i > 0$, and $\times\, {}^\infty(\mathbb{R}_o^{\ell_j}) = {}^\infty B_1$. Gathering similarly on the group - level and using 11) we obtain the required G_i's. To prove 8), a) split up ${}^\infty A_j = {}^\infty(\mathbb{R}_0^{\ell_j}) \times {}^\infty A_j^*$ with *largest* possible ℓ_j. Then ${}^\infty A_j$ is singular iff ${}^\infty A_j^* \neq \{0\}$ and $T_o^i A_j^* = 0$ (use 10)), iff $G_j \neq \mathrm{id}$ (use 13)). 8), b) follows similarly with 1), 11) and 14).

Also theorem 9) follows for $K = \mathbb{R}$:
Let ${}^\infty A \simeq {}^\infty\tilde{A}$, especially $\dim A = \dim \tilde{A}$, therefore $n = m$ (the groups are finite). The irreducible factorisations of G resp. $\tilde{G}$ induce p-irreducible factorisations of ${}^\infty A$, ${}^\infty\tilde{A}$, which are the "same" (1)). By [9] the corresponding groups are conjugate. This proves one direction, the other direction is obvious.

Finally we prove 5) and 6) for $K = \mathbb{R}$:
We have $G \subset \max^\infty G \subset \max^\omega G$ and ${}^\omega(\mathbb{R}^n)/G \simeq {}^\omega(\mathbb{R}^n)/\max^\omega G$, because G and $\max^\omega G$ have the same invariant analytic functions. By [9]: G and $\max^\omega G$ are conjugate; due to $G \subset \max^\omega G$ and G finite, they are equal then. This proves 5). If G is compact, we may suppose G is orthogonal, then G, hence also $\max^\omega G$ keeps invariant the function x^2 on $\mathbb{R}^n$. Then also $\max^\omega G$ is orthogonal, and obviously also closed, hence compact. The same holds for $\max^\infty G$; but due to the proof of 11) we even have $\max^\infty G = \max^\omega G$.

3. FACTORING IN THE COMPLEX CASE $K = \mathbb{C}_o$

Let $G \subset GL(n, \mathbb{C})$ be a finite group without reflections. Again due to the unique decomposition theorems we may restrict at first to $N = \omega^*$ and the following situation:

15) Let ${}^{\omega*}(\mathbb{C}^n)/G \simeq {}^{\omega*}A_1 \times {}^{\omega*}A_2$. Then there exist $G_i \subset GL(n_i, \mathbb{C})$, without reflections, with $n = n_1 + n_2$, $G = G_1 \times G_2$ up to conjugation and ${}^{\omega*}A_i \times {}^{\omega*}(\mathbb{C}^n)/G_i$ for $i = 1, 2$.

<u>Proof</u>. In the following we omit the index ω^*; for ex. A_i means ${}^{\omega*}A_i$. As in Sec. 2 let $q = (q_1, q_2): \mathbb{C}_o^n \to A_1 \times A_2$ be a (holomorphic) quotient map, $q(0) = 0$, $A_1^* := q_2^{-1}(0)$, $A_2^* := q_1^{-1}(0)$ and their

tangent vector spaces $T_o A_1^*$, $T_o A_2^*$ are G-invariant. q is a finitely branched covering map; the branching germ $S \subset \mathbb{C}_o^n$ is of codimension ≥ 2 (: G has no reflections). Also $\bar{q}_i := q_i \mid A_i^* \to A_i \times \{0\}$ is a branched covering and $A_i^*/G \simeq A_i$. Especially $\dim A_i^* = \dim A_i$ everywhere.

a) If $x \in A_1$ is nonsingular, then each $y \in q_1^{-1}(x) \subset A_1^*$ is nonsingular for A_1^*, and $\bar{q}_1$ at these points is a local bimorphism. A similar result holds for A_2 and $A_1 \times A_2$.

B e c a u s e. Through any $v \in T_x A_1 \setminus \{0\}$ passes a field V on the germ A_{1x}. V can be considered as a field on $(A_1 \times A_2)_{(x,0)}$, which is constant in the A_2-variable and zero in the A_2-component. V can be lifted therefore to a field V^* on $\mathbb{C}_y^n$ for $q(y) = (x,0)$ (see the proof of 13´)). V^* is everywhere different from zero, its integral curves push down to integral curves of V, hence to regular curves on $A_1 \times$ const. Therefore: $V^* \mid A_1^*$ is tangent to A_1^* and moreover $\dim_y A_1^* = \dim_x A_1 \leq \dim T_y^i A_1^*$, $T\bar{q}_1(y)\colon T_y^i A_1^* \to T_x A_1 = T_x^i A_1$ is surjective (: $v \in T_x A_1 \setminus \{0\}$ was chosen arbitrarily). 10), e) now gives: A_1^* is a manifold at y. Then $\dim T_y^i A_1^* = \dim T_y A_1^* = \dim T_x A_1$, hence $T\bar{q}_1(y)$ is bijective, and $\bar{q}_1$ is a bimorphism at y.

b) There exist holomorphic retractions $r_i\colon \mathbb{C}_o^n \to A_i^*$, respecting G-orbits. The A_i^* are therefore G-invariant complex manifolds, and the induced mappings $r_i \mid (T_o A_1^*)_o\colon (T_o A_1^*)_o \to A_i^*$ are biholomorphic and respect G-orbits.

B e c a u s e. A point $x = (x_1, x_2) \in A_1 \times A_2$ is regular iff each $x_i \in A_i$ is regular. Using a) we see, that in regular points the projection $A_1 \times A_2 \to A_1$ can be lifted locally and can then be extended holomorphically to a map $\mathbb{C}_o^n \setminus S \to A_1^*$, then to a map $r_1\colon \mathbb{C}_o \to A_1^*$ (: codim S ≥ 2, Riemannscher Hebbarkeitssatz):

$$\begin{array}{ccc}
\mathbb{C}_o^n & \longrightarrow & \mathbb{C}_o^n/G = A_1 \times A_2 \\
\Big\downarrow r_1 & & \Big\downarrow \text{projection} \\
A_1^* & \longrightarrow & A_1^*/G = A_1
\end{array}$$

The diagram commutes, therefore $r_1 \mid A_1^* = g \mid A_1^*$ for some $g \in G$: first locally over a regular point x of A_1 (a)), then globally by the identity-theorem. Changing from r_1 to $g^{-1} \circ r_1$ if necessary we may assume $r_1 \mid A_1^* = \text{id}$. In a similar way we find r_2. b) now follows with [11].

c) $V_i/G \simeq A_i$ for $V_i := (T_o A_i^*)_o$.

<u>Because</u>. $r_i \mid V_i : V_i \to A_i^*$ are biholomorphic; V_i, A_i^* are G-invariant, $r_i \mid V_i$ respect G-orbits. Therefore $V_i/G \simeq A_i^*/G \simeq A_i$.

d) Let $G_i := G \mid V_i$. Then $V_1 \oplus V_2 = \mathbb{R}_0^n$, $G = G_1 \times G_2$ (up to conjugation), and 15) follows.

<u>Because</u>. $A_1^* = r_1(\mathbb{C}_o^n)$, $A_2^* = r_1^{-1}(0)$. Then A_1^*, A_2^* are transversal at $\{0\} = A_1^* \cap A_2^*$. This gives $V_1 + V_2 = \mathbb{R}_0^n$, $V_1 \cap V_2 = \{0\}$, $V_1 \oplus V_2 = \mathbb{R}_0^n$. The quotient mappings $q: \mathbb{C}_o^n \to \mathbb{C}_o^n/G = A_1 \times A_2$ and $p: V_1 \times V_2 \to (V_1/G_1) \times (V_2/G_2) = A_1 \times A_2$ (: 11)) are not branched outside of sets S of codimension ≥ 2 (a)). Biholomorphisms on $\mathbb{C}_o^n$, that push down by q to id, are exactly those given by the $g \in G$ (this is obvious locally over regular points of $A_1 \times A_2$, then also globally). A similar result holds for $V_1 \times V_2$ and p (using a)). Because of $\operatorname{codim} S \geq 2$, the identity $A_1 \times A_2 \leftrightarrow A_1 \times A_2$ can be lifted to mappings $\phi: \mathbb{C}_o^n \to V_1 \times V_2$, $\psi: V_1 \times V_2 \to \mathbb{C}_o^n$ with $p \circ \phi = q$, $q \circ \psi = p$:

$$\begin{array}{ccc}
\mathbb{C}_o^n & \underset{\psi}{\overset{\phi}{\rightleftarrows}} & V_1 \times V_2 \\
\downarrow q & & \downarrow p \\
\mathbb{C}_o^n/G & & (V_1/G_1) \times (V_2/G_2) \\
\parallel & & \parallel \\
A_1 \times A_2 & \longleftrightarrow & A_1 \times A_2
\end{array}$$

If follows: $\psi \circ \phi \in G$ and $\phi \circ \psi \in G_1 \times G_2$. So ϕ is biholomorphic. In the same way: $\phi^{-1} \circ g \circ \phi \in G_1 \times G_2$, hence $T\phi^{-1}(0) \circ g \circ T\phi(0) \in G_1 \times G_2$ for each g G; similarly $T\phi(0) \circ g \circ T\phi^{-1}(0) \in G$ for each $g \in G$.

The theorems 7), 8) now follow for $K = \mathbb{C}_o$:
For "7), b)$\curvearrowright$a)" see 11) and the def. of ${}^\infty A$ in this case (bottom p. 2). "a)$\curvearrowright$b)": $A = \mathbb{C}_o^n/G$ is an irreducible complex analytic germ. Let ${}^\infty A = {}^\infty B_1 \times \ldots \times {}^\infty B_r$ and ${}^{\omega*}A = {}^{\omega*}C_1 \times \ldots \times {}^{\omega*}C_s$ be p-irreducible factorisations. By [14] or [2] we have $s = r$ and up to numbering ${}^\infty B_i \simeq {}^\infty C_i$. Using 15) by induction we obtain up to conjugation $G = G_1 \times \ldots \times G_r$ and ${}^{\omega*}(\mathbb{C}_o^n)/G_i \simeq {}^{\omega*}B_i$, hence ${}^\infty(\mathbb{C}_o^n)/G_i \simeq {}^\infty B_i \simeq {}^\infty C_i$. Thus "a)$\curvearrowright$b)" is proved for p-irreducible factorisations. In the general situation use 1´) and proceed as in the proof for $K = \mathbb{R}$ in Sec. 2. To prove 8), a) note, that ${}^\infty A_i$ is nonsingular iff ${}^{\omega*}A_i$ is nonsingular (for ex. [11]), iff $\mathbb{C}^{n_i}/\mathrm{id} = \mathbb{C}^{n_i}/G_i$, iff $G_i = \mathrm{id}$ (see the proof of 15) above). 8), b) follows similarly.

Also theorem 9) follows for $K = \mathbb{C}$:
Start with p-irreducible factorisations according to 1):

$${}^{\omega*}A_1 \times \ldots \times {}^{\omega*}A_r \simeq {}^{\omega*}A, \quad {}^{\omega*}\tilde{A}_1 \times \ldots \times {}^{\omega*}\tilde{A}_s \simeq {}^{\omega*}\tilde{A},$$

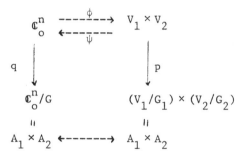

and without restriction again: the A_i, $\tilde{A}_i$ are complex analytic. Then $r = s$, ${}^{\infty}A_1 \times \ldots \times {}^{\infty}A_r \simeq {}^{\infty}A \simeq {}^{\infty}\tilde{A} \simeq {}^{\infty}\tilde{A}_1 \times \ldots \times {}^{\infty}\tilde{A}_r$ are p-irreducible factorisations (3)), and we have up to numbering ${}^{\infty}A_i \simeq {}^{\infty}\tilde{A}_i$ (1)), even ${}^{\omega^*}A_i \simeq {}^{\omega^*}\tilde{A}_i$ or ${}^{\omega^*}\bar{A}_i \simeq {}^{\omega^*}\tilde{A}_i$ (3)). Here $\bar{A}_i \subset \mathbb{C}_o^n$ is the image of A_i under $z \to \bar{z}$. To the p-irreducible factorisations of ${}^{\omega^*}A$ resp. ${}^{\omega^*}\tilde{A}$ correspond irreducible factorisations $G \simeq G_1 \times \ldots \times G_r$, $\tilde{G} \simeq \tilde{G}_1 \times \ldots \times \tilde{G}_r$ (8)). Moreover we have up to numbering ${}^{\omega^*}(\mathbb{C}_o^{n_i})/G_i \simeq {}^{\omega^*}A_i$, ${}^{\omega^*}\tilde{A}_i \simeq {}^{\omega^*}(\mathbb{C}_o^{n_i})/\tilde{G}_i$, therefore also ${}^{\omega^*}(\mathbb{C}_o^{n_i})/\bar{G}_i \simeq {}^{\omega^*}\bar{A}_i$. Because: $f: \mathbb{C}_o^{n_i} \to \mathbb{C}_o^s$ is holomorphic iff $\bar{f}: \mathbb{C}^{n_i} \to \mathbb{C}_o^s$ is holomorphic with $\bar{f}: z \to \bar{z} \to f(\bar{z}) \to \overline{f(\bar{z})}$. One also obtains: f is G_i-invariant iff $\bar{f}$ is $\bar{G}_i$-invariant. This finally implies: $q_i: \mathbb{C}_o^{n_i} \to \mathbb{C}_o^{s_i}$ is a quotient map with respect to G_i iff $\bar{q}_i$ is a quotient map with respect to $\bar{G}_i$. And this was to be shown. As in the proof of 15), d) (or with [4] or [6]) we obtain now up to conjugation: $G_i = \tilde{G}_i$ or $\bar{G}_i = \tilde{G}_i$. With this, 9), b), c) are established. The other direction of the statement is obvious.

Finally 5) and 6) hold:
5) The identity $G = \max{}^{\omega^*}G$ follows as in the proof 15), d) above. Or realising first, that $\max{}^{\omega^*}G$ has no reflections (proof 15), d)) one also may use [4] or [6]. In case $\max^{\infty}G$ or $\max^{\omega}G$, G has to be considered as a "real group"; 5) follows then from Sec. 2. 6) follows from 4) and 5).

4. EXAMPLES

The assumptions made for 7) and 8) are necessary:
I) Case $K = \mathbb{R}$: Let $G := \{g \in O(2) \times O(2) \mid \det g = +1\} \subset O(4)$, where $\mathbb{R}^4 = \mathbb{R}^2 \times \mathbb{R}^2$. G is a compact Lie group, which is *not* maximal, because the G-invariant functions are generated by x^2, y^2, where (x,y) are coordinate functions of $\mathbb{R}^2 \times \mathbb{R}^2$. Therefore $\max^{\infty}G = O(2) \times O(2) \neq G$. We have $\mathbb{R}^4/G = \mathbb{R}_+ \times \mathbb{R}_+$; but there is no splitting $\mathbb{R}^4 = V_1 \times V_2$ as in 15), d), because the only G-invariant subspaces of $\mathbb{R}^4$ are $\mathbb{R}^2 \times \{0\}$ and $\{0\} \times \mathbb{R}^2$. And on these spaces G induces already all of $O(2)$. So G can not truely be of the form $G = G_1 \times G_2$. However, for any compact group $G \subset GL(n, \mathbb{R})$ with $\mathbb{R}^n/G \simeq A_1 \times A_2$, there always exist $\mathbb{R}^n = V_1 \oplus V_2$ with V_i G-invariant and $V_i/G \simeq A_i$ (but in general not $G = (G \mid V_1) \times (G \mid V_2)$): Use $\mathbb{R}^n/G = \mathbb{R}^n/\max G$ and apply 7) for $\max G$.

II) Case $K = \mathbb{C}$: Let $G \subset GL(2,C)$ be a Dieder group of at least 6 elements, generated by two reflections. The quotient $\mathbb{C}_o^2/G$ is regular, so $\mathbb{C}_o^2/G = \mathbb{C}_o^2 = \mathbb{C}_o^1 \times \mathbb{C}_o^1$ ([6]). But this product does not lead to a decomposition of $\mathbb{C}_o^2$ into G-invariant subspaces, because there are only trivial such spaces.

This shows that, despite of their formal similarity, the factorisation theorems 7), 8), 9) are fundamentally different in content for the case $K=\mathbb{R}$ or $K=\mathbb{C}$.

References

[1] CARTAN, H.: 'Quotient d'un espace analytique par un groupe d'automorphismes', in: *Algebraic geometry and topology. A symposion in honor of S. Lefschetz.* (Princeton 1954); *Princeton Univ. Press* 1957, 90-102.

[2] EPHRAIM, R.: 'C^∞ and analytic equivalence of singularities' in: *Proc. of the Conference of Complex Analysis 1972, Rice Univ. Studies.*

[3] ———: 'The cartesian product structure of singularities', *Trans. Amer. Math. Soc.* 224 (1976), 299-311.

[4] GOTTSCHLING, E.: 'Invarianten endlicher Gruppen und biholomorphe Abbildungen', *Inv. Math.* 6 (1969), 315-326.

[5] LUNA, D.: 'Fonctions differentiables invariantes sous l'operation d'un groupe reductif', *Ann. Inst. Fourier* 26,1 (1976), 33-49.

[6] PRILL, D.: 'Local classification of quotients of complex manifolds', *Duke Math. J.* 34 (1967), 375-386.

[7] REICHARD, K.: 'Quotienten analytischer und differenzierbarer Räume nach Transformationsgruppen', Preprint Bochum 1978.

[8] ———: 'C^∞-Diffeomorphismen semianalytischer und subanalytischer Mengen', *Composito Math.* 42,3 (1981), 401-416.

[9] ———: 'Lokale Klassifikation von Quotientensingularitäten reeller Mannigfaltigkeiten nach diskreten Gruppen', *Math. Z.* 179 (1982), 287-292.

[10] SCHWARZ, G.W.: 'Smooth functions invariant under the action of a compact Lie group', *Top.* 14 (1975), 63-68.

[11] SPALLEK, K.: 'Über Singularitäten analytischer Mengen', *Math. Ann.* 172 (1967), 249-268.

[12] ———: 'Differenzierbare Räume', *Math. Ann.* 180 (1969), 269-296.

[13] ———: 'Geometrische Bedingungen für die Integrabilität von Vektorfeldern auf Teilmengen im $\mathbb{R}^n$', *Manuscripta Math.* 25 (1978), 147-160.

[14] ———: 'Produktzerlegung und Äquivalenz von Raumkeimen I. Der allgemeine Fall', *Lecture Notes in Math.* 1014 (1983), 78-100 und: 'Produktzerlegung und Äquivalenz von Raumkeimen II. Der komplexe Fall', *Lecture Notes in Math.* 1014 (1983), 101-111.

APPROXIMATION AND EXTENSION OF C^∞ FUNCTIONS DEFINED ON COMPACT SUBSETS OF $\mathbb{C}^n$

Wiesław Pawłucki and Wiesław Pleśniak
Institute of Mathematics
Jagiellonian University
PL-30-059 Kraków, Poland

ABSTRACT. A topological linear embedding into the Fréchet space of rapidly decreasing sequences is constructed (Section 6) on a (fat) compact subset of $\mathbb{R}^n$ whose Siciak's extremal function is Hölder continuous. The existence of such an embedding appears to be equivalent to the existence of a continuous linear extension for C^∞ functions. Thus, the main result of our previous paper [8] is reproved in a different way. Also a 'complex' version of Bernstein's theorem, proved earlier in our paper [7], is given.

INTRODUCTION

By a classical theorem of Bernstein [2], a function $f : I = [a,b] \to \mathbb{R}$ extends to a C^∞ function on $\mathbb{R}$ if and only if the sequence

$$\mathrm{dist}_I(f,P_k) := \inf\{|f-p|(I) : p \in P_k\}, \quad \text{for} \quad k = 0, \ldots ,$$

is rapidly decreasing, P_k denoting the space of all polynomials of degree at most k. This beautiful result can relatively easily be extended to the multidimensional case by replacing I with a bounded domain in $\mathbb{R}^n$ having a cone property (see [1]). In [7] we proved an essentially stronger version of Bernstein's theorem for compact subsets of $\mathbb{R}^n$ admitting polynomial cusps or, more generally, for compact subsets of $\mathbb{R}^n$ whose Siciak's extremal function is Hölder continuous (HCP sets). An important point in proving that result was an n-dimensional version of Markov's inequality for the derivatives of polynomials. Both results, Bernstein's theorem and Markov's inequality permitted us to give in [8] a simple formula involving Lagrange interpolation polynomials with knots in Fekete-Leja extremal points, for a continuous linear operator extending C^∞ functions defined on (fat) HCP compact sets to the whole space $\mathbb{R}^n$.

In this paper, by applying a similar technique we construct a topological linear embedding of the space of C^∞ functions defined on a (fat) HCP compact set in $\mathbb{R}^n$ into the Fréchet space of rapidly decreasing sequences (Section 6). By the result of Tidten [14], the existence of such an embedding is equivalent to the existence of a

J. Ławrynowicz (ed.), Deformations of Mathematical Structures, 283–295.

continuous linear extension for C^∞ functions. Thus, we reprove in a different way the main result of our previous paper [8]. We also give (in Section 2) a 'complex' version of Bernstein's theorem proved earlier in [7].

1. HÖLDER CONTINUITY OF EXTREMAL FUNCTION

Let E be a compact subset of $\mathbb{C}^n$. Define, for $x \in \mathbb{C}^n$,

$$\Phi_E(x) = \sup\{|p(x)|^{1/\deg p};\ p \text{ is a polynomial from } \mathbb{C}^n \text{ to } \mathbb{C} \text{ of degree at least 1 with } \sup|p|(E) \leq 1\}.$$

The function Φ_E was introduced by Siciak [10]. It is called the *extremal function* of E. This function is a multidimensional counterpart of the *Green function* for $\mathbb{C} \setminus \hat{E}$ with a pole at $z = \infty$, where $\hat{E}$ denotes the polynomially convex hull of E, and for this reason it appears a very useful tool in the complex analysis.

In the sequel we shall be interested in compact subsets E of $\mathbb{C}^n$ having the following Hölder continuity property:

(HCP) There exist positive constants M and μ such that

$$\Phi_E(x) \leq 1 + M\delta^\mu \quad \text{as} \quad \operatorname{dist}(x,E) \leq \delta \leq 1.$$

In virtue of the definition, $\Phi_E(x) \geq 1$ everywhere in $\mathbb{C}^n$ and $\Phi_E(x)=1$ if and only if $x \in \hat{E}$. It is known ([20], [12]) that the continuity of Φ_E at every point of E implies its continuity in the whole space $\mathbb{C}^n$. Hence, in particular for every compact subset E of $\mathbb{C}^n$ having HCP (briefly: HCP compact subset), the function Φ_E is continuous in $\mathbb{C}^n$. (A subset of $\mathbb{C}^n$ with the latter property is usually called *L-regular*.) In [7] we have introduced the following family of HCP sets admitting polynomial cusps:

Let E be a subset of the space $\mathbb{K}^n$, where $\mathbb{K} = \mathbb{C}$ or $\mathbb{K} = \mathbb{R}$. E is said to be *uniformly polynomially cuspidal* (briefly: UPC) if there exist constants $M > 0$ and $m > 0$, a positive integer d and a mapping $h : \overline{E} \times [0,1] \to \mathbb{K}^n$ such that, for each $x \in \overline{E}$, $t \mapsto h(x,t)$ is a polynomial map of degree at most d, and

(i) $h(x,t) \in \overline{E}$ and $h(x,0) = x$ for each $x \in \overline{E}$ and $t \in [0,1]$;

(ii) $\operatorname{dist}(h(x,t), \mathbb{K}^n \setminus E) \geq Mt^m$ for $(x,t) \in \overline{E} \times [0,1]$.

1.1. PROPOSITION ([7], Theorem 4.1). *Suppose* E *is a* UPC *compact subset of* $\mathbb{K}^n$. *Then for* $x \in \mathbb{C}^n$

$$\Phi_E(x) \leq 1 + C\delta^{1/2[m]} \quad \textit{as} \quad \operatorname{dist}(x,E) \leq \delta \leq 1,$$

where C *is a positive constant depending only on the constants* M, m *and* d *of the definition of* UPC *and* $[m] := k$ *as* $k - 1 < m \leq k$, *where* $k \in \mathbb{Z}$.

Thus, every UPC compact set in $\mathbb{K}^n$ is an HPC subset of $\mathbb{C}^n$. It can easily be checked that every bounded domain D satisfying the following cone property is UPC: There exists $r>0$ such that for each point $b\in\overline{D}$ we can find a point $a\in D$ such that the convex hull of $\{x;\ |x-a|\leq r\}\cup\{b\}$ is contained in $D\cup\{b\}$. In particular, every bounded convex domain or bounded domain with Lipschitz boundary is UPC. Using Hironaka's rectilinearization theorem and Łojasiewicz's inequality (see [4]) we have proved in [7] that every fat subanalytic compact set in $\mathbb{R}^n$ is UPC as well. (E is said to be *fat* if $E\subset\overline{\operatorname{int} E}$.)

Let us recall the definition of semianalytic and subanalytic sets. A subset E of $\mathbb{R}^n$ is said to be *semianalytic* if for each x in $\mathbb{R}^n$ there exists a neighbourhood U of x and a finite number of real analytic functions f_i, g_i on U such that

$$E\cap U=\bigcup_i\{f_i=0,\ g_i>0\}.$$

E is called *subanalytic* if for each x in $\mathbb{R}^n$ there exists a neighbourhood U of x such that $E\cap U=\pi(A)$, where A is a bounded semianalytic subset of $\mathbb{R}^{n+m}$ for some m and $\pi:\mathbb{R}^{n+m}\to\mathbb{R}^n$ is the natural projection. The class of all subanalytic sets is stable under most of set-theoretical and topological operations. If $n\geq 3$, the class of all subanalytic subsets of $\mathbb{R}^n$ is essentially larger than that of all semianalytic sets, e.g. the projection of a (relatively compact) semianalytic set need not be semianalytic (cf. [5], pp. 133-135), both classes being identical if $n\leq 2$. For basic properties of semianalytic and subanalytic sets we refer the reader to [5] and [4].

The class of all UPC subsets of $\mathbb{R}^n$ is essentially wider than that of (fat) compact subanalytic sets. In [7] we have given some non-trivial examples of sets that are UPC but not subanalytic. Another class of UPC sets in $\mathbb{R}^n$ is yielded by a class of sets introduced in [14]. Following Tidten [14] we say that a closed subset A of $\mathbb{R}^n$ satisfies the *generalized cone condition* (briefly: GCC) if for each compact set K in $\mathbb{R}^n$ there are a parabolic cone

$$S=\{(x_1,\dots,x_n)\in\mathbb{R}^n;\ 0\leq x_1\leq 1,\ x_2^2+\dots+x_n^2\leq x^{2r}\}$$

of order $r\geq 1$ and a family $(f_i)_{i\in I}$ of C^∞ diffeomorphisms $f_i:S\to f_i(S)\subset A$ such that

(a) $K\cap A\subset\overline{\bigcup_{i\in I} f_i(S)}$;

(b) $\sup_{i\in I}|f_i|_k<\infty$ and $\sup_{i\in I}|f_i^{-1}|_k<\infty$ for each $k=0,1,\dots$.

(As usual, if $h=(h^1,\dots,h^n):E\to\mathbb{R}^n$ is a C^∞ mapping and E is compact, we put $|h|_k:=\max\{\sup_{x\in E}|D^\alpha h^i(x)|;\ |\alpha|\leq k,\ 1\leq i\leq n\}$.)

The fact that every compact GCC set is UPC follows from the proposition we have proved in [8]:

1.2. PROPOSITION. *Let* K *be a compact set in* $\mathbb{R}^n$ *and let* $f:\mathbb{R}^n\to\mathbb{R}^n$ *be a* C^∞ *mapping such that for each* $x\in K$, *the Jacobi determinant of* f, $D(x)\neq 0$. *If then* K *is* UPC, *so is the set* $f(K)$.

On the other hand, the set $E=\{(x,y)\in\mathbb{R}^2;\ 0\leq x\leq 1,\ x^{2/3}\leq y\leq 2x^{2/3}\}$ is UPC (since it is semialgebraic - see [5]) but not GCC.

2. POLYNOMIAL APPROXIMATION OF C^∞ FUNCTIONS

In [7] (Theorem 5.1 and Remark 3.2) we have proved the following extension of the classical Bernstein's theorem:

2.1. THEOREM. *Assume that* E *is a* HCP *compact set in* $\mathbb{R}^n$. *A real-valued function* f *defined on* E *is the restriction to* E *of a* C^∞ *function in* $\mathbb{R}^n$ *if and only if for each* $r>0$

$$\lim_{k\to\infty} k^r \operatorname{dist}_E(f,P_k)=0.$$

Here P_k is the linear space of (the restrictions to E of) all polynomials from $\mathbb{R}^n$ to $\mathbb{R}$ of degree at most k, and $\operatorname{dist}_E(f,P_k)=\inf\{\|f-p\|_E;\ p\in P_k\}$, $\|h\|_E$ denoting the uniform norm of h on E. This theorem results from HCP and from the following multidimensional version of a known Markov's inequality:

2.2. PROPOSITION ([7], Theorem 3.1 and Remark 3.2). *Let* E *be a* HCP *compact set in* $\mathbb{C}^n$. *Then there exists a constant* $c>0$ *such that for each polynomial* $p:\mathbb{C}^n\to\mathbb{C}$ *of degree at most* k *and each multi-index* $\alpha\in\mathbb{Z}^n_+$, *we have*

$$\|D^\alpha p\|_E \leq Ck^{c|\alpha|}\ \|p\|_E,$$

where the constant C *depends only on* E.

There is a 'complex' version of the 'if' part of Theorem 2.1.

2.3. THEOREM. *Let* E *be a compact subset of* $\mathbb{C}^n$ *satisfying* HCP *and let* $f:E\to\mathbb{C}$. *Suppose that for each* $s>0$

$$\lim_{k\to\infty} k^s \operatorname{dist}_E(f,P_k^{\mathbb{C}})=0, \tag{2.1}$$

where $P_k^{\mathbb{C}}$ *is the space of the restrictions to* E *of all polynomials from* $\mathbb{C}^n$ *to* $\mathbb{C}$ *of degree at most* k. *Then* f *is holomorphic in* $\operatorname{int}E$ *and extends to a* C^∞ *function on* $\mathbb{R}^{2n}$.

The proof goes on similar lines as that of Theorem 2.1 (see [7], Theorem 5.1). By the assumptions there is a sequence $\{p_k\}$ of polynomials $p_k\in P_k^{\mathbb{C}}$ such that

$$f = \Sigma_{k=1}^{\infty} p_k \tag{2.2}$$

and

$$\lim_{k\to\infty} k^s \|p_k\|_E = 0 \qquad \text{for each} \quad s > 0. \tag{2.3}$$

Set $\varepsilon_k = (1/Mk)^{1/\mu}$, for $k = 1,2,\ldots$, where the constants M and μ are the same as in (HCP). It is known (see e.g. [16], Chap. IV, Lemma 3.3) that we can find positive constants C_α, for $\alpha \in \mathbb{Z}_+^{2n}$, such that for any k there exists a C^∞ function u_k on $\mathbb{R}^{2n}$ satisfying $0 \leq u_k \leq 1$, $u_k = 1$ in a neighbourhood of E and

$$\operatorname{supp} u_k \subset E_k := \{x \in \mathbb{R}^{2n};\ \operatorname{dist}(x,E) \leq \varepsilon_k\}, \tag{2.4}$$

and

$$|D^\alpha u_k(x)| \leq C_\alpha \varepsilon_k^{-|\alpha|}, \quad \text{for all} \quad x \in \mathbb{R}^{2n} \quad \text{and} \quad \alpha \in \mathbb{Z}_+^{2n}. \tag{2.5}$$

Now define

$$g = \Sigma_{k=1}^{\infty} u_k p_k.$$

By (2.4) and (2.3) the series g is uniformly convergent in $\mathbb{R}^{2n}$, since we have

$$\sup_{\mathbb{R}^{2n}} |u_k p_k| \leq \sup_{E_k} |u_k p_k| \leq \sup_{E_k} |p_k| \leq \|p_k\|_E (\sup_{E_k} \Phi_E)^k$$

$$\leq \|p_k\|_E (1 + 1/k)^k \leq e \|p_k\|_E.$$

Moreover, by (2.4), (2.5) and HCP, for each $\alpha \in \mathbb{Z}_+^{2n}$ we have

$$\sup_{\mathbb{R}^{2n}} |D^\alpha (u_k p_k)| \leq \sum_{\beta \leq \alpha} \binom{\alpha}{\beta} \sup_{E_k} |D^{\alpha-\beta} u_k D^\beta p_k| \leq$$

$$\sum_{\beta \leq \alpha} \binom{\alpha}{\beta} C_{\alpha-\beta} \varepsilon_k^{-|\alpha-\beta|} \sup_E \left| i^{|\delta|} \frac{\partial^{|\beta|}}{\partial z^{\gamma+\delta}} p_k(z) \right| (1+1/k)^k \leq M_\alpha k^{q|\alpha|} \|p_k\|_E,$$

where $\beta = (\gamma,\delta) \in \mathbb{Z}_+^n \times \mathbb{Z}_+^n$, $q = \max(1/\mu, c)$ and M_α is independent of k. Hence, by (2.3), g is C^∞ on $\mathbb{R}^{2n}$ and by (2.4), $g = f$ on E.

Now suppose that $n = 1$. Let E be a compact set in the complex plane $\mathbb{C}$ satisfying the following condition:

(r) There exists a constant $r > 0$ such that for each $x \in E$ we can find a continuum $F \subset E$ with $\sup\{|y - z|;\ y,z \in F\} \geq 2r$ such that $x \in F$.

It is known ([11]) that if E satisfies (r) then

$$\Phi_E(x) \leq 1 + M\delta^{1/2}, \quad \text{as} \quad \operatorname{dist}(x,E) \leq \delta \leq 1, \tag{2.6}$$

where M is a positive constant depending only on r. Hence, in particular, we get

2.4. COROLLARY. *If* E *is a plane compact set satisfying* (r), *then every function* $f : E \to \mathbb{C}$ *such that for each* $s > 0$

$$\lim_{k\to\infty} k^s \operatorname{dist}_E(f, P_k^{\mathbb{C}}) = 0$$

extends to a C^∞ *function in* $\mathbb{R}^2$.

Corollary 2.4 is useful e.g. in studying the distribution of the zeros of the polynomials of best approximation to a C^∞ function on E (see [19]).

If $n > 1$ and $E = E_1 \times \dots \times E_n$, where E_i is a plane continuum, then it is known ([10]) that for each $x = (x_1,\dots,x_n) \in \mathbb{C}^n$

$$\Phi_E(x_1,\dots,x_n) = \max_i \Phi_{E_i}(x_i).$$

This property together with (2.6) yield another class of compact sets in $\mathbb{C}^n$ having HCP.

Contrary to the 'real' case, the condition (2.1) is not necessary in order that a function f defined on $E \subset \mathbb{C}^n$ be extendible to a C^∞ function on $\mathbb{R}^{2n}$. Let us consider the following example:

2.5. Example. Let $E = \{z \in \mathbb{C};\ |\operatorname{Re} z| \leq 1,\ \operatorname{Im} z = |\operatorname{Re} z|\}$. Let $f(z) = \bar{z}$. If there existed polynomials $p_k \in P_k^{\mathbb{C}}$, $k = 1,2,\dots,$ such that for each $s > 0$

$$\lim_{k\to\infty} k^s \|f - p_k\|_E = 0,$$

then due to (2.6) and Proposition 2.2 we would have

$$\lim_{E\ni z\to 0} \frac{f(z)-f(0)}{z} = \lim_{E\ni z\to 0} \Big\{\frac{p_1(z)-p_1(0)}{z} + \sum_{k=1}^{\infty} \frac{p_{k+1}(z)-p_k(z)-[p_{k+1}(0)-p_k(0)]}{z}\Big\} = p'(0) + \sum_{k=1}^{\infty} [p'_{k+1}(0) - p'_k(0)],$$

what is impossible.

3. LAGRANGE INTERPOLATION POLYNOMIALS

Let $\kappa : \mathbb{N} \ni j \to \kappa(j) = (\kappa_1(j),\dots,\kappa_n(j)) \in \mathbb{Z}_+^n$ be a one-to-one mapping such that for each j $|\kappa(j)| \leq |\kappa(j+1)|$. Let m_k denote the number of monomials

$$x^\alpha := x_1^{\alpha_1} \dots x_n^{\alpha_n}$$

of degree at most k. We can easily verify that $m_k = \binom{n+k}{k}$. Let

$e_j(x) := x^{\kappa(j)}$ for $j=1,2,\ldots$. The set of monomials $e_1,\ldots,e_{m_k}$ is a basis of the space $P_k^{\mathbb{K}}$ of all polynomials from $\mathbb{K}^n$ to $\mathbb{K}$ of degree at most k.

Let $t^{(k)} = \{t_1,\ldots,t_k\}$ be a system of k points of $\mathbb{K}^n$. Consider the Vandermondian

$$V(t^{(k)}) = V(t_1,\ldots,t_k) := \det[e_j(t_i)],$$

where $i,j \in \{1,\ldots,k\}$. If $V(t^{(k)}) \neq 0$, we define

$$L^{(j)}(x,t^{(k)}) := V(t_1,\ldots,t_{j-1},x,t_{j+1},\ldots,t_k)/V(t^{(k)}).$$

Since $L^{(j)}(t_i,t^{(k)}) = \delta_{ij}$ (Kronecker's symbol), we get the following *Lagrange interpolation formula* (cf. [10], Lemma 2.1):

(LIF) If $p \in P_k^{\mathbb{K}}$ and $t^{(m_k)}$ is a system of m_k points of $\mathbb{K}^n$ such that $V(t^{(m_k)}) \neq 0$, then

$$p(x) = \sum_{j=1}^{m_k} p(t_j) L^{(j)}(x,t^{(m_k)}) \quad \text{for } x \in \mathbb{K}^n.$$

Let E be a compact set in $\mathbb{K}^n$. A system $t^{(k)}$ of k points $t_1,\ldots,t_k$ of E is called a *Fekete-Leja system of extremal points* of E of order k, if $V(t^{(k)}) \geq V(s^{(k)})$ for all systems $s^{(k)} = \{s_1,\ldots,s_k\} \subset E$. Observe that if $t^{(k)}$ is a system of extremal points of E such that $V(t^{(k)}) \neq 0$, then

$$|L^{(j)}(x,t^{(k)})| \leq 1 \quad \text{on } E, \quad \text{for } j=1,\ldots,k. \tag{3.1}$$

The set E is said to be *unisolvent* (see [10]) if, for each k and $p \in P_k^{\mathbb{K}}$, $p=0$ on E implies $p=0$ in $\mathbb{K}^n$. Let E be a unisolvent compact subset of $\mathbb{K}^n$ and let $f: E \to \mathbb{K}$. For each k, let $t^{(m_k)}$ be a system of extremal points of E of order m_k. Define

$$L_k f(x) = \sum_{j=1}^{m_k} f(t_j) L^{(j)}(x,t^{(m_k)}). \tag{3.2}$$

$L_k f$ is called the *Lagrange interpolation polynomial* of f of degree k.

Suppose f is continuous on E. Let p_k be a polynomial of degree at most k such that $\|f-p_k\|_E = \mathrm{dist}_E(f,P_k^{\mathbb{K}})$. Then by LIF, (3.1) and (3.2), we get

$$\begin{aligned}\|f-L_k f\|_E &\leq \|f-p_k\|_E + \|L_k f - L_k(p_k)\|_E \leq (m_k+1)\|f-p_k\|_E \\ &\leq 4k^n \,\mathrm{dist}_E(f, P_k^{\mathbb{K}}).\end{aligned} \tag{3.3}$$

4. C^∞ FUNCTIONS ON COMPACT SETS IN $\mathbb{R}^n$

In this section we assume that E is a fat compact set in $\mathbb{R}^n$. A C^∞ function on E is a function $f: E \to \mathbb{R}$ such that there exists a C^∞ function g on $\mathbb{R}^n$ with $g|E = f$. Let $C^\infty(E)$ be the space of all such functions. Given a compact subset K of $\mathbb{R}^n$ we define, for $k = 0,1,\ldots,$

$$q_{K,k}(f) := \inf\{|g|_K^k;\ g \in C^\infty(\mathbb{R}^n),\ g|E = f\},$$

where

$$|g|_K^k := \max_{|\alpha| \leq k} \|D^\alpha g\|_K.$$

Let τ_1 be the topology defined by the seminorms $q_{K,k}$. Then τ_1 is exactly the quotient topology of the space $C^\infty(\mathbb{R}^n)/I(E)$, where $C^\infty(\mathbb{R}^n)$ is endowed with the natural topology τ_o defined by the seminorms $|\ |_K^k$, and $I(E) = \{g \in C^\infty(\mathbb{R}^n);\ g|E = 0\}$. Since the space $(C^\infty(\mathbb{R}^n), \tau_o)$ is complete and $I(E)$ is a closed subspace of $C^\infty(\mathbb{R}^n)$, the quotient space $C^\infty(\mathbb{R}^n)/I(E)$ is also complete, whence $(C^\infty(E), \tau_1)$ is a Fréchet space.

Denote by J the mapping

$$C^\infty(E) \ni f \to J(f) = (D^\alpha\, g|E)_{\alpha \in \mathbb{Z}_+^n},$$

where $g \in C^\infty(\mathbb{R}^n)$ and $g|E = f$. By Whitney's extension theorem [18], J is a linear bijection of the space $C^\infty(E)$ onto the space $\mathcal{E}(E)$ of C^∞ *Whitney fields*

$$F = (F^\alpha)_{\alpha \in \mathbb{Z}_+^n},$$

where each F^α is a continuous function on E. $\mathcal{E}(E)$ is a Fréchet space with the topology τ_2 defined by the seminorms

$$\|F\|_E^k = |F|_E^k + \sup\{|(R_x^k F)^\alpha(y)|/|x-y|^{k-|\alpha|};\ x,y \in E,\ x \neq y,\ |\alpha| \leq k\}$$

$(k = 0,1,\ldots)$, where

$$|F|_E^k = \sup\{|F^\alpha(x)|;\ x \in E,\ |\alpha| \leq k\}$$

and

$$(R_x^k F)^\alpha(y) = F^\alpha(y) - \sum_{|\beta| \leq k-|\alpha|} (1/\beta!) F^{\alpha+\beta}(x)(y-x)^\beta.$$

If the geodesic distance on a compact set K in $\mathbb{R}^n$ is equivalent to the Euclidean one, the seminorms $\|\ \|_K^k$ and $|\ |_K^k$ are equivalent as well (see [18]). Hence, in particular, if K is a cube containing E, the seminorms $\|\ \|_K^k$ and $|\ |_K^k$ are equivalent. Therefore the linear bijection J is a continuous mapping from $(C^\infty(E), \tau_1)$ to $(\mathcal{E}(E), \tau_2)$ and by Banach's theorem J is an isomorphism.

Now we equip $C^\infty(E)$ with another topology. We set $d_{-1}(f) := \|f\|_E$, $d_o(f) := \mathrm{dist}_E(f, P_o)$ and, for $k = 1,2,\ldots$,

$$d_k(f) := \sup_{l \geq 1} l^k \,\mathrm{dist}_E(f, P_l).$$

By Jackson's theorem (see e.g. [15]), each d_k is a seminorm on $C^\infty(E)$. Denote by τ_3 the topology in $C^\infty(E)$ defined by the system of the seminorms d_k $(k = -1, 0, \ldots)$. Then, by Bernstein's theorem (Theorem 2.1), if E is a fat compact set in $\mathbb{R}^n$ with HCP, the space $(C^\infty(E), \tau_3)$ is a Fréchet space. Moreover, following the proof of Proposition 2.2 in [8] (applying both HCP and Markov's inequality) we show that the topologies τ_1 and τ_3 coincide.

Finally, let τ_4 denote the topology for $C^\infty(E)$ defined by the seminorms

$$|f|_E^k = \sup\{|D^\alpha f(x)|;\ x \in \mathrm{int}\, E,\ |\alpha| \leq k\}.$$

Assume E has the following *Whitney extension property*:

(WEP) For every C^∞ function f on $\mathrm{int}\, E$, whose partial derivatives of all orders are uniformly continuous on $\mathrm{int}\, E$, there exists a C^∞ function g in $\mathbb{R}^n$ such that $g = f$ on $\mathrm{int}\, E$.

Then, the space $(C^\infty(E), \tau_4)$ is complete and by Banach's theorem the topologies τ_2 (and consequently τ_1) and τ_4 coincide. In particular, if E is a fat subanalytic compact set in $\mathbb{R}^n$ then all the topologies τ_1, τ_2, τ_3, and τ_4 coincide (see [8], Section 2).

5. EXTENSION OF C^∞ FUNCTIONS

Whitney's extension theorem [18] provides a continuous linear operator extending C^k functions (k finite) defined on closed subsets E of $\mathbb{R}^n$. For C^∞ functions such an operator does not, in general, exist even if E is fat (see e.g. Remark 5.3 below). However, Mityagin [6] and Seeley [9] showed the existence of an extension operator if E is a half-space of $\mathbb{R}^n$. Stein [13] proved that such an operator exists if E is the closure of a Lipschitz domain in $\mathbb{R}^n$ of class Lip 1. Stein's result was then extended by Bierstone [3] to the case of a domain with Lipschitz boundary of any order. He also proved in [3], using Hironaka's desingularization theorem, that such an operator exists if E is a fat closed subanalytic set in $\mathbb{R}^n$. Recently, Wachta [17] has constructed an extension operator for fat subanalytic sets in $\mathbb{R}^n$ without making use of Hironaka's desingularization theorem. In [8] we have constructed an extension operator for a family of subsets of $\mathbb{R}^n$ that includes all types of sets mentioned above. Our result reads as follows (see [8], Theorem 4.1 and the remarks preceding Theorem 1.5):

5.1. <u>THEOREM</u>. *Let* E *be a fat* HCP *compact set in* $\mathbb{R}^n$. *Let* u_k *be the* C^∞ *functions defined in the proof of Theorem 2.4. Given a function* $f \in C^\infty(E)$, *let* $L_k f$ *be the Lagrange interpolation polynomial defined by* (3.2). *Then the operator*

$$Lf = u_1 L_1 f + \sum_{k=1}^{\infty} u_k (L_{k+1} f - L_k f)$$

is a continuous linear operator from the space $(C^\infty(E), \tau_3)$ *to the space* $C^\infty(\mathbb{R}^n)$ *endowed with the natural topology* τ_0.

5.2. Remark. In [14] Tidten showed (by different methods) the existence of an extension operator for the class of GCC sets (see Section 1). As it has been mentioned in Section 1, this class is strictly contained in the class of fat HCP sets.

5.3. Remark. Tidten [14] proved that there is no extension operator in the case when $E = \{(x,y) \in R^2;\ 0 \leq x \leq 1,\ 0 \leq y \leq \phi(x)\}$, where ϕ is a C^∞ function infinitely flat at 0, and $\phi(x) > 0$ as $x > 0$. Hence, by Theorem 5.1, the set E is an example of an L-regular set without HCP.

6. EMBEDDING OF C^∞ FUNCTIONS INTO THE SPACE OF RAPIDLY DECREASING SEQUENCES

In his work on the existence of an extension operator for C^∞ functions defined on a GCC subset E of $\mathbb{R}^n$, Tidten shows that the space $C^\infty(E)$ can be embedded (by a topological linear embedding) into the Fréchet space S of all rapidly decreasing sequences $x = (x_j)$, endowed with the system of the seminorms

$$|x|_r := \sup_j |j^r x_j| \qquad \text{for } r = 0,1,\ldots .$$

Using the Lagrange interpolation polynomials we shall construct such an embedding in the case that E is a fat HCP compact subset of $\mathbb{R}^n$.

Let E be a unisolvent compact set in $\mathbb{R}^n$ and $f: E \to \mathbb{R}$. Fix a point $t_o \in E$ and put $L_o f(x) \equiv f(t_o)$. For each $k = 1,2,\ldots$, let

$$t^{(m_k)} = \{t_1^k, \ldots, t_{m_k}^k\} \subset E$$

be a Fekete-Leja system of extremal points of E of order m_k (see Section 3). (We recall that $m_k = \binom{n+k}{k}$.) Denote by $L_k f$ the Lagrange interpolation polynomial of f of degree k with knots in the points of the set $t^{(m_k)}$. Define a sequence $\{\phi_p(f)\}_{p=1}^{\infty}$ as follows:

$$\phi_1(f) = f(t_o)$$

and

$$\phi_j(f) = (L_{k+1} f - L_k f)(t_{j-M_k}^{k+1}),$$

for $M_k < j \leq M_{k+1}$, $k = 0,1,\ldots,$ where $M_k = m_o + \ldots + m_k$.

6.1. THEOREM. *If* E *is a fat* HCP *compact subset of* $\mathbb{R}^n$ *then, for each* $f \in C^\infty(E)$, *the sequence*

$$\phi(f) = \{\phi_j(f)\}_{j=1}^{\infty}$$

belongs to S *and the mapping* ϕ *is a topological linear embedding of the space* $(C^\infty(E), \tau_3)$ *into the space* S.

Proof. It is clear that ϕ is linear. To prove that ϕ is continuous observe that for each $r = 0,1,\ldots,$ and $M_k < j \leq M_{k+1}$, by (3.3) we get

$$|j^r\phi_j(f)| \leq M_{k+1}^r \|L_{k+1}f - L_kf\|_E \leq Ck^{(n+1)r+n} \operatorname{dist}_E(f, P_k)$$
$$\leq Cd_{(n+1)r+n}(f),$$

if $k \geq 1$, where C is a positive constant depending only on n and r. By (3.1), we also have

$$|\phi_j(f)| \leq 2m_1\ d_{-1}(f)$$

for $1 \leq j \leq M_1$. Hence $\{\phi(f)\} \in S$ and, moreover, ϕ is continuous.

Suppose now that $\phi(f) = 0$. Then, by the definition of $\phi(f)$ and by LIF (see Section 3), $L_kf = 0$ for each $k \geq 0$. Hence, since by (3.3) $L_kf \to f$, we get $f = 0$. Thus, ϕ is an injection. We shall prove that ϕ is a homeomorphism onto its image. For this it is sufficient to show that for each $r = 0,1,\ldots,$ there exist a positive integer s and a constant $M > 0$ such that for each $f \in C^\infty(E)$

$$d_r(f) \leq M|\phi(f)|_s.$$

Write $q_k = L_kf - L_{k-1}f$, if $k > 1$, and $q_1 = L_1f$. Since $f = \Sigma_{k=1}^\infty q_k$ on E, we have

$$l^r \operatorname{dist}_E(f,P_l) \leq l^r \|\sum_{k=l+1}^{\infty} q_k\|_E \leq \sum_{k=l+1}^{\infty} k^r \|q_k\|_E \leq \frac{\pi^2}{6} \sup_{k\geq 1} k^{r+2} \|q_k\|_E.$$

On the other hand, by LIF we have

$$\|q_k\|_E = \sup_{x\in E} |\sum_{i=1}^{m_k} q_k(t_i^k) L^{(i)}(x, t^{(m_k)})| \leq m_k \max_i |q_k(t_i^k)|.$$

Hence we get

$$d_r(t) \leq M \sup_{k\geq 1} k^{r+n+2} \max_{M_{k-1}<j\leq M_k} |\phi_j(f)| \leq M|\phi(f)|_{r+n+2},$$

where $M = 2^n\pi^2/6$, since for each $k \geq 1$ we have $k \leq j$, if $M_{k-1} < j \leq M_k$. The proof is complete.

6.2. Remark. By [14], Folgerung 2.4, the existence of an extension operator from $C^\infty(E)$ to $C^\infty(\mathbb{R}^n)$ is equivalent to the existence of an embedding of $C^\infty(E)$ into the space S. Thus, Theorem 6.1 yields also another proof of the existence of an extension operator for HCP compact subsets of $\mathbb{R}^n$ that we have shown in Theorem 5.1.

References

[1] BAOUENDI, M.S. and C. GOULAOUIC: 'Approximation polynomiale de fonctions C^∞ et analytiques', *Ann. Inst. Fourier, Grenoble* 21 (1971), 149-173.

[2] BERNSTEIN, S.N.: *Collected Works.I*, Akad. Nauk SSSR, Moskva 1952 (Russian).

[3] BIERSTONE, E.: 'Extension of Whitney fields from subanalytic sets', *Invent. Math.* 46 (1978), 277-300.

[4] HIRONAKA, H.: 'Introduction to real-analytic sets and real-analytic maps', *Istituto Matematico "L. Tonelli"*, Pisa 1973.

[5] ŁOJASIEWICZ, S.: 'Ensembles semi-analytiques', *Inst. Hautes Etudes Sci.*, Bures-sur-Yvette 1964.

[6] MITYAGIN, B.: 'Approximate dimension and bases in nuclear spaces', *Russian Math. Surveys* 16/4 (1961), 59-128 (see also *Uspekhi Mat. Nauk* 16/4 (1961), 63-132).

[7] PAWŁUCKI, W. and PLEŚNIAK, W.: 'Markov's inequality and C^∞ functions on sets with polynomial cusps', *Math. Ann.* 275(3) (1986), 467-480.

[8] —— and ——: 'Extension of C^∞ functions from sets with polynomial cusps', *Studia Math.* 88 (1988), 83-91.

[9] SEELEY, R.T.: 'Extension of C^∞ functions defined on a half space', *Proc. Amer. Math. Soc.* 15 (1964), 625-626.

[10] SICIAK, J.: 'On some extremal functions and their applications in the theory of analytic functions of several complex variables', *Trans. Amer. Math. Soc.* 105(2) (1962), 322-357.

[11] ——: 'Degree of convergence of some sequences in the conformal mapping theory', *Colloq. Math.* 16 (1967), 49-59.

[12] ——: 'Extremal plurisubharmonic functions in $\mathbb{C}^n$', *Ann. Pol. Math.* 39 (1981), 175-211.

[13] STEIN, E.M.: *Singular integrals and differentiability properties of functions*, University Press, Princeton 1970.

[14] TIDTEN, M.: 'Fortsetzungen von C^∞-Funktionen, welche auf einer abgeschlossenen Menge in $\mathbb{R}^n$ definiert sind', *Manuscripta Math.* 27 (1979), 291-312.

[15] TIMAN, A.F.: *Theory of approximation of functions of a real variable*, Pergamon Press, Oxford 1963.

[16] TOUGERON, J.Cl.: *Idéaux de fonctions différentiables*, Springer, Berlin-Heidelberg-New York 1972.

[17] WACHTA, K.: 'Prolongement de fonctions définies sur les ensembles sous-analytiques', Uniwersytet Jagielloński, Kraków 1986, preprint.

[18] WHITNEY, H.: 'Analytic extension of differentiable functions defined in closed sets', *Trans. Amer. Math. Soc.* 36 (1934), 63-89.

[19] WÓJCIK, A.: 'On the distribution of zeros of the polynomials of uniform best approximation', *Monatsh. Math.*, to appear.

[20] ZAKHARYUTA, V.P.: 'Extremal plurisubharmonic functions, orthogonal polynomials, and the Bernstein-Walsh theorem for analytic functions of several complex variables', *Ann. Pol. Math.* 33 (1976), 137-148 (Russian).

NEW EXISTENCE THEOREMS AND EVALUATION FORMULAS FOR ANALYTIC FEYNMAN INTEGRALS

Robert Horton Cameron
School of Mathematics
University of Minnesota
Minneapolis, MN 55455, USA

and David Arne Storvick
School of Mathematics
University of Minnesota
Minneapolis, MN 55455, USA

ABSTRACT. Existence theorems and evaluation formulas are obtained for the analytic Feynman integral corresponding to functionals of the form given below in (1.1).

1. INTRODUCTION

In this paper we show the existence of the analytic Feynman integral (see [3] and [4]) for functionals of the form

$$F(\vec{x}) = G(\vec{x})\ \Psi(\vec{x}(b)). \tag{1.1}$$

Here $\vec{x}$ is an element of ν-dimensional Wiener space, $C^{\nu}[a,b] = \times_{j=1}^{\nu} C[a,b]$, G is a functional on C^{ν} which is expressable as the Fourier transform of a measure of finite variation on $L_2^{\nu}[a,b]$ and Ψ is of the form $\Psi = \Psi_1 + \Psi_2$ where $\Psi \in L_1^{\nu} \equiv L_1(\mathbb{R}^{\nu})$ and Ψ_2 is a Fourier transform of a measure of bounded variation over $\mathbb{R}^{\nu}$.

In the applications of the Feynman integral to quantum theory, functionals of the type

$$\exp\{\int_a^b \theta(s, \vec{x}(s))\, ds\}\Psi(\vec{x}(b)) \tag{1.2}$$

are often employed, with Ψ corresponding to the initial condition associated with Schroedinger's equation. See Theorem 6.1 of [6], Sec. 7 of [9], Theorem 4 of [12] and also [1], [2], [3], [7], [8] and [13]. It has been shown in [6] and [10] that such functionals (1.2) are sequentially and analytically Feynman integrable when Ψ is the Fourier transform of a measure of bounded variation on $\mathbb{R}^{\nu}$ and θ satisfies a similar condition. However this condition restricts Ψ to be bounded and continuous, so that this Ψ could be applied in connection with the Schroedinger equation only when the initial condition for the Schroedinger equation was bounded and continuous.

We prove the existence of the analytic Feynman integral for functionals of the type (1.2) for which Ψ need not be bounded or con-

J. Ławrynowicz (ed.), Deformations of Mathematical Structures, 297–308.

tinuous. We also establish formulas for the evaluation of such integrals.

The existence theorems and formulas for calculations of analytic Feynman integrals given in this paper parallel those given for the sequential Feynman integral in an earlier paper [7]. In addition, the present paper shows the equality of the analytic and sequential Feynman integrals all under appropriate hypotheses.

2. EXISTENCE THEOREMS FOR THE ANALYTIC FEYNMAN INTEGRAL

We now establish the existence of the analytic Feynman integral for functionals of the form $G(\vec{x})\Psi(\vec{x}(b))$ where $G \in S[L_2^\nu]$ and $\Psi \in L_1^\nu$. The analytic Feynman integral and the space S introduced in [3] are defined below.

Notation: $C[a,b] \equiv \{x(\cdot) \mid x \text{ continuous on } [a,b],\ x(a)=0\}$.

Definition. Let F be a functional such that the ν-dimensional Wiener integral

$$J(\lambda) \equiv \int_{C^\nu[a,b]} F({}^{-1/2}\vec{x})\,d\vec{x} \tag{2.1}$$

exists for all real $\lambda > 0$. If there exists a function $J^*(\lambda)$ analytic in the half-plane $\operatorname{Re}\lambda > 0$ such that $J^*(\lambda) = J(\lambda)$ for all real $\lambda > 0$, then we define J^* to be the *analytic Wiener integral of* F *over* $C^\nu[a,b]$ *with parameter* λ, and for $\operatorname{Re}\lambda > 0$ we write

$$\int_{C^\nu[a,b]}^{\text{anw}_\lambda} F(\vec{x})\,d\vec{x} \equiv J^*(\lambda). \tag{2.2}$$

Definition. Let q be a real parameter $(q \neq 0)$ and let F be a functional whose analytic Wiener integral exists for $\operatorname{Re}\lambda > 0$. Then if the following limit exists, we call it the *analytic Feynman integral of* F *over* $C^\nu[a,b]$ with parameter q, and we write

$$\int_{C^\nu[a,b]}^{\text{anf}_q} F(\vec{x})\,d\vec{x} \equiv \lim_{\substack{\lambda \to -iq \\ \operatorname{Re}\lambda > 0}} \int_{C^\nu[a,b]}^{\text{anw}_\lambda} F(\vec{x})\,d\vec{x}. \tag{2.3}$$

Terminology. We shall say that two functionals $F(\vec{x})$ and $G(\vec{x})$ are equal s-almost everywhere if for each $\rho > 0$ the equation $F(\rho\vec{x}) = G(\rho\vec{x})$ holds for almost all $\vec{x} \in C^\nu[a,b]$, in other words, if $F(\vec{x}) = G(\vec{x})$ except for a scale-invariant null set. We denote the equivalence relation between functionals by $F \approx G$. (Our measure in C^ν is Wiener measure.)

The definition of S also involves the P.W.Z. (Paley-Wiener-Zygmund) integral [11] which is defined as follows.

D e f i n i t i o n. Let $\phi_1, \phi_2, \ldots$ be a C.O.N. (complete orthonormal) set of real functions of bounded variation on $[a,b]$. Let $v \in L_2[a,b]$ and $v_n(t) = \Sigma_{j=1}^{n} \phi_j(x) \int_a^b v(s)\phi_j(s)\,ds$. Then the P.W.Z. integral is defined by

$$\int_a^b v(s)\,d\tilde{x}(s) \equiv {}^{\{\phi\}}\!\int_a^b v(s)\,d\tilde{x}(s) \equiv \lim_{n\to\infty} \int_a^b v_n(s)\,dx(s)$$

for all $x \in C[a,b]$ for which the above limit exists.

N o t e. It was shown in [11] that this integral exists for almost all $x \in C[a,b]$ and is essentially independent of the choice of ϕ_1, $\phi_2, \ldots$ Moreover if v is of bounded variation, it is essentially equivalent to the Riemann-Stieltjes integral. Clearly "almost all" may be replaced by "s-almost all" in this statement.

D e f i n i t i o n. Let $D[a,b]$ be the class of elements $x \in C[a,b]$ such that x is absolutely continuous on $[a,b]$ and $x' \in L_2[a,b]$. Let $D^\nu = \times_1^\nu D$.

D e f i n i t i o n. Let $M \equiv M(L_2^\nu[a,b])$ be the class of complex measures of finite variation defined on $B(L_2^\nu)$, the Borel measurable subsets of $L_2^\nu[a,b]$. We set $\|\mu\| = \text{var}\,\mu$. (In this paper, L_2 always means real L_2).

D e f i n i t i o n. Let $S \equiv S(L_2^\nu)$ be the space of functionals expressible in the form

$$F(\vec{x}) \equiv \int_{L_2^\nu} \exp\{ i \sum_{j=1}^{\nu} \int_a^b v_j(x)\,d\tilde{x}_j(t)\}\,d\mu(\vec{v})$$

for s-almost all $\vec{x} \in C^\nu[a,b]$, where $\mu \in M_\nu \equiv M[L_2^\nu]$. (Note, it is assumed that the P.W.Z. integral $\int_a^b v_j(t)\,d\tilde{x}_j(t)$ is based on the same C.O.N. sequence $\{\phi_n\}$ for all choices of $\vec{v}$ and $\vec{x}$).

D e f i n i t i o n. The functional F defined on a subset of C^ν that contains D^ν is said to be an element of $\hat{S} \equiv \hat{S}(L_2^\nu)$ if there exists a measure $\mu \in M$ such that for $\vec{x} \in D^\nu$,

$$F(\vec{x}) \equiv \int_{L_2^\nu} \exp\{ i \sum_{j=1}^{\nu} \int_a^b v_j(t)\left(\frac{dx_j(t)}{dt}\right) dt\}\,d\mu(\vec{v}).$$

N o t a t i o n. If $F(\vec{x}) = G(\vec{x})$ s-almost everywhere on $C^\nu[a,b]$ and also for *every* $\vec{x} \in D^\nu[a,b]$, we shall write $F(\vec{x}) \overset{\approx}{=} G(\vec{x})$.

We now define the space S^* which we introduced in [6].

D e f i n i t i o n. We say $F \in S^* \equiv S^*[L_2^\nu]$ iff there exists a $\mu \in M$ such that

$$F(\vec{x}) \overset{\approx}{=} \int_{L_2^\nu} \exp\{ i \sum_{j=1}^{\nu} \int_a^b v_j(t)\tilde{d}x_j(t)\} d\mu(\vec{v}). \tag{2.4}$$

It was shown in Lemma 2.6 of [6] that

$$S^* \underset{\neq}{\subset} S \cap \hat{S}. \tag{2.5}$$

In order to prove existence theorems for the analytic Feynman integral of functionals of the form $G(\vec{x})\Psi(\vec{x}(b))$, we first prove the following lemma.

LEMMA 1. *If* $g \in L_2[a,b]$, $\Psi \in L_1(\mathbb{R})$, c *is a positive number and* γ *is a complex number, then the following Wiener integral exists and its value is given by*

$$\int_{C[a,b]} \exp\{\gamma \int_a^b g(t)\tilde{d}x(t)\}\Psi(c\, x(b))dx \tag{2.6}$$

$$= \frac{1}{c\sqrt{2\pi(b-a)}} \exp\{\frac{\gamma^2}{2} \int_a^b [g(s)]^2 ds - \frac{\gamma^2}{2(b-a)} [\int_a^b g(s)ds]^2\}$$

$$\int_{-\infty}^{\infty} \Psi(u) \exp\{ \frac{-u^2}{2c^2(b-a)} + \frac{\gamma\, u \int_a^b g(s)ds}{c(b-a)}\} du.$$

Proof of Lemma 1. (Assume without loss of generality that g is not equivalent to a constant). Let

$$\phi_1(t) \equiv \frac{1}{\sqrt{b-a}}, \quad \phi_2(t) \equiv \frac{1}{\alpha} [g(t) - \phi_1(t) \int_a^b \phi_1(s)g(s)ds],$$

where $\alpha = \{\int_a^b [g(t)]^2 dt - (b-a)^{-1} [\int_a^b g(s)ds]^2\}^{1/2}$. Clearly ϕ_1 and ϕ_2 are normal and orthogonal on $[a,b]$. Now

$$g(t) = \alpha\, \phi_2(t) + \phi_1(t) \int_a^b \phi_1(s)g(s)ds = \alpha\, \phi_2(t) + \beta\, \phi_1(t),$$

where

$$\beta = \int_a^b \phi_1(s)g(s)ds = (b-a)^{-1/2} \int_a^b g(s)ds,$$

so that

$$\int_a^b g(t)\tilde{d}x(t) \approx \alpha \int_a^b \phi_2(t)\tilde{d}x(t) + \beta \int_a^b \phi_1(t)dx(t)$$

and

$$x(b) = \int_a^b dx(t) = (b-a)^{1/2} \int_a^b \phi_1(t)dx(t).$$

Thus by the Paley-Wiener-Zygmund theorem, the left member of (2.6) equals

$$\int_{C[a,b]} \exp\{\gamma\alpha\int_a^b \phi_2(t)d\tilde{x}(t) + \gamma\beta\int_a^b \phi_1(t)dx(t)\}$$

$$\Psi[c(b-a)^{1/2}\int_a^b \phi_1(t)dx(t)]dx$$

$$= \frac{1}{2\pi}\int_{-\infty}^{\infty}\int_{-\infty}^{\infty} \exp\{\gamma\alpha u_2 + \gamma\beta u_1\}\Psi[c(b-a)^{1/2}u_1]$$

$$\exp\{\frac{-u_1^2}{2} - \frac{u_2^2}{2}\}du_1du_2$$

$$= \frac{1}{2\pi}[\int_{-\infty}^{\infty} \exp\{\gamma\alpha u_2 - \frac{u_2^2}{2}\}du_2]\int_{-\infty}^{\infty}\Psi(c(b-a)^{1/2}u_1)\exp\{\frac{-u_1^2}{2} + \gamma\beta u_1\}du_1$$

$$= \frac{1}{c\sqrt{2\pi(b-a)}}\exp\{\frac{\gamma^2}{2}\int_a^b [g(t)]^2dt - \frac{\gamma^2}{2(b-a)}[\int_a^b g(s)ds]^2\}$$

$$\int_{-\infty}^{\infty}\Psi(v)\exp\{\frac{-v^2}{2c^2(b-a)} + \frac{\gamma v\int_a^b g(s)ds}{c(b-a)}\}dv,$$

and the lemma is proved.

THEOREM 1. *Let*

$$F(\vec{x}) \approx G(\vec{x})\Psi(\vec{x}(b)),$$

where $G \in S[L_2^\nu]$ *and* $\Psi \in L_1^\nu \equiv L_1(\mathbb{R}^\nu)$. *Then for each real* $q \neq 0$, F *is analytic Feynman integrable; and if*

$$G(\vec{x}) \approx \int_{L_2^\nu} \exp\{i\sum_{j=1}^{\nu}\int_a^b v_j(t)d\tilde{x}_j(t)\}d\mu(\vec{v}),$$

where μ *is a complex measure of bounded variation on* L_2^ν, *then*

$$\int_{C^\nu[a,b]}^{\mathrm{anf}_q} F(\vec{x})d\vec{x} = \left(\frac{q}{2i(b-a)}\right)^{\nu/2}\int_{L_2^\nu}\int_{\mathbb{R}^\nu}$$

$$\exp\{\frac{i}{2q(b-a)}\sum_{j=1}^{\nu}[(q\xi_j + \int_a^b v_j(s)ds)^2 - (b-a)\int_a^b [v_j(s)]^2ds]\}\Psi(\vec{\xi})d\vec{\xi}\,d\mu(\vec{v}). \tag{2.7}$$

P r o o f. Let λ be a real positive number (for the present). We begin by computing the Wiener integral,

$$\int_{C^\nu[a,b]} F(\lambda^{-1/2}\vec{x})d\vec{x} = \int_{C^\nu[a,b]} \int_{L_2^\nu} \exp\{ i\, \lambda^{-1/2} \sum_{j=1}^{\nu} \int_a^b v_j(s)\tilde{d}x_j(s)\}$$

$$d\mu(\vec{v})\Psi(\lambda^{-1/2}\vec{x}(b))d\vec{x} \tag{2.8}$$

$$= \int_{L_2^\nu} [\int_{C^\nu[a,b]} \exp\{i\, \lambda^{-1/2} \sum_{j=1}^{\nu} \int_a^b v_j(s)\tilde{d}x_j(s)\}\Psi(\lambda^{-1/2}\vec{x}(b))d\vec{x}]d\mu(\vec{v})$$

$$\equiv \int_{L_2^\nu} I\, d\mu(\vec{v}),$$

where

$$I \equiv \int_{C^\nu[a,b]} \exp\{ i\, \lambda^{-1/2} \sum_{j=1}^{\nu} \int_a^b v_j(s)\tilde{d}x_j(s)\}\Psi(\lambda^{-1/2}\vec{x}(b))d\vec{x}$$

$$= \int_{C[a,b]} \overset{(\nu)}{\cdots} \int_{C[a,b]} \exp\{ i\, \lambda^{-1/2} \sum_{j=1}^{\nu} \int_a^b v_j(s)\tilde{d}x_j(s)\}$$

$$\Psi(\lambda^{-1/2}\vec{x}(b))dx_1 \ldots dx_\nu .$$

The integral I can be computed by applying Lemma 1 successively to each of the ν integrals over $C[a,b]$. Thus

$$I = [\frac{\lambda}{2\pi(b-a)}]^{1/2}\exp\{\frac{-1}{2\lambda} \int_a^b [v_1(s)]^2 ds + \frac{1}{2\lambda(b-a)} [\int_a^b v_1(s)ds]^2\}$$

$$\int_{C[a,b]} \overset{(\nu-1)}{\cdots} \int_{C[a,b]} \int_{-\infty}^{\infty} \exp\{i\, \lambda^{-1/2} \sum_{j=2}^{\nu} \int_a^b v_j(s)\tilde{d}x_j(s)\}$$

$$\Psi(u_1, \lambda^{-1/2} x_2(b), \ldots, \lambda^{-1/2} x_\nu(b))$$

$$\exp\{\frac{-\lambda u_1^2}{2(b-a)} + i\, \frac{(\int_a^b v_1(s)ds)}{(b-a)} u_1\} du_1\, dx_2 \ldots dx_\nu \tag{2.9}$$

$$= [\frac{\lambda}{2\pi(b-a)}]^{\nu/2} \exp\{ \frac{-1}{2\lambda} \sum_{j=1}^{\nu} \int_a^b [v_j(s)]^2 ds +$$

$$+ \frac{1}{2\lambda(b-a)} \sum_{j=1}^{\nu} [\int_a^b v_j(s)ds]^2\} \int_{-\infty}^{\infty} \overset{(\nu)}{\cdots} \int_{-\infty}^{\infty} \Psi(u_1, u_2, \ldots, u_\nu)$$

$$\exp\{\frac{-\lambda \sum_{j=1}^{\nu} u_j^2}{2(b-a)} + i\, \frac{\sum_{j=1}^{\nu} (u_j \int_a^b v_j(s)ds)}{(b-a)}\} du_1 \ldots du_\nu .$$

We now substitute the value of I from equation (2.9) into equation (2.8) and obtain

$$\int_{C^\nu[a,b]} F(\lambda^{-1/2}\vec{x})\,d\vec{x} = \left[\frac{\lambda}{2\pi(b-a)}\right]^{\nu/2}$$

$$\int_{L_2^\nu} \exp\left\{\frac{-1}{2\lambda}\sum_{j=1}^{\nu}\left(\int_a^b [v_j(s)]^2\,ds - \frac{1}{(b-a)}\left[\int_a^b v_j(s)\,ds\right]^2\right)\right\}$$

$$\int_{-\infty}^{\infty}\cdots\int_{-\infty}^{\infty}\Psi(u_1,\ldots,u_\nu)\exp\left\{\frac{-\lambda\sum_{j=1}^{\nu}u_j^2}{2(b-a)} + \frac{i\sum_{j=1}^{\nu}\left(u_j\int_a^b v_j(s)\,ds\right)}{(b-a)}\right\}$$

$$du_1\ldots du_\nu\,d\mu(\vec{v}).$$

Using the Schwarz inequality in the first exponent we now observe that the exponentials above are bounded in absolute value by unity and since $\Psi\in L_1$, and μ is a complex measure of bounded variation, it follows that the right hand member is analytic in λ for $\mathrm{Re}\,\lambda>0$. We now pass to the limit as $\lambda\to -iq$, $\mathrm{Re}\,\lambda>0$, and obtain the formula given in equation (2.7) and the theorem is proved.

The following corollary to Theorem 1 shows the equality of the analytic and sequential Feynman integrals for an important subclass of the functionals considered in Theorem 1. The sequential Feynman integral was defined in [6]. The corollary follows from the fact that the formulas for these integrals in Theorem 1 and Theorem 1 of [7] are identical.

COROLLARY 1 to Theorem 1. *Let*

$$F(\vec{x}) \approx G(\vec{x})\Psi(\vec{x}(b)),$$

where $G\in S^*[L_2^\nu]$ *and* $\Psi\in L_1^\nu$. *Then for each real* $q\neq 0$, F *is both sequentially Feynman integrable and analytically Feynman integrable and we have*

$$\int^{sf_q} F(\vec{x})\,d\vec{x} = \int_{C^\nu}^{anf_q} F(\vec{x})\,d\vec{x}.$$

Definition. Let J^ν be the set of functions Ψ defined on $\mathbb{R}^\nu$ by

$$\Psi(\vec{r}) = \int_{\mathbb{R}^\nu}\exp\left\{i\sum_{j=1}^{\nu} r_j t_j\right\}d\rho(\vec{t}),$$

where ρ is a complex Borel measure of bounded variation on $\mathbb{R}^\nu$.

Definition. Let K^ν be the set of functions of the form

$$\Psi_1(\cdot)+\Psi_2(\cdot),$$

where $\Psi_1\in L_1^\nu$ and $\Psi_2\in J^\nu$.

Note: In [7] we have given several examples to illustrate the scope of K^ν. In particular the function $\Psi(u) \equiv u^{-2/3}$ is in K but not in either L_1^1 or J^1.

THEOREM 2. *Let* $G \in S(L_2^\nu)$ *and* $\Psi \in J^\nu$ *and* $q \neq 0$ *be a real number and let*

$$F(\vec{x}) \approx G(\vec{x})\Psi(\vec{x}(b)). \tag{2.10}$$

Then F *is analytic Feynman integrable; and if* ρ *is a measure of* B.V. *on* $\mathbb{R}^\nu$ *such that*

$$\Psi(\vec{u}) \equiv \int_{\mathbb{R}^\nu} \exp\{ i \sum_{j=1}^{\nu} u_j t_j\} d\rho(\vec{t}), \tag{2.11}$$

and

$$G(\vec{x}) \approx \int_{L_2^\nu} \exp\{ i \sum_{j=1}^{\nu} \int_a^b v_j(t)\, d\tilde{x}_j(t)\} d\mu(\vec{v}), \qquad \mu \in M, \tag{2.12}$$

we have

$$\int_{C^\nu[a,b]}^{\text{anf}_q} F(\vec{x})d\vec{x} = \int_{L_2^\nu}\int_{\mathbb{R}^\nu} \exp\{\frac{1}{2qi} \sum_{j=1}^{\nu} \int_a^b [v_j(s) + t_j]^2 ds\} d\rho(\vec{t})d\mu(\vec{v}). \tag{2.13}$$

Proof. By Lemma 2.8 of [6], there exists $G^* \in S^*$ such that $G^* \approx G$ on $C^\nu[a,b]$. By the proof of this lemma, G and G^* are both based on the same $\mu \in M$ and we have

$$G^*(\vec{x}) \overset{\approx}{=} \int_{L_2^\nu} \exp\{ i \sum_{j=1}^{\nu} \int_a^b v_j(t)d\tilde{x}_j(t)\} d\mu(\vec{v}).$$

By (2.5), $S^* \subset \hat{S}$, so $G^* \in \hat{S}$. By the definition of S'' in [3], $\Psi(x(b)) \in S'' \subset S'$ and by Lemma $\hat{2}$.6 of [6], $\Psi(x(b)) \in S^* \subset \hat{S}$. By Lemma 2.2 and 2.5 of [6], S^* and $\hat{S}$ are Banach algebras. Let $F^*(\vec{x}) \equiv G^*(\vec{x})\Psi(\vec{x}(b))$ whenever the right member exists. Since S^* is a Banach algebra, we have $F^* \in S^* \subset \hat{S}$.
By Theorem 2 of [7],

$$\int^{\text{sf}_q} F^*(\vec{x})d\vec{x} = \int^{\text{sf}_q} G^*(\vec{x})\Psi(\vec{x}(b))d\vec{x}$$

$$= \int_{L_2^\nu}\int_{\mathbb{R}^\nu} \exp\{\frac{1}{2qi} \sum_{j=1}^{\nu} \int_a^b [v_j(s) + t_j]^2 ds\} d\rho(\vec{t})d\mu(\vec{v}).$$

By Theorem 3.1 of [6], we have

$$\int_{C^\nu}^{\text{anf}_q} F^*(\vec{x})d\vec{x} = \int^{\text{sf}_q} F^*(\vec{x})d\vec{x}.$$

Now since $G^* \approx G$, $F^* \approx F$, their Wiener integrals are equal, and

$$\int_{C^\nu} F^*(\lambda^{-1/2}\vec{x})\,d\vec{x} = \int_{C^\nu} F(\lambda^{-1/2}\vec{x})\,d\vec{x}$$

for all positive real λ. Hence their analytic extensions are equal and

$$\int_{C^\nu}^{\mathrm{anf}_q} F(\vec{x})\,d\vec{x} = \int_{C^\nu}^{\mathrm{anf}_q} F^*(\vec{x})\,d\vec{x}.$$

COROLLARY to Theorem 2. *Let*

$$F(\vec{x}) \stackrel{\approx}{=} G(\vec{x})\Psi(\vec{x}(b)),$$

where $G \in S^*[L_2^\nu]$ *and* $\Psi \in J^\nu$. *Then for each real* $q \neq 0$, F *is both sequentially and analytically Feynman integrable and we have*

$$\int^{\mathrm{sf}_q} F(\vec{x})\,d\vec{x} = \int_{C^\nu}^{\mathrm{anf}_q} F(\vec{x})\,d\vec{x}.$$

THEOREM 3. *Let* $F(\vec{x}) \approx G(\vec{x})\Psi(\vec{x}(b))$ *where* $G \in S(L_2^\nu)$ *and* $\Psi \in K^\nu$. *Then for each real* $q \neq 0$, F *is analytically Feynman integrable. Moreover if* $\Psi = \Psi_1 + \Psi_2$ *where* $\Psi_1 \in L_1^\nu$ *and* $\Psi_2 \in J^\nu$, *with* Ψ_2 *given by* (2.11), *we have*

$$\int_{C^\nu[a,b]}^{\mathrm{anf}_q} F(\vec{x})\,d\vec{x} = \left(\frac{q}{2\pi i(b-a)}\right)^{\nu/2} \int_{L_2^\nu}\int_{\mathbb{R}^\nu} \exp\left\{\frac{i}{2q(b-a)}\right.$$

$$\left.\sum_{j=1}^{\nu}\left[\left(q\xi_j + \int_a^b v_j(s)\,ds\right)^2 - (b-a)\int_a^b [v_j(s)]^2 ds\right]\right\}\Psi_1(\vec{\xi})\,d\vec{\xi}\,d\mu(\vec{v})$$

$$+ \int_{L_2^\nu}\int_{\mathbb{R}^\nu} \exp\left\{\frac{1}{2qi}\sum_{j=1}^{\nu}\int_a^b [v_j(s)+t_j]^2 ds\right\} d\rho(\vec{t})\,d\mu(\vec{v}).$$

Proof. This theorem follows from Theorem 1 and 2, and the linearity of the analytic Feynman integral.

COROLLARY 1 to Theorem 3. *Let* $F(\vec{x}) \stackrel{\approx}{=} G(\vec{x})\Psi(\vec{x}(b))$ *where* $G \in S^*(L_2^\nu)$ *and* $\Psi \in K^\nu$. *Then for each real* $q \neq 0$, F *is sequentially and analytically Feynman integrable and*

$$\int^{\mathrm{sf}_q} F(\vec{x})\,d\vec{x} = \int_{C^\nu[a,b]}^{\mathrm{anf}_q} F(\vec{x})\,d\vec{x}.$$

Since $S^* \subset \hat{S} \cap S$, this corollary follows immediately from Theorem 3 of [7] and Theorem 3 above.

Definition. Let H be the class of families of measures $\sigma(\cdot,\cdot)$ such that for each $s \in [a,b]$, $\sigma(s,\cdot)$ is a measure of bounded variation on $\mathbb{R}^\nu$ where for each Borel set $E \subset \mathbb{R}^\nu$, $\sigma(s,E)$ is measurable in s on $[a,b]$, and $\|\sigma(s,\cdot)\| \equiv \operatorname{var}\sigma(s,\cdot)$ is bounded in s on $[a,b]$.

THEOREM 4. *Let* $\Psi \in K^\nu$ *and let* Ψ_1 *and* Ψ_2 *and* ρ *be as in Theorem 3. Let*

$$\theta(s,\xi) = \int_{\mathbb{R}^\nu} \exp\{i \sum_{j=1}^{\nu} \xi_j u_j\} d_{\vec{u}}\sigma(s,\vec{u}),$$

where $\sigma \in H$. *Then the functional*

$$\exp\{\int_a^b \theta(s,\vec{x}(s))ds\}$$

is an element of S^* *and the following analytic and sequential Feynman integrals exist and the equation holds:*

$$\int_{C^\nu[a,b]}^{\text{anf}_q} \exp\{\int_a^b \theta(s,\vec{x}(s))ds\}\Psi(\vec{x}(b))d\vec{x} =$$

$$= \int^{\text{sf}_q} \exp\{\int_a^b \theta(s,\vec{x}(s))ds\}\Psi(\vec{x}(b))d\vec{x} = \sum_{n=0}^{\infty} \Omega_n$$

where

$$\Omega_n \equiv \left(\frac{q}{2\pi i(b-a)}\right)^{\nu/2} \int_{\Delta_n} \int_{\mathbb{R}^{\nu n}} \int_{\mathbb{R}^\nu} \exp\{\frac{1}{2q(b-a)} \sum_{j=1}^{\nu} [(q\xi_j + \sum_{k=1}^{n} u_{j,k}(s_k - a))^2 - (b-a)\sum_{k=1}^{n}[\sum_{\ell=k}^{n} u_{j,\ell}]^2(s_k - s_{k-1})]\}$$

$$\Psi_1(\vec{\xi})d\vec{\xi}\, d\sigma(s_1,\vec{u}_1)\ldots d\sigma(s_n,\vec{u}_n)d\vec{s}$$

$$+ \int_{\Delta_n} \int_{\mathbb{R}^{\nu n}} \int_{\mathbb{R}^\nu} \exp\{\frac{1}{2qi} \sum_{j=1}^{\nu} [\sum_{k=1}^{n} (\sum_{\ell=1}^{n} u_{j,\ell})^2 (s_k - s_{k-1})$$

$$+ 2t_j \sum_{k=1}^{n} u_{j,k}(s_k - a) + t_j^2(b-a)]\}d\rho(\vec{t})d\sigma(s_1,\vec{u}_1)\ldots d\sigma(s_n,\vec{u}_n)d\vec{s}$$

where $\Delta_n = \{\vec{s}: a = s_0 < s_1 < s_2 < \ldots < s_n \leq b\}$.

Proof of Theorem 4. Since $\theta(s,\vec{u})$ is the Fourier transform of a measure σ of bounded variation, it is continuous in $\vec{u}$ for each $s \in [a,b]$. Thus for each $s \in [a,b]$, $\theta(s,\vec{x}(s))$ is continuous in the uniform topology for $\vec{x} \in C^\nu[a,b]$. Since $\|\sigma(s,\cdot)\|$ is bounded in s on $[a,b]$, $\theta(s,\vec{u})$ is bounded on $[a,b] \times \mathbb{R}^\nu$. By bounded convergence, $\exp\{\int_a^b \theta(s,\vec{x}(s))ds\}$ is continuous in the uniform topology for $\vec{x} \in C^\nu[a,b]$. Let

$$G(\vec{x}) \equiv \exp\{\int_a^b \theta(s,\vec{x}(s))ds\} \quad \text{for } \vec{x} \in C^\nu[a,b].$$

By Theorem 8 of [7], $G \in \hat{S}$, and by Corollary 2 of Theorem 2.1 of [6], there exists a functional $G^* \in S^*$ such that $G^*(\vec{x}) = G(\vec{x})$ for $\vec{x} \in D^\nu$. By Theorem 2.1 of [6], G^* is continuous with respect to binary polygonal approximation s-almost everywhere on C^ν. Consequently $G^*(\vec{x}) = G(\vec{x})$ s-almost everywhere on C^ν and therefore $G^* \overset{s}{=} G$. Hence $G \in S^*$.

By Corollary 1 to Theorem 3, $G\Psi$ is sequentially and analytically Feynman integrable and

$$\int^{sf_q} G(\vec{x})\Psi(\vec{x}(b))d\vec{x} = \int_{C^\nu[a,b]}^{anf_q} G(\vec{x})\Psi(\vec{x}(b))d\vec{x}.$$

By Theorem 8 of [7], we have

$$\int_{C^\nu[a,b]}^{anf_q} \exp\{\int_a^b \theta(s,\vec{x}(s))ds\}\Psi(\vec{x}(b))d\vec{x} = \sum_{n=0}^{\infty} \Omega_n .$$

Thus Theorem 4 is proved.

References

[1] ALBEVERIO, A. and R. HOEGH-KROHN: *Mathematical theory of Feynman path integrals*, Lecture Notes in Mathematics 523, Springer-Verlag, Heidelberg-Berlin-New York 1976.

[2] —— and ——: 'Feynman path integrals and the corresponding method of stationary phase, Feynman path integrals', *Lecture Notes in Physics* **106**, Springer-Verlag, Berlin-Heidelberg-New York 1979, 3-57.

[3] CAMERON, R.H. and D.A. STORVICK: 'Some Banach algebras of analytic Feynman integrable functionals', in: *Proceedings, Analytic Functions, Kozubnik 1979*, ed. by J. Ławrynowicz, Lecture Notes in Mathematics 798, Springer-Verlag, Berlin-Heidelberg-New York 1980, 18-67.

[4] —— and ——: 'Analytic Feynman integral solutions of an integral equation related to the Schrödinger equation', *J. Analyse Math.* **38** (1980), 34-66.

[5] —— and ——: 'A new translation theorem for the analytic Feynman integral', *Rev. Roum. Math. Pures et Appl.* **27**,**9** (1982), 937-944.

[6] —— and ——: 'A simple definition of the Feynman integral, with applications', *Mem. Amer. Math. Soc.* **46**,**288** (1983).

[7] —— and ——: 'New existence theorems and evaluation formulas for sequential Feynman integrals', *Proc. London Math Soc.* (**3**) **52** (1986), 557-581.

[8] ELWORTHY, D. and A. TRUMAN: 'A Cameron-Martin formula for Feynman

integrals', in: *Mathematical problems in theoretical physics*, Lecture Notes in Physics 153, Springer-Verlag, Berlin-Heidelberg-New York 1981, 288-294.

[9] FEYNMAN, R.P.: 'Space-time approach to non-relativistic quantum mechanics', *Rev. Mod. Phys.* 20 (1948), 367-387.

[10] JOHNSON, G.W. and D. SKOUG: 'Notes on the Feynman integral, III: the Schrödinger equation', *Pacific J. Math.* 105 (1983), 321-358.

[11] PALEY, R.E.A.C., N. WIENER and A. ZYGMUND: 'Notes on random functions', *Math. Z.* 37 (1933), 647-688.

[12] TRUMAN, A.: 'The Feynman maps and the Wiener integral', *J. Math. Phys.* 19,8 (1978).

[13] ——: 'The polygonal path formulation of the Feynman path integral' in: *Feynman path integrals*, Lecture Notes in Physics 106, Springer-Verlag, Berlin-Heidelberg-New York 1979, 73-102.

ON THE CONSTRUCTION OF POTENTIAL VECTORS AND GENERALIZED POTENTIAL VECTORS DEPENDING ON TIME BY A CONTRACTION PRINCIPLE

Wolfgang Tutschke
Sektion Mathematik
Universität Halle
DDR-4020 Halle, GDR

ABSTRACT. Let L be the first order differential operator acting on $u = u(t,x)$, $x = (x_1,x_2,x_3)$, $t \in \mathbb{R}$, $x \in G$, G being a bounded domain in $\mathbb{R}^3$. Consider the initial value problem $(\partial/\partial t)u = Lu$, $u(0,x) = u_o(x)$, where $u_o = u_o(x)$ is a given generalized potential vector field in G. It is proved that the problem is solvable globally in the cone $\{(t,x) : x \in G,\ 0 \leq t < \eta\ \mathrm{dist}(x,\partial G)\}$ by a contraction principle provided that the slope η is sufficiently small. The method relies upon W. Walter's ideas (1985). After introducing a Banach space of generalized potential vectors depending on time, Nagumo's lemma is proved for generalized potential vectors.

0. INTRODUCTION

Nagumo's approach [1] to the Cauchy-Kowalewski theorem has led, in connection with the concept of a scale of Banach spaces, to abstract versions of the Cauchy-Kowalewski theorem (to consider the linear case cf., for instance, F. Trèves' book [6], whereas nonlinear abstract Cauchy-Kowalewski theorems are proved, for instance, in L. Nirenberg's paper [2] and in his book [3], and in T. Nishida's paper [4]). By applying an abstract Cauchy-Kowalewski theorem to the scale of Banach spaces of holomorphic functions, we obtain the classical Cauchy-Kowalewski theorem again. Replacing the scale of Banach spaces of holomorphic functions by a scale of generalized analytic functions, we get generalizations of the Cauchy-Kowalewski theorem to the case of generalized analytic functions, i.e. initial value problems with generalized analytic initial functions can be solved instead of holomorphic ones. For example, a Cauchy-Kowalewski theorem for generalized analytic functions in the sense of I.N. Vekua was formulated in [7]. The method of scales of Banach spaces is also applicable to initial value problems for which the initial function is a solution of a higher-dimensional generalization of the Cauchy-Riemann system: potential vectors as initial functions are investigated in [8], whereas the paper [9] deals with the case of generalized potential vectors.

J. Ławrynowicz (ed.), Deformations of Mathematical Structures, 309–317.

On the other hand, starting also from Nagumo's considerations, W. Walter [11] has recently proved the Cauchy-Kowalewski theorem by using a contraction principle. When rewriting the initial value problem as integral equation, the corresponding integral operator turns out to be contractive in sufficiently small time intervals. The underlying Banach space is equipped with a modified maximum norm, where the modulus is multiplied with a factor vanishing at the boundary of the domain in which the space-like variable varies.

W. Walter's idea is always applicable if the Banach space under consideration can be equipped with the maximum norm. In the present paper this will be demonstrated in the cases of both potential vectors and generalized potential vectors as initial vectors (in the papers [8] and [9] such initial value problems have been reduced to operator equations in scales of Banach spaces). Finally, notice that also initial value problems in abstract scales can be proved by a contraction principle [10].

1. INITIAL VALUE PROBLEMS FOR VECTORS IN $\mathbb{R}^3$ DEPENDING ON TIME

Let L be the first order differential operator

$$L\,u = \Big(\sum_{i,j} A_{ij}^{(1)} \frac{\partial u_i}{\partial x_j} + \sum_i B_i^{(1)} u_i + D^{(1)}, \ldots \Big),$$

where $u = u(t,x)$, $x = (x_1,x_2,x_3)$, $A_{ij}^{(k)} = A_{ij}^{(k)}(t,x)$ and so on. The solutions to the first order system

$$\begin{aligned} &\operatorname{div} u + (a,u) = 0, \\ &\operatorname{rot} u + [u \times b] = 0, \end{aligned} \tag{1}$$

where $a = (a_1,a_2,a_3)$ and $b = (b_1,b_2,b_3)$ are constant vectors, are called generalized potential vectors (cf. [5]). Suppose, further, that the differential operator L is associated with the differential operator l defined by the left-hand sides of (1), i.e. L transforms the set of all solutions to (1) into itself (see [9]; as for the special case of potential vectors we refer to [8]).

Now, let G be a given bounded domain in $\mathbb{R}^3$. Denote by $d(x)$ the distance of $x \in G$ from the boundary of G. Let

$$M = \{(t,x) : x \in G,\ 0 \leq t < \eta\, d(x)\}$$

be the cone with base G and slope η. Assume that the coefficients of L are defined in M for every $\eta \leq \eta_o$. Consider the initial value problem

$$\partial u/\partial t = L\,u, \tag{2}$$

$$u(0,x) = u_o(x), \tag{3}$$

where $u_o = u_o(x)$ is a given generalized potential vector in G. Then the following theorem is true:

THEOREM. *The initial value problem* (2), (3) *is globally solvable in the cone* M *by a contraction principle provided that the slope* η *of the cone* M *is sufficiently small.*

Making use of the scale of Banach spaces of generalized potential vectors defined in a family of subdomains of G, the existence of the solution of the initial value problem (2), (3) has already been proved in [9], whereas the case of potential vectors has been investigated in [8]. In the present paper the above-mentioned theorem will be proved by using W. Walter's ideas [11].

2. A BANACH SPACE OF GENERALIZED POTENTIAL VECTORS DEPENDING ON TIME

Define a modified distance

$$d(t,x) = d(x) - \frac{t}{\eta}$$

for points belonging to the cone M. Take any point $x_0 \in G$ and any t with $0 \leq t < \eta\, d(x)$ and let r be any number satisfying the inequality

$$r < d(t,x_0).$$

Denote the Euclidean distance of two points x and y in $\mathbb{R}^3$ by r_{xy}. Further, the (closed) ball centred at x_0 with radius r is denoted by $K_r(x_0)$ and its boundary by $S_r(x_0)$. For all points (t,y) with $r_{x_0 y} \leq r$ we have

$$d(y) \geqq d(x_0) - r$$

and, consequently,

$$d(t,y) = d(y) - \frac{t}{\eta} \geqq d(x_0) - \frac{t}{\eta} - r \geqq d(t,x_0) - r > 0. \tag{4}$$

Therefore, all points (t,y) with $y \in K_r(x_0)$ belong to M.

Now, let the vector $u = u(t,x)$ be a solution to the system (1) for every t. Without any loss of generality we may assume that $a = b$ (see [5]). Then $u = u(t,x)$ is also a solution of the second order equation

$$\Delta u - |a|^2 \mu = 0,$$

where $|a|^2 = a_1^2 + a_2^2 + a_3^2$. Taking into consideration that $(-4\pi\, r_{xy})^{-1}$ defines a fundamental solution to the Laplace equation with singularity at y, in $K_r(x_0)$ the vector $u = u(t,x)$ can be represented by

$$\mu(t,x) = \frac{|a|^2}{4\pi} \iiint_{K_r(x_0)} r_{xy}^{-1}\, u(t,y)\,dy + \tilde{u}(t,x), \tag{5}$$

where $\tilde{u} = \tilde{u}(t,x)$ turns out to be a solution to the Laplace equation.

Denote the set of all solutions $u = u(t,x)$ defined in M by $R(M)$. The subset of all those $u(t,x)$, for which

$$|u(t,x)|d^p(t,x),$$

is bounded and denoted by $R_*(M)$, where p is a fixed positive number (the case $p = 0$ could also be included by analogous considerations) and $|u(t,x)|$ means the maximum of all $|u_j(t,x)|$. Obviously,

$$\|u(t,x)\|_* = \sup_M |u(t,x)|d^p(t,x) \tag{6}$$

is a norm in $R_*(M)$. This definition implies immediately that at each point (t,x) of M the solution $u = u(t,x)$ can be estimated by

$$|u(t,x)| \leq \frac{\|u(t,x)\|_*}{d^p(t,x)} . \tag{7}$$

Now, we shall show that $R_*(M)$, equipped with the modified maximum norm (6), is complete. Let K be any compact subset of G and δ any positive number. Then, define $\tilde{K}$ by

$$\tilde{K} = \{(t,x) : x \in K,\ 0 \leq t \leq \eta(d(x) - \delta)\}.$$

For each point (t,x) belonging to $\tilde{K}$ we have

$$d(t,x) \geq \delta$$

and, therefore,

$$|u(t,x)| \leq \frac{\|u(t,x)\|_*}{\delta^p} .$$

This implies that a fundamental sequence in $R_*(M)$ converges uniformly in $\tilde{K}$.

On the other hand, for each point (t',x') of M we can find a compact subset K of G and a positive δ such that (t',x') belongs to $\tilde{K}$. Therefore, a fundamental sequence possesses a continuous limit function in M. Notice that $\tilde{K}$ can be chosen in such a way that the arbitrarily chosen point (t',x') is an inner point of the intersection of $\tilde{K}$ with the hyperplane defined by $t = t'$. It remains to prove that the limit function is a generalized potential vector. For this end we deduce an integral representation for the derivatives which also be applied by proving Nagumo's lemma for generalized potential vectors. Replacing $\tilde{\mu}(t,x)$ in (5) by the Poisson integral

$$\frac{1}{4\pi r} \iint_{S_r(x_0)} r_{xx'}^{-3}(r^2 - r_{x_0 x}^2)\mu(t,x')d\mu,$$

where $d\mu$ is the area element of the sphere $S_r(x_0)$, and differentiating (5) with respect to x_j, we obtain

$$\frac{\partial u}{\partial x_j}(t,x) = -\frac{|a|^2}{4\pi} \iiint_{K_r(x_o)} r_{xy}^{-3}(x_j - y_j)u(t,y)\,dy$$
$$-\frac{1}{4\pi r} \iint_{S_r(x_o)} [3r_{xx'}^{-5}(x_j - x_j')(r^2 - r_{x_o x}^2) + 2r_{xx'}^{-3}(x_j - x_{oj})]u(t,x')\,d\mu. \tag{8}$$

In view of E. Schmidt's inequality the modulus of the first term may be estimated by

$$|a|^2\, r \sup_{y\in K_r(x_o)} |u(t,y)|.$$

Restricting x to the smaller ball $K_{r'}(x_o)$, where $r' < r$, we get for the modulus of the second integral in (8) the upper bound

$$\left(3\frac{r^3}{(r-r')^4} + 2\frac{r'\,r}{(r-r')^3}\right) \sup_{x'\in S_r(x_o)} |u(t,x')|$$

since $r_{xx'} \geq r - r'$ in $K_{r'}(x_o)$. Therefore, the following lemma holds:

LEMMA 1. *In the smaller ball* $K_{r'}(x_o)$, $r' < r$, *the first order derivatives of a generalized potential vector can be estimated by*

$$\left|\frac{\partial u}{\partial x_j}(t,x)\right| \leq \left(|a|^2 r + \frac{3r^3}{(r-r')^4} + \frac{2r'r}{(r-r')^3}\right) \sup_{y\in K_r(x_o)} |u(t,y)|. \tag{9}$$

Notice that for $r' = 0$, i.e. $x = x_o$, this estimate passes into

$$\left|\frac{\partial u}{\partial x_j}(t,x_o)\right| \leq \left(|a|^2 r + \frac{3}{r}\right) \sup_{y\in K_r(x_o)} |u(t,y)|. \tag{10}$$

Now, take any fundamental sequence $\{u_n(t,x)\}_{n=1,2,\ldots}$ in $R_*(M)$. Applying the estimate (9) to $u_n - u_m$ we can see that

$$\left|\frac{\partial u_n}{\partial x_j} - \frac{\partial u_m}{\partial x_j}\right|$$

is uniformly small in $K_{r'}(x_o)$ provided that n and m are sufficiently large. Hence, the limit $u = u(t,x)$ of the $u_n = u_n(t,x)$ is a generalized potential vector as well.

3. NAGUMO'S LEMMA FOR GENERALIZED POTENTIAL VECTORS

Let $u = u(t,x)$ be any element belonging to $R_*(M)$. Take again an arbitrary point (t,x_o) of M. Specify the radius r, introduced at the beginning of Section 2, to

$$r = \frac{1}{p+1}\, d(t,x_o). \tag{11}$$

Then, in view of (4), all points (t,y) with $y \in K_r(x_o)$ satisfy the inequality

$$d(t,y) \geq \frac{p}{p+1}\, d(t,x_o).$$

Taking into account the inequality (7) we, consequently, get

$$\sup_{y \in K_r(x_o)} |u(t,y)| \leq \left(1+\frac{1}{p}\right)^p \frac{\|u(t,x)\|_*}{d^p(t,x_o)}.$$

The choice (11), the a-priori estimate (10) and the latter estimate lead to

$$\left|\frac{\partial u}{\partial x_j}(t,x_o)\right| \leq \left(|a|^2 \frac{d(t,x_o)}{p+1} + \frac{3(p+1)}{d(t,x_o)}\right)\left(1+\frac{1}{p}\right)^p \frac{\|u(t,x)\|_*}{d^p(t,x_o)}.$$

Since G is bounded, we have $d(t,x) \leq d(x) \leq d_o$ and, consequently,

$$1 \leq \frac{d_o}{d(t,x_o)}. \tag{12}$$

Therefore, the following generalization of Nagumo's lemma to the case of generalized potential vectors is true:

LEMMA 2. *Suppose that* $u = u(t,x)$ *belongs to* $R_*(M)$. *Then the first order derivatives can be estimated by*

$$\left|\frac{\partial u}{\partial u_j}(t,x_o)\right| \leq \left(|a|^2 \frac{d_o^2}{p+1} + 3(p+1)\right)\left(1+\frac{1}{p}\right)^p \frac{\|u(t,x)\|_*}{d^{p+1}(t,x_o)},$$

where (t,x_o) *is an arbitrary point belonging to* M.

For the sake of brevity we denote the factor

$$\left(|a|^2 \frac{d_o^2}{p+1} + 3(p+1)\right)\left(p+\frac{1}{p}\right)^p$$

by C_p. In the case of ordinary potential vectors the constant C_p is equal to

$$3(p+1)\left(1+\frac{1}{p}\right)^p.$$

4. PROOF OF THE CAUCHY-KOWALEWSKI THEOREM FOR POTENTIAL VECTORS AND GENERALIZED POTENTIAL VECTORS

First, we wish to note that the initial value problem (2), (3) is equivalent to the integral equation

$$\begin{aligned} u(t,x) &= u_o(x) + \int_0^t (Lu)(\tau,x)d\tau \\ &= u_o(x) + \int_0^t \Big(\sum_{i,j} A_{ij}(\tau,x) \frac{\partial u_i}{\partial x_j}(\tau,x) + \sum_i B_i(\tau,x)u_i(\tau,x) + D(\tau,x)\Big)d\tau, \end{aligned} \tag{13}$$

where $A_{ij} = (A_{ij}^{(1)}, A_{ij}^{(2)}, A_{ij}^{(3)})$ and so on.

Assume that the operator L is associated with the operator 1 defined by the left-hand sides of the system (1). In the papers [8] and [9], mentioned already in Section 1, there are given sufficient conditions for L guaranteeing that L is associated with 1. Moreover, the coefficients

$$A_{ij}^{(k)}, \; B_i^{(k)}, \; D^{(k)},$$

and the initial function $u_o = u_o(x)$ have to satisfy the following inequalities:

$$\begin{aligned} &|A_{ij}^{(k)}(t,x)| \leq A_*, \quad |B_i^{(k)}(t,x)| \leq \frac{B_*}{d(t,x)}, \\ &|D^{(k)}(t,x)| \leq \frac{D_*}{d^{p+1}(t,x)}, \quad |\mu_o(x)| \leq \frac{E_*}{d^p(x)}, \end{aligned} \tag{14}$$

where A_*, B_*, D_*, E_* are given constants. In view of (12) the second and third conditions (14) are satisfied, especially if the coefficients $B_i^{(k)}$ and $D^{(k)}$ are bounded. Analogously, the fourth condition (14) is satisfied if onlu μ_o is bounded.

Now, consider the operator that is defined by the right-hand side of (13):

$$U(t,x) = u_o(x) + \int_0^t (Lu)^0(\tau,x)d\tau. \tag{15}$$

Since L is associated with the operator defining potential vectors and generalized potential vectors, respectively, the operator (15) maps $R(M)$ into itself (see [8] for the case of potential vectors and [9] for generalized potential vectors). Taking into consideration that

$$\int_0^t \frac{d\tau}{d^{p+1}(\tau,x)} < \frac{\eta}{p} \frac{1}{d^p(t,x)} \tag{16}$$

(see [11]), the assumptions (14) imply the following lemma:

LEMMA 3. *The operator* (15) *maps* $R_*(M)$ *into itself.*

Now, we shall look for sufficient conditions under which the operator (15) is contractive. Take only two elements $\mu = \mu(t,x)$ and $v = v(t,x)$ belonging to $R_*(M)$ and denote the images by $U(t,x)$ and $V(t,x)$, respectively. Then we have

$$U(t,x) - V(t,x) = \int_0^t \Big(\sum_{i,j} A_{ij}(\tau,x) \Big(\frac{\partial u_i}{\partial x_j}(\tau,x) - \frac{\partial u_i}{\partial x_j}(\tau,x)\Big)$$

$$+ \sum_i B_i(\tau,x)(u_i(\tau,x) - v_i(\tau,x))\Big)d\tau .$$

Again, taking into account the inequalities (14), (16) and Nagumo's lemma, we obtain

$$|U(t,x) - V(t,x)| \leq \frac{\eta}{p}(9A_* C_p + 3B_*) \frac{\|u(t,x) - v(t,x)\|_*}{d^p(t,x)} .$$

Thus, the operator (15) is contractive, provided that

$$3\,\frac{\eta}{p}\,(3A_* C_p + B_*) < 1 .$$

By virtue of the definition of C_p the following lemma is proved:

LEMMA 4. *Assume that*

$$\eta < \left(3\left|9A_*(1 + \frac{1}{p})^{p+1} + 3A_*|a|^2 \frac{d_o^2}{p(p+1)} + \frac{B_*}{p}\right|\right)^{-1} .$$

Then the operator (15) *is contractive.*

This means that the solution to the initial value problem (2), (3) may be constructed globally by a contraction principle if η is small enough.

References

[1] NAGUMO, M.: 'Über das Anfangswertproblem partieller Differentialgleichungen', *Japan Journ. Math.* 18 (1941), 41-47.

[2] NIRENBERG, L.: 'An abstract form of the nonlinear Cauchy-Kowalewski theorem', *Journ. Diff. Geom.* 6 (1972), 561-576.

[3] ——: *Topics in nonlinear functional analysis*, New York 1974 (Russian transl. Moscow 1977).

[4] NISHIDA, T.: 'A note on Nirenberg's theorem as an abstract form of the nonlinear Cauchy-Kowalewski theorem in a scale of Banach spaces', *Journ. Diff. Geom.* 12 (1977), 629-633.

[5] OBOLASHVILI, E.I.: 'Space-like analogon of generalised analytical functions', *Soob. Akad. Nauk GrSSR* 73,1 (1974), 21-24.

[6] TREVES, F.: *Basic linear partial differential equations*, New York-San Francisco-London 1975.

[7] TUTSCHKE, W.: 'An initial value problem for generalised analytical functions depending on time (generalised Cauchy-Kowalewska theorem)', *Dokl. Akad. Nauk SSSR* 262,5 (1982), 1081-1082; English transl.: *Soviet Math. Dokl.* 25,1 (1982), 201-205.

[8] ——: 'Potential vectors depending on time', *Trudy Symp. 75-let. I.N. Vekua*, Tbilisi 1982, in print.

[9] ——: 'A solution of the Cauchy problem for some classes of functions being a generalisation of the spaces of generalised analytic functions' in: *Mathematical structures. Computational mathematics. Mathematical modelling.*, vol. 2, 85-89, Sofia 1984.

[10] ——: 'An abstract nonlinear Cauchy-Kowalewski theorem and its proof by a contraction principle. II', Yugoslavian Conference on Complex Analysis, Budva 1986, in print.

[11] WALTER, W.: 'An elementary proof of the Cauchy-Kowalewski theorem', *Amer. Math. Monthly* 92 (1985), 115-126.

SYMBOLIC CALCULUS APPLIED TO CONVEX FUNCTIONS AND ASSOCIATED DIFFUSIONS

Masami Okada
Department of Mathematics
College of General Education
Tôhoku University
Sendai 980, Japan

ABSTRACT. We prove classical inequalities for generalized second-order differential operators in order to study some potential theoretic properties of convex functions and associated diffusions.

0. NOTATION AND DEFINITION

Let $D \subset \mathbb{R}^n$ be a domain with Lipschitz boundary and let a_{ij}, $i,j = 1,2,\ldots,n$, be Borel measures on D such that $a_{ij} = a_{ji}$ and $\Sigma a_{ij} \xi_i \xi_j$ is a non-negative Borel measure on D for any $\xi = (\xi_1,\ldots,\xi_n) \in \mathbb{R}^n$. We associate to (a_{ij}) the $(n-1)$-form w with measure coefficients whose values are in the space of differential operator of first order:

$$w = 1/(n-2)! \sum_{\sigma \in \sigma_n} a_{\sigma(1)\sigma(2)} (* \, dx_{\sigma(1)}) \frac{\partial}{\partial x_{\sigma(2)}}$$

$$= \sum_{k,l=1}^{n} a_{kl} (* \, dx_k) \frac{\partial}{\partial x_l}, \tag{1}$$

where $(*dx_k) = (-1)^{k+1} dx_1 \wedge dx_2 \wedge \ldots \overset{k}{\vee} \ldots \wedge dx_n$ and σ_n is the permutation group of $\{1,2,\ldots,n\}$.

The rule of calculus is summarized for convenience in the following formulae:

$$dw := \sum_{k,l} \frac{\partial a_{kl}}{\partial x_k} \frac{\partial}{\partial x_l} dV, \qquad dV = dx_1 \wedge \ldots \wedge dx_n, \tag{2}$$

where derivatives are understood in the distribution sense. Here we assume that $\Sigma \, \partial a_{kl}/\partial x_k$ is a Borel measure for any $l = 1,2,\ldots,n$. Furthermore, for any $\phi \in C_o^2(D)$,

$$w\phi = \Sigma \, a_{kl} \frac{\partial \phi}{\partial x_l} (* \, dx_k) \tag{3}$$

J. Ławrynowicz (ed.), Deformations of Mathematical Structures, 319–329.

and

$$w \wedge d\phi = (-1)^{n-1} \Sigma\, a_{kl} \frac{\partial^2 \phi}{\partial x_k\, \partial x_l}\, dV. \tag{4}$$

Now, let L and ε be defined by

$$L\,\phi = \Sigma \frac{\partial}{\partial x_k}\left(a_{kl} \frac{\partial \phi}{\partial x_l}\right) \tag{5}$$

and

$$\varepsilon(\phi,\phi) = \int_D \Sigma\, a_{kl} \frac{\partial \phi}{\partial x_k} \frac{\partial \phi}{\partial x_l}\, dV. \tag{6}$$

Therefore the following equalities hold easily:

$$L\,\phi\, dV = d(w\phi) = (dw)\phi + (-1)^{n-1}\, w \wedge d\phi \tag{7}$$

and

$$\varepsilon(\phi,\phi) = \int_D d\phi \wedge w\phi = -\int_D \phi\, L\,\phi\, dV. \tag{8}$$

1. POINCARÉ INEQUALITY AND TRACE INEQUALITY

For convenience, $dw\phi$ will denote $(dw)\phi$ from now on. In this chapter, we fix a non-negative function $u \in C^1(D) \cap L^\infty(D)$ such that $(-1)^{n-1} w \wedge du$ is a non-negative volume form. Next, we put

$$A = \int_D \phi^2\, du \wedge wu, \quad I = \int_D \phi^2 (-1)^{n-1}\, w \wedge du, \quad J_u = \int_D \phi^2 (-1)^{n-1}\, uw \wedge du. \tag{9}$$

Then we have

<u>LEMMA 1</u>.

$$A + J_u \leq \int_{\partial D} \phi^2 u\, wu + 2\,||u||\, A^{\frac{1}{2}}\, \varepsilon^{\frac{1}{2}} - \int_D \phi^2 u\, dwu, \tag{10}$$

$$A + J_{\tilde{u}} \leq \int_{\partial D} \phi^2 \tilde{u}\, wu + 2\,||\tilde{u}||\, A^{\frac{1}{2}}\, \varepsilon^{\frac{1}{2}} - \int_D \phi^2 \tilde{u}\, dwu, \tag{11}$$

where $||u|| = \sup_{z \in D} u(z)$ *and* $\tilde{u} = u - ||\tilde{u}||$.

<u>Proof</u>. We easily get

$$\phi^2 du \wedge wu = d(\phi^2 u\; wu) - 2u\,\phi\, d\phi \wedge wu - \phi^2 u\, dwu - (-1)^{n-1}\, \phi^2 u \wedge du. \tag{12}$$

Then we apply Stokes' formula to the first term and Schwarz's inequality to the second one, so that

$$\left|\int_D u\phi \, d\phi \wedge wu\right| \leq \|u\| \, A^{\frac{1}{2}} \varepsilon^{\frac{1}{2}}. \tag{13}$$

(see [10]). The proof of (11) is now immediate.

Next, let us decompose the surface element wu on ∂D into the two non-negative parts: $wu = (wu)_+ - (wu)_-$. We notice that there are cases where $(wu)_- = 0$. For example, take the case where u is the defining function of D, i.e. $D = \{u < 1\}$.

Now, let $F_\pm$ denote $\int_{\partial D}(wu)_\pm$. Then we have

LEMMA 2.

$$F_\pm \leq F_\mp + \int_D \phi \, dwu \pm I + 2A^{\frac{1}{2}} \varepsilon^{\frac{1}{2}}. \tag{14}$$

Proof. It sufficies to note that

$$d(\phi^2 wu) = 2\phi \, d\phi \wedge wu + \phi^2 dwu + \phi^2(-1)^{n-1} w \wedge du. \tag{15}$$

Let the similar decomposition of dwu be $dwu = (dwu)_+ - (dwu)_-$ and suppose that there exists $0 \leq \delta < 1$ such that

$$u(dwu)_- \leq \delta\{du \wedge wu + (-1)^{n-1} uw \wedge du\}. \tag{H_1}$$

Then we have

THEOREM 1.

$$I \leq (3 - 2\delta)(1 - \delta)^{-2}(\|u\| + 1)F_+ + 5(1 - \delta)^{-2}(\|u\| + 1)^2 \varepsilon. \tag{16}$$

Proof. On account of (H_1) we induce from (10)

$$A + J_u \leq \int_{\partial D} \phi^2 u(wu)_+ + 2\|u\| \, A^{\frac{1}{2}} \varepsilon^{\frac{1}{2}} + \delta(A + J_u).$$

Since A and J_u are non-negative,

$$A \leq (1 - \delta)^{-1} \|u\| \{F_+ + 2A^{\frac{1}{2}} \varepsilon^{\frac{1}{2}}\}, \tag{17}$$

$$J_u \leq (1 - \delta)^{-1} \|u\| \{F_+ + 2A^{\frac{1}{2}} \varepsilon^{\frac{1}{2}}\}. \tag{18}$$

Therefore, by elementary calculus, it follows from (17) that

$$A \leq 2(1 - \delta)^{-2} \|u\| \{2\|u\| \varepsilon + (1 - \delta) F_+\}. \tag{19}$$

Thus, by (18),

$$J_u \leq (3 - 2\delta)(1 - \delta)^{-2} \|u\| F_+ + 5(1 - \delta)^{-2} \|u\|^2 \varepsilon. \tag{20}$$

Now it suffices to replace u with $u + 1$, noting that $I \leq J_{u+1}$.

COROLLARY 1. *A generalized Poincaré inequality holds under the assumption* (H_1) *for* $\phi \in C_0^\infty(D)$.

Proof. Note that $F_+ = 0$.

COROLLARY 2. *We suppose that* (H_1) *is satisfied and* $(wu)_+ = 0$. *Then we can conclude that* $w \wedge du = 0$.

Proof. It suffices to apply (16) to $\phi \equiv 1$.

This means that the value of wu cannot be assigned arbitrarily on the boundary ∂D.

Now, let us proceed to a generalized trace inequality.

THEOREM 2. *Assume that there exists* $0 \leqq \gamma < 1$ *and the positive numbers* N_1, N_2 *and* N_3 *such that*

$$(\|u\| - u)(dwu)_+ \leqq \gamma du \wedge wu + N_1 \|u\| (-1)^{n-1} \, w \wedge du,$$
$$(dwu)_+ \leqq N_2 \|u\|^{-1} \, du \wedge wu + N_3(-1)^{n-1} \, w \wedge du. \qquad (H_2)$$

Then we get

$$F_+ \leqq C_3 \, F_- + C_4 \|u\| \, \varepsilon + C_5 \, I, \qquad (21)$$

where $C_3 = 2(N_2+1)(1-\gamma)^{-1} + 1$, $C_4 = 4(N_2+1)(1-\gamma)^{-1} + 1$, *and* $C_5 = 2(N_2+1)(N_1+1)(1-\gamma)^{-1} + (N_3+1)$.

Remark. An estimate of F_- by F_+ and ε holds as in the proof of Theorem 1 under a similar condition (see (H_1)).

Proof of Theorem 2. In view of (11), we have

$$A \leqq \|u\| \, F_- + 2\|u\| \, A^{\frac{1}{2}} \, \varepsilon^{\frac{1}{2}} - \int_D \tilde{u} \, \phi^2 (dwu)_+ .$$

Therefore, by the assumption (H_2),

$$A \leqq \|u\| \, F_- + 2\|u\| \, A^{\frac{1}{2}} \, \varepsilon^{\frac{1}{2}} + (N_1+1)\|u\| \, I + \gamma A.$$

Thus, by elementary calculus

$$A \leqq 4(1-\gamma)^2 \, \|u\|^2 \, \varepsilon + 2(1-\gamma)^{-1} \, \|u\| \, F_- + 2(1-\gamma)^{-1}(N_1+1)\|u\| \, I \qquad (22).$$

On the other hand, by (14) and (H_2) we have

$$F_+ \leqq F_- + N_2 \|u\|^{-1} \, A + (N_3+1) I + 2A^{\frac{1}{2}} \, \varepsilon^{\frac{1}{2}}.$$

Then the inequality holds by elementary calculus.

2. ASSOCIATED CAPACITY

In this chapter we shall show that a capacity is naturally associated to w. Put $|\phi| = \varepsilon(\phi,\phi)^{\frac{1}{2}}$ and let $\bar{\varepsilon}$ be the closure of $C_o^\infty(D)$ with respect to the norm $|\ |$. To rely on the Hilbert space based on the potential theory of Beurling-Deny, $\bar{\varepsilon}$ is required to be imbedded in $L^2(m,D)$ with m as an appropriate Borel measure. This property will be guaranteed by the following closability (see [5] for the general reference to this chapter):

(C) If a Cauchy sequence $\{\phi_n\} \subset C_o^\infty(D)$ with respect to $|\ |$ tends to zero in the L^2-norm, then $|\phi_n|$ tends to zero as well.

Then, if we define m_1 by $m_1 = m_o + \Sigma[dw\,x_j]$, where $m_o = \Sigma\, d\,x_j \wedge w\,x_j$ and $[dw\,x_j] = (dw\,x_j)_+ + (dw\,x_j)_-$, we have the following proposition:

(<u>Remark.</u> If $u(x) = \Sigma\, x_j^2 \equiv v^o(x)$, then $(-1)^{n-1}\, w \wedge dv^o = 2m_o$.)

<u>PROPOSITION 1</u>. *Under the above hypothesis, the Dirichlet form* ε *is* C_o^∞*-closable in* $L^2(m_1,D)$.

<u>Proof</u>. We repeat a similar argument as in [10] for the sake of completeness. Let $\{\phi_k\} \subset C_o^\infty(D)$ be Cauchy with respect to $|\ |$ satisfying $\|\phi_k\|_{L^2(m_1)} \to 0$ as $k \to \infty$. Then

$$\varepsilon(\phi_k - \phi_l,\ \phi_k - \phi_l) = \sup\{\varepsilon(\phi_k - \phi_l, \psi)\ |\ \psi \in C_o^\infty(D),\ |\psi| \leqq 1\}$$

$$= \sup_\psi\{-\textstyle\int_D(\phi_k - \phi_l)\,dw\,\psi - \int_D(\phi_k - \phi_l)(-1)^{n-1}\, w \wedge d\psi\}.$$

Therefore, when $l \to +\infty$, we have by hypothesis

$$\textstyle\int_D(\phi_k - \phi_l)\,dw\psi \to \int_D \psi_k\, dw\psi \qquad \text{and} \qquad \int_D(\phi_k - \phi_l)\, w \wedge d\psi \to \int_D \psi_k\, w \wedge d\psi.$$

Hence

$$E(\phi_k, \phi_k) = \sup_\psi\{-\textstyle\int_D \phi_k\, dw\,\psi - \int_D \phi_k (-1)^{n-1}\, w \wedge d\psi\} \to 0 \quad \text{as} \quad k \to +\infty.$$

Now, consequently, a capacitary function C is naturally defined for compact sets $K \subset D$ by

$$C(K) = \inf\{\varepsilon(\phi,\phi) + \|\phi\|^2_{L^2(m_1)}\ |\ \phi \in C_o^\infty(D),\ \phi \geqq 1 \ \text{on}\ K\}. \tag{23}$$

<u>COROLLARY 2</u>. C *is extended to be a Choquet capacity.*

3. ASSOCIATED DIFFUSION

From now on, we shall add the assumption that the volume form m_1 is dense in D, namely $m_1(U) > 0$ for any open set $U \subset D$. Then, under the assumption of Proposition 1, there exists a diffusion (X_t^w, ζ, P_x^w) associated to (w, m_1) on account of the general theory of [5].

Next, we shall give a criterion in order to show that our diffusion is transient. We assume that D is a bounded set contained in $\{(x_1, x_2, \ldots, x_n) \in \mathbb{R}^n;\ x_j \geqq 1,\ j = 1,2,\ldots,n\}$.

PROPOSITION 2. *If there exist* $0 \leqq \delta < 1$ *and* $c > 0$ *such that*

$$\Sigma(dw\ x_k)_+ \leqq c\ m_o \tag{24}$$

and

$$\Sigma(dw\ x_k)_- \leqq \delta\ m_o, \tag{25}$$

then (X_t^w, ζ, P_x^w) *is transient.*

Proof. Let $A_j = \int_D \phi^2\, dx_j \wedge wx_j$. Then, as in the proof of Theorem 1,

$$A_j \leqq \int_{\partial D} \phi^2\, x_j (wx_j)_+ + 2\|x_j\|\, A_j^{\frac{1}{2}}\, \varepsilon^{\frac{1}{2}} + \delta A_j,$$

since the equality $w \wedge dx_j = 0$ implies $J_{x_j} = 0$. Thus, for $\phi \in C_o^\infty(D)$, (19) gives us

$$A_j \leqq 4(1-\delta)^{-2}\, \|x_j\|^2\, \varepsilon. \tag{26}$$

Therefore, by summing up both sides, we get

$$\int_D \phi^2\, dm_o \leqq 4(1-\delta)^{-2}\ \Sigma\, \|x_j\|^2\, \varepsilon.$$

On the other hand, by the hypotheses (24) and (25), $m_1 \leqq (1+c+\delta)m_o$, so that

$$\int_D \phi^2\, dm_1 \leqq 4(1-\delta)^{-2}(1+c+\delta)\Sigma\, \|x_j\|^2\, \varepsilon.$$

We can see that this yields our conclusion.

Now, let us show that our diffusions satisfy a kind of symmetric property. Let D be the unit ball of $\mathbb{R}^n$ and S_+ the semi-sphere $\{x \in S^{n-1},\ x_1 \geqq 0\}$.

PROPOSITION 3. *Let* (H_1) *be assumed. If* $w = 0$ *on* S_+, *i.e. if there is no diffusing on* S_+, *then* $w \equiv 0$ *in* D.

Proof. We take $u(x) \equiv u_N(x) = (x_1 + N)^2 + \Sigma_2^n x_j^2$ with a sufficiently large positive constant N. Then, since $(wu_N)_+^- \to 0$ as $N \to +\infty$, which is easy to show, Corollary of Chapter 1 implies that $w \wedge d\|x\|^2 = w \wedge du_N \to 0$ as $N \to \infty$. And hence $w = 0$.

4. CONVEX FUNCTIONS

First of all, we recall some non-trivial examples of convex functions:

E 1. $v_1(x,y) = (x^2+y^2)/y, \quad D = \{(x,y) \in \mathbb{R}^2, \ y>0\},$

E 2. $v_2(x,y) = \sup \left\{ \begin{array}{l} v(x,y) \mid v \ \text{ convex in } \ \mathbb{R}^2, \quad v \leq 1 \ \text{ on } \ K \\ \sup v(x,y) - (x^2+y^2)^{\frac{1}{2}} < +\infty \end{array} \right\}$

(cf. [13]),

E 3. $v_3(x,y) = \begin{cases} \{(1-y)^2+x^2\}/(1-y), & x+y \geq 1, \\ 2x, & x+y<1, \end{cases} \quad D = \text{unit disk},$

E 4. $v_4(x,y) = \max(y, 1-y, \min(1, 1-x)), \quad D = \mathbb{R}^2 \setminus \{(0,t) \mid 0 \leq t \leq 1\}$

(cf. [3]).

Secondly, we notice that, as it is well known, most of potential theoretic properties of convex functions are closely related to those of plurisubharmonic functions. Some of them are easier to show for a convex case and others are not because of the lack of the conjugate operator d^c. Here let us discuss rather briefly some properties relevant to those from the preceding chapters.

Let $v^{(k)}$, $k=2,3,\ldots,n$, be convex functions defined on D. We suppose that $v^{(k)}$ are of the class C^2. This hypothesis is not essential. Then we define (a_{kl}), $k,l=1,2,\ldots,n$, by

$$a_{kl} = \Sigma \operatorname{sgn}\sigma \ \operatorname{sgn}\tau \ v^{(2)}_{\sigma(2)\tau(2)}, \ldots, v^{(n)}_{\sigma(n)\tau(n)}, \tag{27}$$

where the summation is taken over the set $\sigma,\tau \in \sigma_n$ such that $\sigma(1)=k$ and $\tau(1)=l$ and where $v_{ij}^{(k)}$ denote $\partial^2 v^{(k)}/\partial x_i \, \partial x_j$.

R e m a r k. If $v^{(k)}$ are convex functions which are not smooth, a_{kl} are defined as Borel measures, as it is well known (cf. [12]).

LEMMA 3.

$$a_{kl} = a_{lk} \tag{28}$$

and $\Sigma a_{kl} \xi_k \xi_l$ *is a non-negative Borel measure for any* $\xi \in \mathbb{R}^n$.

P r o o f. The first equality is a direct consequence of the definition. The second statement follows from the next equality. Let $v(x) = (\Sigma \xi_j x_j)^2$. Then we have

$$\Sigma\, a_{kl}\, \xi_k\, \xi_l = \sum_{\sigma\, \sigma_n} \operatorname{sgn} \sigma \; dv_{\sigma(1)} \wedge dv^{(2)}_{\sigma(2)} \wedge \ldots \wedge dv^{(n)}_{\sigma(n)}, \tag{29}$$

which is a non-negative volume form, since v is also a convex function.

LEMMA 4. *Let* w *be defined by* (1) *with* (27). *Then*

$$d\,w = 0. \tag{30}$$

Proof. By elementary calculus we have for any k

$$\sum_{l} \frac{\partial a_{kl}}{\partial x_l} = \sum_{\sigma(1)=1} \operatorname{sgn} \sigma (\sum_{\tau} \operatorname{sgn} \tau \, \frac{\partial}{\partial x_{\tau(1)}} \, (v^{(2)}_{\sigma(2)\tau(2)}, \ldots, v^{(n)}_{\sigma(n)\tau(n)})) \tag{31}$$

which is shown to be equal to zero by a combinatorial argument.

Lemma 4 shows that our differential operator-valued (n-1)-form w, induced from convex functions, has an interesting property (30) which simplifies our arguments of the preceding chapters. For example, for $\phi \in C^\infty_o(D)$, we have

$$\int_D \phi^2 (-1)^n \; w \wedge du \leqq \text{const}(\|u\| + 1)\; \varepsilon(\phi,\phi), \tag{32}$$

$$\begin{aligned} L\,\phi\, dV &= \Sigma\, a_{kl} \frac{\partial^2 \phi}{\partial x_k\, \partial x_l} \; dV = (-1)^{n-1} \; w \wedge d\,\phi \\ &= \sum_{\sigma\in\sigma_n} \operatorname{sgn}\sigma \;\; d\phi_{\sigma(1)} \wedge dv^{(2)}_{\sigma(2)} \wedge \ldots \wedge dv^{(n)}_{\sigma(n)}, \end{aligned} \tag{33}$$

$$m_o = m_1 = \tfrac{1}{2}(-1)^{n-1} \; w \wedge dv^o = \sum_{\sigma\in\sigma_n} \operatorname{sgn} dx_{\sigma(1)} \wedge dv^{(2)}_{\sigma(2)} \wedge \ldots \wedge dv^{(n)}_{\sigma(n)}. \tag{34}$$

Let us proceed to study the induced form w from Chapter 2. On account of (16) and (30), $C(K)$ is equivalent to $C(K, v^{(2)},\ldots,v^{(n)})$ defined by

$$C(K,\; v^{(2)},\ldots,v^{(n)}) = \inf\{\varepsilon(\phi,\phi) \,|\, \phi \in C^\infty_o(D), \;\; \phi \geqq 1 \;\; \text{on} \;\; K\}.$$

PROPOSITION 4.

$$\inf_{\phi} \sup_{v^\bullet} \varepsilon(\phi,\phi) \sim \sup_{v^\bullet} \inf_{\phi} \varepsilon(\phi,\phi),$$

where inf *is taken over the set* $\{\phi | \phi \in C^\infty_o(D), \;\; \phi \geqq 1$ *on* $K\}$ *and* sup *is taken over* $\{(v^{(2)},\ldots,v^{(n)}) | v^{(k)}$ *convex on* D, $0 < v^{(k)} < 1$, $k = 2,\ldots,n\}$.

P r o o f. Analogously as in [11], we notice that the above quantity is also equivalent to

$$\sup_{v^\bullet} \int_K \operatorname{sgn}\sigma\, dv^{(1)}_{\sigma(1)} \wedge \ldots \wedge dv^{(n)}_{\sigma(n)}.$$

5. AN EXAMPLE OF DIFFUSIONS ASSOCIATED TO CONVEX FUNCTIONS

First of all, we recall some facts connected with diffusions associated to convex functions. Let $v^{(k)}$, $k = 2,\ldots,n$, be given convex functions. We define L by (33) and set $m = m_o$ of (34). Our hypothesis is the following:

$$m \text{ is dense in } D. \tag{H_3}$$

Then we deduce from Proposition 1 and (32) that the associated second-order differential operator defined by

$$L^{(m)}\phi = (\Sigma \operatorname{sgn}\sigma\, d\phi_{\sigma(1)} \wedge dv^{(2)}_{\sigma(2)} \wedge \ldots \wedge dv^{(n)}_{\sigma(n)})/m \tag{35}$$

has a Friedrichs' extension on $L^2(m)$ which generates a Markovian semigroup. Therefore, it follows from [5] that there corresponds to $L^{(m)}$ a diffusion, which is shown to be transient from (32).

Now, let us proceed to study an example of our diffusions associated with a degenerated convex function. We assume that a smooth convex function v satisfies the following condition:

$$dv_{s_1} \wedge \ldots \wedge dv_{s_{p+1}} = 0 \tag{H_4}$$

for any subset $\{s_1,\ldots,s_{p+1}\} \subset \{1,\ldots,n\}$ with $s_1 < \ldots < s_{p+1}$, but there exists $\{r_1,\ldots,r_p\}$ with $r_1 < \ldots < r_p$ such that

$$dv_{r_1} \wedge \ldots \wedge dv_{r_p} \neq 0.$$

Then, we can define m and $L^{(m)}$ by choosing $v^{(2)} = \ldots = v^{(n-p)} = \frac{1}{2}\|x\|^2$ in (H_3) and (35):

$$m = \Sigma \operatorname{sgn}\sigma\, dx_{\sigma(1)} \wedge \ldots \wedge dx_{\sigma(n-p)} \wedge dv_{\sigma(n-p+1)} \wedge \ldots \wedge dv_{\sigma(n)},$$

$$L^{(m)}\phi = (\Sigma \operatorname{sgn}\sigma\, d\phi_{\sigma(1)} \wedge dx_{\sigma(2)} \wedge \ldots \wedge dx_{\sigma(n-p)} \wedge dv_{\sigma(n-p+1)} \wedge \ldots \wedge dv_{\sigma(n)}) / m.$$

Then we have

THEOREM 3. *Let* (x_t, ζ, P_x) *be the diffusion associated with* v *given above. Then, starting at each point of* D, *it remains to diffuse it on each* $(n-p)$*-dimensional leaf passing through the point.*

Proof. We reduce our question to the corresponding one in $\mathbb{C}^n$. Let π be a mapping from $\mathbb{C}^n$ to $\mathbb{R}^n$ defined by $\pi(z_1,\dots,z_n) = (\log|z_1|,\dots,\log|z_n|)$. We suppose that D is a bounded domain so that $\pi^*v = v(\log|z_1|,\dots,\log|z_n|)$ is pluriharmonic on $\pi^{-1}(D)$.

Then we have

LEMMA 5.

$$(dd^c\ \pi^*\ v)^P \neq 0 \quad \text{and} \quad (dd^c\ \pi^*\ v)^{P+1} = 0 \tag{36}$$

for any v *satisfying* (H_4).

Proof. Since $dd^c\,\pi^*v = \Sigma\,\pi^*v_{kl}\,\sqrt{-1}(dZ_k/2Z_k)\wedge(d\bar{Z}_l/2\bar{Z}_l)$, we have by elementary calculus

$$(dd^c\pi^*v)^P = \Sigma\pi^*(v_{k_1l_1},\dots,v_{k_pl_p})\sqrt{-1}\,\frac{dZ_{k_1}}{2Z_{k_1}}\wedge\frac{d\bar{Z}_{l_1}}{2\bar{Z}_{l_1}}\wedge\dots$$

$$\dots\wedge\sqrt{-1}\,\frac{dZ_{k_p}}{2Z_{k_p}}\wedge\frac{d\bar{Z}_{l_p}}{2\bar{Z}_{l_p}},$$

where the summation is taken over all subsets $\{k_1,\dots,k_p\}$, $\{l_1,\dots,l_p\}$ of $\{1,2,\dots,n\}$. Thus, if $\Sigma_{r,s}$ means the summation over all subsets of P elements $r_1 < r_2 < \dots < r_p$ and $s_1 < s_2 < \dots < s_p$ of $\{1,2,\dots,n\}$, we get

$$(dd^c\,\pi^*v)^P = \pi^*(\Sigma_{r,s}\,\Sigma_{\sigma\,\tau}\,\mathrm{sgn}\,\sigma\,\mathrm{sgn}\,\tau\,v_{s_{\sigma(1)}r_{\tau(1)}},\dots,v_{s_{\sigma(P)}r_{\sigma(P)}})$$

$$\cdot\sqrt{-1}\,\frac{dZ_{s_1}}{2Z_{s_1}}\wedge\frac{d\bar{Z}_{r_1}}{d\bar{Z}_{r_1}}\wedge\dots\wedge\sqrt{-1}\,\frac{dZ_{s_p}}{2Z_{s_p}}\wedge\frac{d\bar{Z}_{r_p}}{2\bar{Z}_{r_p}},$$

where σ, τ are permutations of $\{1,\dots,p\}$.

Thus, the coefficient of the latter term is

$$P!\ \pi^*(\Sigma\,\mathrm{sgn}\,\sigma\,v_{s_{\sigma(1)}r_1},\dots,v_{s_{\sigma(p)}r_p}).$$

Therefore our conclusion holds since $dv_{r_1}\wedge\dots\wedge dv_{r_p}$ has the coefficient $\Sigma\,\mathrm{sgn}\,\sigma\,v_{s_{\sigma(1)}r_1}\wedge\dots\wedge v_{s_{\sigma(p)}r_p}$ with respect to the form $dx_{s_1}\wedge\dots\wedge dx_{s_p}$.

Secondly, we set

$$\theta = (dd^c\pi^*v^o)^{n-p-1}\wedge(dd^c\pi^*v)^P \quad \text{and} \quad m_2 = (dd^c\pi^*v^o)\wedge\theta.$$

Then, we know that there exists a conformal diffusion Z_t corresponding to the pair (θ,m) (see [6]). Next, we can see that the projection of

Z_t by π to D is no other than our diffusion X_t associated with $L^{(m)}$ due to a standard stochastic analysis (cf. [4]).

Finally, we can use Theorem 4 of [9], which says that Z_t diffuses on each leaf of complex dimension $n-p$. Thus, the proof is completed by taking into account the dimension in view of the definition of π.

References

[1] BEDFORD, E. and M. KALKA: 'Foliations and complex Monge-Ampère equations', *Comm. Pure Appl. Math.* 30 (1977), 543-570.

[2] —— and B.A. TAYLOR: 'Variational properties of the complex Monge-Ampère equation II. Intrinsic norm', *Amer. J. Math.* 101 (1979), 1131-1166.

[3] CEGRELL, U.: 'On the domains of existence for plurisubharmonic functions', *Banach Center Publications* 11 (1983), 33-37.

[4] DYNKIN, E.B.: *Markov processes*, vols. 1, 2, Springer-Verlag, Berlin 1965.

[5] FUKUSHIMA, M.: *Dirichlet forms and Markov processes*, Kodansha and North-Holland, 1980.

[6] —— and M. OKADA: 'On conformal martingale diffusions and pluripolar sets', *J. Functional Anal.* 55 (1984), 377-388.

[7] GAVEAU, B. and J. ŁAWRYNOWICZ: 'Intégrale de Dirichlet sur une variété complexe I', in: *Lect. Notes in Math.* 919 , Springer 1982, pp. 131-167.

[8] KALINA, J. and ——: 'Foliations and the generalized complex Monge-Ampère equations', *Banach Center Publications* 11 (1983), 111-119.

[9] ŁAWRYNOWICZ, J. and M. OKADA: 'Canonical diffusion and foliation involving the complex hessian', *Bull. Polish Acad. Sci. Math.* 34 (1986), 661-667.

[10] OKADA, M.: 'Espaces de Dirichlet généraux en analyse complexe', *J. Functional Anal.* 46 (1982), 396-410.

[11] ——: 'Sur une capacité définie par la forme de Dirichlet associée aux fonctions plurisousharmoniques', *Tôhoku Math. J.* 35 (1983), 513-517.

[12] RAUCH, J. and B.A. TAYLOR: 'The Dirichlet problem for the multidimensional Monge-Ampère equation', *Rocky Mountain J. Math.* 7 (1977), 345-364.

[13] SICIAK, J.: 'Extremal plurisubharmonic functions and capacities in $\mathbb{C}^n$, *Lect. Notes Sophia Univ.* 14 (1982).

LAGRANGIAN FOR THE SO-CALLED NON-POTENTIAL SYSTEMS: THE CASE OF MAGNETIC MONOPOLES

Guy Laville
Mathématiques, L.A. 213 du C.N.R.S.
Université Pierre et Marie Curie
F-75252 Paris Cédex 05, France

ABSTRACT and INTRODUCTION. We study two distinct cases. Firstly, we construct a lagrangian associated with the Newton equations $m\ddot{x}_k = f_k(x)$, $1 \le k \le 3$, where f_k are components of a force which depends only on the position (without the classical hypothesis of integrability for the force). For clarifying the idea, here we take the frame-work of the Newtonian mechanics. Secondly, we consider the quantum electrodynamics in the case where the field is generated by the magnetic and electric charges. Mathematically speaking, we perceive that we have to give up exterior differential calculus which is totally unsuitable here. We propose to treat these questions with "interior" differential calculus.

1. CLASSICAL MECHANICS FOR A MATERIAL POINT IN THREE DIMENSIONS

Let $\sigma_0, \sigma_1, \sigma_2, \sigma_3$ be a basis for the Clifford algebra associated with the Euclidean space $\mathbb{R}^3$. The Newton equations take the form

$$m\ddot{x}_k = f_k(x), \quad 1 \le k \le 3. \tag{1}$$

Classically, we look for a potential V such that

$$dV = f_1 dx^1 + f_2 dx^2 + f_3 dx^3; \tag{2}$$

it exists only if the differential form at the right-hand side of (2) is closed. To integrate the non-closed differential form, we have to write it in the framework of a "Cliffordian differential form". The equalities (1) are equivalent to

$$\sum_{k=1}^{3} m\ddot{x}_k \sigma_k = \sum_{k=1}^{3} f_k \sigma_k. \tag{3}$$

The spin does not appear in our considerations because we deal here with the classical mechanics.

Let U be the Clifford function (represented by a 2×2-matrix) such that

J. Ławrynowicz (ed.), Deformations of Mathematical Structures, 331–337.

$$\sum_{h=1}^{3} \sigma_h \frac{\partial U}{\partial x_h} = \sum_{k=1}^{3} f_k \sigma_k . \tag{4}$$

To find U we take the laplacian of U. From (4) it follows that

$$\Delta U = \sum_{h=1}^{3} \sigma_h \frac{\partial}{\partial x_h}\left(\sum_{k=1}^{3} f_k \sigma_k\right) = \sum_{h<k} \sigma_p \left(\frac{\partial f_k}{\partial x_h} - \frac{\partial f_h}{\partial x_k}\right) + \sum_{h=1}^{3} \frac{\partial f_h}{\partial x_h} \tag{5}$$

with $\{h, k, p\}$ being an even permutation of $\{1, 2, 3\}$. Suppose that the first derivatives of f tend to 0 at infinity. Therefore we can set

$$U = \sum_{h<k} \sigma_p \, G * \left(\frac{\partial f_k}{\partial x_h} - \frac{\partial f_h}{\partial x_k}\right) + G * \sum_{h=1}^{3} \frac{\partial f_h}{\partial x_h}, \tag{6}$$

where G is a Green function for the laplacian. It is easy to verify that

$$\sum_{h=1}^{3} \sigma_h \frac{\partial U}{\partial x_h} = \sum_{h=1}^{3} \sigma_h \frac{\partial}{\partial x_h}\left[G * \sum_{h=1}^{3} \sigma_p \frac{\partial}{\partial x_p}\left(\sum_{k=1}^{3} f_k \sigma_k\right)\right] = \Delta G * \sum_{k=1}^{3} f_k \sigma_k$$

$$= \sum_{k=1}^{3} f_k \sigma_k ,$$

so (4) is fulfilled indeed. The classical case is when the first term at the right-hand side of (6) is 0. We have arrived at the following

THEOREM 1. *Define the lagrangian*

$$L = \tfrac{1}{2} m \sum_{j=1}^{3} \dot{x}_j^2 + U \tag{7}$$

with U *given by* (6). *Then we have the Lagrange equation*

$$\sum_{k=1}^{3} \sigma_k \left| \frac{d}{dt}\left(\frac{\partial L}{\partial \dot{x}_k}\right) - \frac{\partial L}{\partial x_k} \right| = 0, \tag{8}$$

which is equivalent to the Newton equations (1).

It is easy to investigate the invariance properties of the equations (8). Let $q_j = \phi_j(x_1, x_2, x_3)$, $j = 1, 2, 3$, be the new variables such that the Clifford structure is preserved, i.e.,

$$\overrightarrow{\text{grad}}\, \phi_j \cdot \overrightarrow{\text{grad}}\, \phi_k = \delta_{jk} \quad \text{and} \quad \overrightarrow{\text{grad}}\, \phi_j \wedge \overrightarrow{\text{grad}}\, \phi_k = \overrightarrow{\text{grad}}\, \phi_\ell$$

with $\{j, k, \ell\}$ being an even permutation of $\{1, 2, 3\}$. Then

$$\sum_i \sigma_i \left[\frac{d}{dt}\left(\frac{\partial L}{\partial \dot{x}_i}\right) - \frac{\partial L}{\partial x_i}\right] = \sum_{i,j} \sigma_i \left[\frac{d}{dt}\left(\frac{\partial L}{\partial \dot{q}_j}\frac{\partial \dot{q}_j}{\partial x_i} + \frac{\partial L}{\partial q_j}\frac{\partial q_j}{\partial \dot{x}_i}\right) - \frac{\partial L}{\partial \dot{q}_j}\frac{\partial \dot{q}_j}{\partial x_i}\right.$$

$$\left. - \frac{\partial L}{\partial q_j}\frac{\partial q_j}{\partial x_i}\right] = \sum_{i,j} \sigma_i \left[\frac{d}{dt}\left(\frac{\partial L}{\partial \dot{q}_j}\frac{\partial \phi_j}{\partial x_i}\right) - \frac{\partial L}{\partial q_j}\frac{\partial \phi_j}{\partial x_i}\right],$$

and hence

$$\sum_i \sigma_i \left[\frac{d}{dt}\left(\frac{\partial L}{\partial \dot{x}_i}\right) - \frac{\partial L}{\partial \dot{x}_i} \right] = \sum_j \left(\sum_i \sigma_i \frac{\partial \phi_j}{\partial x_i} \right) \left(\frac{d}{dt} \frac{\partial L}{\partial \dot{q}_j} - \frac{\partial L}{\partial q_j} \right).$$

Now, we change the basis of the Clifford algebra:

$$\tilde{\sigma}_j = \sum_i \sigma_i \frac{\partial \phi_j}{\partial x_i}, \quad \tilde{\sigma}_j \tilde{\sigma}_k = \sum_i \frac{\partial \phi_j}{\partial x_i} \sum_i \sigma_i \frac{\partial \phi_k}{\partial x_j} = (\overrightarrow{\mathrm{grad}}\, \phi_j, \overrightarrow{\mathrm{grad}}\, \phi_k) + \sum_h \sigma_h \overrightarrow{\mathrm{grad}}\, \phi_h$$

with $\{j, k, h\}$ being an even permutation of $\{1, 2, 3\}$. If $j = k$, the second term at the right-hand side of the latter relation has to be omitted which gives the right commutation relations.

2. CONSERVATION LAWS

From the lagrangian, we can obtain some relations generalizing the well-known conservation laws. For instance, we can consider the energy of one particle. Write L in the form

$$L = L^0 + \sigma_1\sigma_2 L^{12} + \sigma_2\sigma_3 L^{23} + \sigma_3\sigma_1 L^{31} = L^0 + \sigma_3 L^{12} + \sigma_1 L^{23} + \sigma_2 L^{31},$$

where, according to (8), L^0 and L^{jk} fulfil four real equations

$$\frac{\partial L^{12}}{\partial x_3} + \frac{\partial L^{23}}{\partial x_1} + \frac{\partial L^{31}}{\partial x_2} = 0, \qquad \frac{d}{dt}\frac{\partial L^0}{\partial \dot{x}_2} - \frac{\partial L^0}{\partial x_2} + \frac{\partial L^{12}}{\partial x_1} - \frac{\partial L^{23}}{\partial x_3} = 0,$$

$$\frac{d}{dt}\frac{\partial L^0}{\partial \dot{x}_1} - \frac{\partial L^0}{\partial x_1} + \frac{\partial L^{31}}{\partial x_3} - \frac{\partial L^{12}}{\partial x_2} = 0, \quad \frac{d}{dt}\frac{\partial L^0}{\partial \dot{x}_3} - \frac{\partial L^0}{\partial x_3} + \frac{\partial L^{23}}{\partial x_2} - \frac{\partial L^{31}}{\partial x_1} = 0.$$

If the lagrangian L does not explicitly depend on time, then

$$\frac{dL^0}{dt} = \sum_k \left(\frac{\partial L^0}{\partial x_k}\dot{x}_k + \frac{\partial L^0}{\partial \dot{x}_k}\ddot{x}_k \right) = \sum_k \dot{x}_k \frac{d}{dt}\frac{\partial L^0}{\partial \dot{x}_k} + \ddot{x}_k \frac{\partial L^0}{\partial \dot{x}_k} + \dot{x}_1 \frac{\partial L^{31}}{\partial x_3} - \dot{x}_3 \frac{\partial L^{31}}{\partial x_1}$$

$$+ \dot{x}_2 \frac{\partial L^{12}}{\partial x_1} - \dot{x}_1 \frac{\partial L^{12}}{\partial x_2} + \dot{x}_3 \frac{\partial L^{23}}{\partial x_2} - \dot{x}_2 \frac{\partial L^{23}}{\partial x_3}$$

and, consequently,

$$\frac{d}{dt}\left(\sum_k \dot{x}_k \frac{\partial L^0}{\partial \dot{x}_k} - L^0 \right) = \dot{x}_3 \frac{\partial L^{31}}{\partial x_1} - \dot{x}_1 \frac{\partial L^{31}}{\partial x_3} + \dot{x}_1 \frac{\partial L^{12}}{\partial x_2} - \dot{x}_2 \frac{\partial L^{12}}{\partial x_1}$$

$$+ \dot{x}_2 \frac{\partial L^{23}}{\partial x_3} - \dot{x}_3 \frac{\partial L^{23}}{\partial x_2}.$$

In the classical case where all $L^{jk} = 0$ we get the usual conservation law for the energy.

3. QUANTIZATION

The quantization of the preceding system is straightforward. For example, by the classical Hamiltonian Method we get $p_i = (\partial/\partial\dot{x}_i)L$. The hamiltonian is $H = \sum_{i=1}^{3} p_i \dot{q}_i - L$, i.e.,

$$H = \frac{1}{2m} \sum_{i=1}^{3} p_i^2 - U, \tag{9}$$

and the corresponding modified Schrödinger equation reads

$$\left(\frac{1}{2m}\Delta - U\right)\Phi = ih\frac{\partial\Phi}{\partial t}, \tag{10}$$

where Φ is a two-component wave function. The equation resembles the Pauli equation.

4. QUANTUM ELECTRODYNAMICS WITH MAGNETIC MONOPOLES

The idea of considering a potential for the electromagnetic field with more than 4 components is not new at all (cf. [1]), but the associated difficulties (cf. [6]) disappear only within the framework of the Clifford analysis.

Without monopoles, the potential A_μ, $1 < \mu < 4$, is determined by the electromagnetic field $F_{\mu\nu}$:

$$F_{\mu\nu} = \nabla_\mu A_\nu - \nabla_\nu A_\mu, \tag{11}$$

where

$$(\nabla_1, \nabla_2, \nabla_3, \nabla_4) = \left(-\frac{\partial}{\partial x_1}, -\frac{\partial}{\partial x_2}, -\frac{\partial}{\partial x_3}, \frac{\partial}{\partial x_4}\right).$$

It is well-known how to write the system (11) in terms of the exterior differential calculus. Yet, this is possible only if the form is closed, so it is impossible if there exist magnetic monopoles. If we take into account these poles, there are, loosely speaking, three methods which are based on: (i) cutting the space by the Dirac string, i.e. considering a singular potential (cf. [2]), (ii) covering the space, i.e. changing its topology (cf. [8]), and (iii) considering the path dependence of the potential (cf. [4]), respectively. All these methods more or less modify the topology. Here we do not modify the topology, but the algebra.

Let $\gamma^1, \gamma^2, \gamma^3, \gamma^4$ denote the four Dirac matrices, e the electric charge, and g the magnetic charge. If there is no magnetic charge, the system (11) can be written as

$$\sum_\mu \gamma^\mu \nabla_\mu \left(\sum_\nu \gamma^\nu A_\nu\right) = \sum_{\mu<\nu} \gamma^\mu \gamma^\nu F_{\mu\nu}. \tag{12}$$

In the case of electric and magnetic charges, let $\gamma^5 = i\gamma^1\gamma^2\gamma^3\gamma^4$. We are looking for a potential

$$A = \sum_\mu \gamma^\mu (A_\mu + \gamma^5 B_\mu) \tag{13}$$

such that

$$\sum_{\mu} \gamma^{\mu} \nabla_{\mu} A = \sum_{\mu<\nu} \gamma^{\mu} \gamma^{\nu} F_{\mu\nu}. \tag{14}$$

Let us construct explicitly the 8 component-potential A. Let $(j_1, \ldots, j_4)$ be the 4-vector electric current and $(g_1, \ldots, g_4)$ the 4-vector magnetic current. When introducing the current matrix

$$J = \sum_{\mu} \gamma^{\mu} (j_{\mu} + \gamma^5 g_{\mu}) \tag{15}$$

the Maxwell equations can be written as

$$\sum_{\mu} \gamma^{\mu} \nabla_{\mu} \Big(\sum_{\nu<\rho} \gamma^{\mu} \gamma^{\rho} F_{\nu\rho} \Big) = 4\pi J. \tag{16}$$

The potential A has to satisfy the condition

$$\sum_{\alpha} \gamma^{\alpha} \nabla_{\alpha} \sum_{\mu} \gamma^{\mu} \nabla_{\mu} A = \sum_{\alpha} \gamma^{\alpha} \nabla_{\alpha} \Big(\sum_{\mu<\nu} \gamma^{\mu} \gamma^{\nu} F_{\mu\nu} \Big),$$

i.e.,

$$\Box A = 4\pi J, \tag{17}$$

where $\Box$ is the d'Alembertian (the wave equation). If j_μ and g_μ are well-behaving at infinity and regular, we have $A_\mu = N * j_\mu$ and $B_\mu = N * g_\mu$, where N is a fundamental solution of $\Box$. The conservation law for currents: $\nabla_\mu j_\mu = 0$ and $\nabla_\mu g_\mu = 0$ (the Lorentz scalar product) implies the Lorentz condition on the gauge A:

$$\sum_{\mu} \nabla_{\mu} A_{\mu} = N * \sum_{\mu} \nabla_{\mu} j_{\mu} = 0 \quad \text{and} \quad \sum_{\mu} \nabla_{\mu} B_{\mu} = N * \sum_{\mu} \nabla_{\mu} g_{\mu} = 0.$$

Now it is easy to verify the equality (14):

$$\sum_{\mu} \gamma^{\mu} \nabla_{\mu} A = \sum_{\mu} \gamma^{\mu} \nabla_{\mu} (N * J) = \sum_{\mu} \gamma^{\mu} \nabla_{\mu} \Big[N * \sum_{\beta} \gamma^{\beta} \nabla_{\beta} \sum_{\nu<\rho} \gamma^{\nu} \gamma^{\rho} F_{\nu\rho} \Big]$$

$$= \Box N * \sum_{\nu<\rho} \gamma^{\nu} \gamma^{\rho} F_{\nu\rho} = \sum_{\nu<\rho} \gamma^{\nu} \gamma^{\rho} F_{\nu\rho},$$

so we have arrived at

THEOREM 2. *The following lagrangian describes the quantum electromagnetism with magnetic as well as electric charges:*

$$\begin{aligned} L = & -\tfrac{1}{2} \nabla_{\nu} A_{\mu} \nabla_{\nu} A_{\mu} - \tfrac{1}{2} \nabla_{\nu} B_{\mu} + \tfrac{1}{2} i [\bar{\psi} \gamma^{\mu} \nabla_{\mu} \psi - \nabla_{\mu} \bar{\psi} \gamma^{\mu} \psi] - m \bar{\psi} \psi \\ & - \psi \gamma^{\mu} (eA_{\mu} + g\gamma^5 B_{\mu}) \psi \quad \textit{(summation on the repeated indices)}. \end{aligned} \tag{18}$$

The wave function ψ describes a particle with spin $\frac{1}{2}$, electric charge e, and magnetic charge g. It is important to notice that we have got out singular potential and then got rid of the condition of quantification of the electric charge which was the original idea of Dirac (cf. [2] and the following studies on the magnetic charge). We can conclude that even if somebody could find a magnetic monopole, it would not really explain the electric charge quantification (cf. [7]).

The Euler equations of the lagrangian (18) give the modified Dirac

equation
$$\sum_\mu \gamma^\mu (i\nabla_\mu - eA_\mu - g\gamma^5 B_\mu)\psi = m\psi \qquad (19)$$
(the Dirac equation with Dirac monopoles). The same lagrangian also yields
$$A_\mu = e\psi\gamma^\mu\psi \quad \text{and} \quad B_\mu = g\psi\gamma^\mu\gamma^5\psi, \qquad (20)$$
i.e. the Maxwell equations. The "minimal coupling" for the field is here $p_\mu \mapsto p_\mu - eA_\mu - g\gamma^5 B_\mu$. Notice that, contrary to Schwinger's approach [6], the electric and magnetic charges are carried here by the same particle of mass m and spin $\frac{1}{2}$. The electrodynamics with the classical perturbation method seems to have difficulties if g is great, so it may be worth to think of the method sketched in [5].

Consider now the case where there is no magnetic monopole. Let θ be an arbitrary differentiable real function of the four space-time coordinates. When changing ψ to $\psi\exp(-\gamma^5\theta)$ in the Dirac equation
$$\sum_\mu \gamma^\mu (i\nabla_\mu - eA_\mu)\psi = m\psi, \qquad (21)$$
we get an equivalent equation
$$\sum_\mu \gamma^\mu (i\nabla_\mu - eA_\mu)\psi\exp(-\gamma^5\theta),$$
i.e.,
$$-\sum_\mu \exp(-\gamma^5\theta)\gamma^\mu (i\nabla^\mu - eA_\mu - i\gamma^5\nabla_\mu\theta)\psi = m\exp(-\gamma^5\theta)\psi,$$
i.e.,
$$\sum_\mu \gamma^\mu (i\nabla_\mu - eA_\mu - i\gamma^5\nabla_\mu\theta)\psi = -m\psi. \qquad (22)$$
The "local invariance" consists here in substituting $i\nabla_\mu\theta$ by the four components B_μ which, in general. are not the partial derivatives of a scalar function. As previously known, the interaction involving magnetic monopoles appears as a symmetry breaking.

5. INVARIANT PROPERTIES

The Lorentz covariance is utilized here since the very beginning because of the use of the Dirac matrices. Now, we observe the following duality. The transformation of ψ at the end of Section 4 leaves all the results unchanged. The transformation $\psi' = \psi\exp(-\frac{1}{2}\gamma^5\theta)$ changes a current component to
$$\begin{aligned}\overline{\psi'}\gamma^\mu(e+g\gamma^5)\psi &= \overline{\psi}\exp(\tfrac{1}{2}\gamma^5\theta)\gamma^\mu(e+g\gamma^5)\exp(-\tfrac{1}{2}\gamma^5\theta)\psi\\ &= \overline{\psi}\gamma^\mu(\cos\theta - \gamma^5\sin\theta)(e+g\gamma^5)\psi\\ &= \overline{\psi}\gamma^\mu[(e\cos\theta + g\sin\theta) + \gamma^5(-e\sin\theta + g\cos\theta)]\psi.\end{aligned}$$
Hence $e' = e\cos\theta + g\sin\theta$ and $g' = e\sin\theta + g\cos\theta$, so we have arrived at the classical duality transformation between the electricity and magnetism.

Finally we turn our attention to the gauge invariance. Let Λ be a matrix of the form

$$\Lambda = u + \sum_{\mu<\nu} \gamma^\mu \gamma^\nu v_{\mu\nu} + \gamma^5 w,$$

where u, $v_{\mu\nu}$, and w are 8 functions which satisfy the conditions $\Box u = \Box v_{\mu\nu} = \Box w = 0$. Then we get another gauge by taking $A + \sum_\mu \gamma^\mu \nabla_\mu \Lambda$.

6. CONCLUSION

The classical argument of Dirac showing that if there exists only one magnetic monopole then the electric charge is quantized, is probably specious because it is possible to construct an electromagnetic theory with magnetic and electric charges with, a priori, arbitrary values. Yet, the reasoning of A.S. Goldhaber (cf. [6]) is not overthrown since he does not utilize the potential.

I wish to thank Professor E. Kerner for the remarks concerning this paper.

References

[1] CABIBBO, N. and E. FERRARI: 'Quantum electrodynamics with Dirac monopoles', *Nuovo Cimento* **23** (1962), 1147-1154.

[2] DIRAC, P.A.M.: 'Quantized singularities in the electromagnetic field', *Phys. Rev.* **74** (1948), 817-830; cf. also *Proc. Roy. Soc. A* **133** (1931), 60-74.

[3] GAMBINI, R., S. SALAMO, and A. TRIAS: 'Self-dual covariant Lagrangian formulation of electromagnetism with magnetic charges', *Lett. Nuovo Cimento* **27** (1980), 385-388.

[4] ——— and A. TRIAS: 'Path dependent quantum formulation of electromagnetism with magnetic charges', *J. Math. Phys.* **21** (1980), 1539-1545.

[5] LAVILLE, G.: 'Sur une famille de solutions de l'équation de Dirac', *C.R. Acad. Sci. Paris Série I* **296** (1983), 1029-1032.

[6] SCHWINGER, J.: 'Magnetic charge and quantum field theory', *Phys. Rev.* **144** (1966), 1087-1093.

[7] ———: 'Source and magnetic charge', *ibid.* **173** (1968), 1536-1544.

[8] WU, T.T. and C.N. YANG: 'Dirac's monopole without strings: classical Lagrangian theory', *ibid. D* **14** (1976), 437-445; cf. also *ibid. D* **12** (1975), 3845-3853.

EXAMPLES OF DEFORMATIONS OF ALMOST HERMITIAN STRUCTURES

Stancho Dimiev and Kalin Petrov
Institute of Mathematics
Bulgarian Academy of Sciences
BG-1090 Sofia, Bulgaria

ABSTRACT. The possibilities of developing some deformation theory of almost Hermitian structures and its interconnection with almost holomorphic isometries are examined. Some examples of deformations on the manifold of Thurston-Abbena are given.

INTRODUCTION

The manifold of Thurston-Abbena is the simplest compact nilpotent 4-dimensional manifold related with nil-geometry. In connection with this it seems interesting to develop complex function theory on this manifold and on other manifolds of this kind. On the other hand, the manifolds mentioned have recently attracted the attention of a number of specialists in differential geometry [1], [2], [3]. The reason is that these manifolds admit no Kähler structures but can be equipped with an almost Kähler one. The non-existence of Kähler structures is due to Thurston (the famous example of a compact symplectic manifold which admits no Kähler metrics [4]). The structure of nil-manifold has been mentioned by J. Bresin and the almost Kähler structure was introduced by E. Abbena [1].

The question of the generalization of the classical deformation theory of Kähler manifolds ([5], [6]) appears naturally since the generalized Kähler geometry (almost Kähler, nearly Kähler manifolds etc.) can be considered a generalization of the classical Kähler geometry (Kähler manifolds). In this paper we investigate some deformations of almost Hermitian manifolds of the simplest kind as well as their interconnection with almost holomorphic isometries. The notion of deformation of almost Hermitian structure is defined via the Griffits' theory of deformations of G-structures. Some explicit examples on the manifold of Thurston-Abbena are given.

1. RECALLING THE GENERAL NOTION OF DEFORMATION OF A G-STRUCTURE

In this chapter we recall some fundamental notions of the Griffits' paper [7] adapted for our purposes.

J. Ławrynowicz (ed.), Deformations of Mathematical Structures, 339–348.

1.1. The Notion of G-Structure

Let M be a smooth manifold covered by the local coordinate systems (charts) $(U_j, x_j, \mathbb{R}^n)$, i.e. $x_j = (x_j^1, \ldots, x_j^n)$, $n = \dim M$, and let G be a closed linear subgroup of the general linear group $GL(n, \mathbb{R})$. A G-structure is defined on M by a system σ of frames $\{\sigma_j\}$ ($\sigma_j = \sigma_j(p)$ being a frame on U_j, $p \in U_j$) and smooth maps $g_{jk} : U_j \cap U_k \to G$ such that $\sigma_k(p) = \sigma_j(p) g_{jk}(p)$ for each point $p \in U_j \cap U_k$. The G-structure $\sigma = \{\sigma_j\}$ is called an integrable one if there exist local coordinate systems $(U_j, x_j^!, \mathbb{R}^n)$ such that

$$\sigma_j(p) = \frac{\partial}{\partial x_j}\Big|_p$$

for each point $p \in U_j$ (if necessary the covering U_j may be refined as usual).

Let us denote by $\hat{\sigma}_j(x_j)$ a matrix of smooth functions on U_j such that

$$\sigma_j(p) = \frac{\partial}{\partial x_j}\Big|_p \hat{\sigma}_j(x_j), \quad p \in U_j, \quad x_j = x_j(p).$$

1.2. The Sheaf $\Gamma_G[t]$ and the Exponential Map $\exp\theta\,(p,t)$

Let D be an interval on real axis $\mathbb{R}$, $0 \in D$, and U a domain of a local coordinate system on M. We shall consider maps $f : U \times D \to U$, such that $f(p,0) = p$, i.e. the map f is the identity for $t = 0$. We suppose that each $f \times 1$ is a bi-map, i.e. $f \times 1 : U \times D \to U \times D$ is an invertible map.

By $\Gamma_G[t]$ we denote the sheaf of non-abelian groups on M defined by the germs of local bi-maps f, depending on t, $t \in D$, and such that

$$f_{*,(p,t)}\,\sigma_j = \sigma_j \circ f \quad \text{for each point } p \in U_j,$$

where $f_{*,(p,t)}$ is a tangent map to $F_t(p) = f(p,t)$. In fact, the above condition means that

$$D_x f(x_j,t)\,\hat{\sigma}_j(x_j) = \hat{\sigma}_j(f(x_j,t)),$$

where $D_x f(x_j,t)$ denotes the derivative of the local expression of $f(x_j,t)$ related to the variable x_j.

Let $\theta = \theta(t)$ be a vector field defined locally on M and depending on the parameter t. Then, in a neighbourhood U of every point in which θ is defined, there exists a unique local bi-map $f : U \times D \to U$ (or a family $f(t)$ of local bi-maps) such that $f(p,0) = p$, $p \in U$, and

$$\frac{df(p,t)}{dt} = \theta(f(p,t),t).$$

This map is denoted by $\exp\theta\,(p,t)$. In this case θ does not depend on t and we write $f(p,t) = \exp\theta\,(p)$.

Let us notice that in local coordinates the map f is determined as $f(x_j,0)=x_j$, $D_t f(x_j,t)=\theta_j(f(x_j,t),t)$, where $D_t f(x_j,t)$ denotes the derivative related to t and

$$\theta(p,t)=\frac{\partial}{\partial x_j}\Big|_p\ \theta_j(x_j,t).$$

By Θ_G we denote the sheaf of germes of local vector fields such that $\exp t\,\theta\in\Gamma_G[t]$.

1.3. Definition of Deformation of a G-Structure

Let $\sigma=\{\sigma_j\}$, $\sigma_j=\sigma_j(p)$, be a G-structure on M and $\{\sigma_j(p,t)\}$ a system of frames depending smoothly on the parameter $t\in D$, D being an open interval of $\mathbb{R}$, $0\in\mathbb{R}$, i.e.

$$\theta_j(p,t)=\frac{\partial}{\partial x_j}\Big|_p\ \hat{\sigma}_j(x_j,t),$$

where $\hat{\sigma}_j(x_j,t)$ is a matrix of smooth functions on $U_j\times D$. We also assume that $\sigma_k(p,t)=\sigma_j(p,t)g_{jk}(p,t)$, where $g_{jk}:(U_j\cap U_k)\times D\to G$ is a smooth map and $\sigma_j(p,0)=\sigma_j(p)$.

We say that the system of frames $\{\sigma_j(p,t)\}$ defines a deformation $\sigma(t)$ of the given G-structure $\sigma=\{\sigma_j\}$ if there exist smooth local bi-maps $\gamma_j:U_j\times D\to U_j$ and $\phi_j:U_j\times D\to U_j$, such that $(\gamma_j\times 1)^{-1}=\phi_j\times 1$ and the following condition (A) is satisfied:

$$\gamma_{j*,(p,t)}\ \sigma_j(p,0)=\sigma_j(\gamma_j(p,t),t)\ g_j(p,t), \tag{A}$$

where $g_j:U_j\times D\to G$.

In local coordinates the condition (A) is as follows:

$$D_y\gamma_j(y_j,t)\ \hat{\sigma}_j(y_j,0)=\hat{\sigma}_j(x_j,t)g(x_j,t),$$

where $y_j=\phi_j(x_j,t)$.

Let us consider the manifold $\mathcal{M}=M\times D$ and the projection $\tilde{\omega}:\mathcal{M}\to D$, $\tilde{\omega}(p,t)=t$. If M is covered by the set of local coordinate systems $(U_j,x_j,\mathbb{R}^n)$, then $\mathcal{M}$ is covered by the following set of local coordinate systems: $(U_j\times D,\ x_j\times 1,\ \mathbb{R}^{n+1})$. We shall also use the local coordinate systems $(U_j\times D,\ y_j,\ \mathbb{R}^n)$ on $\mathcal{M}$, where $y_j=\phi_j(x_j,t)$, $t_j=t$ and $x_j=\gamma_j(y_j,t)$, $t=t_j$.

Let

$$\sigma_j(p,t)=\frac{\partial}{\partial y_j}\Big|_{(p,t)}\ \hat{\sigma}_j(y_j,t)$$

exist in the coordinates (y_j,t). It is clear that $\hat{\sigma}_j(y_j,t)=D_x\phi_j(x_j,t)\hat{\sigma}_j(x_j,t)$. Now the condition (A) is equivalent to the following condition:

$$\hat{\sigma}_j(y_j,t) = \hat{\sigma}_j(y_j,0)\ g_j(y_j,t). \tag{B}$$

Let us notice that for a fixed $t \in D$ the frames $\{\sigma_j(p,t)\}$ define a G-structure on the fibre $M_t = \tilde{\omega}^{-1}(t)$. Sometimes the coordinates $(x) = (x_j)$ are called coordinates of the first kind or Euler-type coordinates and the coordinates $(y) = (y_j)$ - coordinates of the second kind or Lagrange-type coordinates.

Finally, we recall the notion of the equivalence of deformations. Let $\sigma(t)$, $t \in D$, and $\sigma'(t')$, $t' \in D$, be two deformations of σ. They are called equivalents if there exist a family of bi-maps $\psi : M \times D \to M$ and a bi-map $\tilde{\psi} : D \to D'$, such that

$$\psi(t)_{*,p}\ \sigma(t)(p) = \sigma'(\tilde{\psi}(t))(\psi(p,t)).$$

The deformation $\sigma(t) = \{\sigma_j(t)\}$ is called trivial if it is equivalent to the constant deformation $\sigma_j(t) = \sigma_j(0)$.

1.4. Almost Complex and Almost Hermitian Structures as G-structures

Let S be the matrix of the so-called standard almost complex structure $S^2 = -1$,

$$S = \begin{bmatrix} 0 & -1 & & & \\ 1 & 0 & & & \\ & & \ddots & & \\ & & & 0 & -1 \\ & & & 1 & 0 \end{bmatrix}.$$

By $AH(2\mu, \mathbb{R})$ we denote the subgroup of $GL(2\mu, \mathbb{R})$ composed by the matrices $A \in GL(2\mu, \mathbb{R})$ which commute with the matrix S.

Let M be a smooth manifold of real dimension 2μ. If $\sigma = \{\sigma_j(p)\}$ is a $AH(2\mu, \mathbb{R})$-structure on M and $\sigma(t) = \{\sigma_j(p,t)\}$ is a deformation of this structure, then on each fibre M_t of $\tilde{\omega} : M \to D$ a complex structure J_t appears, which is defined as follows:

$$\sigma_j(p,t) \to \sigma_j(p,t).S : TM_t \to TM_t.$$

In coordinates (y_j) we shall have

$$\frac{\partial}{\partial y_j}\Big|_{(p,t)} \to \frac{\partial}{\partial y_j}\Big|_{(p,t)}\ \hat{\sigma}_j(y_j,t).S.\hat{\sigma}_j^{-1}(y_j,t).$$

From the condition (A) we obtain

$$D_y\gamma_j(y_j,t)\ \hat{\sigma}_j(y_j,0) = \hat{\sigma}_j(x_j,t)\ g_j(x_j,t),$$

where $g_j : U_j \times D \to G$. Thus, in coordinates (y_j) we have $\hat{\sigma}_j(y_j,t) = \hat{\sigma}_j(y_j,0).g_j(y_j,t)$ and therefore $J_j(y_j,t) = J_j(y_j,0)$, where

$$J_j\left(\frac{\partial}{\partial y_j}\right) = \frac{\partial}{\partial y_j} J_j(y_j,t), \quad (p,t) \Longleftrightarrow (x_j,t) \Longleftrightarrow (y_j,t).$$

Now, let G be the orthogonal subgroup of $GL(2\mu, \mathbb{R})$, i.e. G is the group of matrices A such that ${}^{\tau}AA = 1$, ${}^{\tau}A$ being the transposed matrix of A, 1 - the unit matrix. We shall write it as ordinary $G = O(2\mu)$. Having in mind the notations introduced above, we get that on each fibre M_t a Riemann metric can be defined if we take $\sigma_j(p,t)$ as an orthogonal frame. If we set

$$g_j(y_j,\tau) = g\left({}^{\tau}\left(\frac{\partial}{\partial y_j}\right), \frac{\partial}{\partial y_j}\right)$$

we obtain the matrix of g in the base $\partial/\partial y_j$. By the condition (B) we have $g_j(y_j,t) = g_j(y_j,0)$.

Finally, let (M,g,J) be an almost Hermitian manifold with a Hermitian metric g, i.e. $g(JX,JY) = g(X,Y)$ for each couple of vector fields on M. By the definition the fundamental 2-form Ω is $\Omega(X,Y) = g(X,JY)$. Thus, in the base $\partial/\partial y_j$ it has a matrix $\Omega_j(y_j,t)$. By the condition (B) we can obtain $\Omega_j(y_j,t) = \Omega_j(y_j,0)$. We get a smooth family of almost Hermitian manifolds (M_t,g_t,J_t), $t \in D$.

2. COMMON DEFORMATIONS OF G-STRUCTURES

We shall consider some auxiliary notion of common deformations of G-structures.

2.1. Abstract Setting

Let A be a set of indices and $\{G^\alpha : \alpha \in A\}$ a set of subgroups of $GL(n, \mathbb{R})$. We shall consider sets of G-structures $\{\sigma^\alpha : \alpha \in A\}$, where σ^α is a G^α-structure. By definition, A common deformation of the set $\{\sigma^\alpha : \alpha \in A\}$ means every set of deformations $\{\sigma^\alpha(t) : \alpha \in A\}$ which is defined by unique couple of maps $(\gamma,\phi) : \gamma_j : U_j \times D \to U_j$ and $\phi_j : U_j \times D \to U_j$ (see 1.3.).

All σ^α have the same (y)-coordinates on $\mathcal{M} = M \times D$. Let $\{f_{ij}\}$ be the transition functions for the (y)-coordinates, i.e.

$$y_i = f_{ij}(y_j,t) \quad \text{on} \quad U_i \cap U_j.$$

<u>PROPOSITION 1</u>. *If* M *is a compact manifold, we have a bijective map between the set of common deformations of* $\{\sigma^\alpha : \alpha \in A\}$ *and the intersection* $\bigcap_\alpha H^1(M,\Gamma_{G^\alpha}[t])$. *In other words, the germes of common deformations may be identified with the elements of the intersection written above.*

<u>Proof</u>. We have the 1-cochain $\{f_{ij}\} \in C^1(\{U_i\}, \Gamma_{G^\alpha}[t])$. Obviously, it belongs to $Z^1(\{U_i\}, \Gamma_{G^\alpha}[t])$ for each $\alpha \in A$. Therefore $\{f_{ij}\}$

determine an element of $\bigcap_\alpha H^1(M, \Gamma_{G^\alpha}[t])$. If $\{f_{ij}\} = 0$, each deformation is trivial. Conversely, when $\gamma = \{\gamma_{ij}\} \in Z^1(\{U_i\}, \Gamma_{G^\alpha}[t])$ for each $\alpha \in A$, then in (y)-coordinates we have deformations if we set $f_{ij}(y_j, t) = \gamma_{ij}(p,t)$. The proposition is proved.

Having in mind that

$$H^1(M, \Gamma_G[t]) \subset \bigcap_\alpha H^1(M, \Gamma_{G^\alpha}[t]), \quad \text{where } G = \bigcap_\alpha G^\alpha,$$

we obtain that a common deformation exists if the cohomological group $H^1(M, \Gamma_G[t])$ is non-empty.

Now, let us take the sheaf morphism $r : \Gamma_G[t] \to \Theta_G$, which sends the germs of $f(t)$ into the germs defined by $df(t)/dt|_{t=0}$. This morphism induces the map $f : H^1(M, \Gamma_G[t]) \to H^1(M, \Theta_G)$. If $\{\gamma\} \in H^1(M, \Gamma_G[t])$ is a 1-cochain of deformation, then $\{r_\gamma\} \in H^1(M, \Theta_G)$ is an infinitesimal deformation corresponding to $\{\gamma\}$. Thus we have the following

PROPOSITION 2. *If* M *is compact, we have that*

$$\{rf_{ij}\} \in \bigcap_\alpha H^1(M, \Theta_G\alpha)$$

is an infinitesimal deformation determined by the transition functions $\{f_{ij}\}$ *of the* (y)*-coordinates of the common deformation of* $\{\sigma^\alpha : \alpha \in A\}$.

Proof. Indeed, we have

$$\{rf_{ij}\} = \frac{\partial}{\partial y_j} \left. \frac{df_{ij}(y_j, t)}{dt} \right|_{t=0} \in H^1(M, \Theta_G\alpha)$$

for each $\alpha \in A$. Then, if we take $f_{ij}(y_j, t) = \gamma_{ij}(p,t)$ in (y)-coordinates, we obtain deformations.

2.2. Almost Holomorphic Isometries

Let (M,g,J) be an almost Hermitian manifold, J being its almost complex structure and g a Hermitian metric on M, i.e. $g(JX,JY) = g(X,Y)$ for each couple of vector fileds on M. On (M,g,J) we have three fundamental groups of transformations: the group $AH_J(M)$ of all almost holomorphic transformations, the group $ASp_J(M)$ of all almost symplectic transformations, and the group $O(M)$ of all orthogonal transformations.

PROPOSITION 3. *The following three equalities hold on* M:

$$AH_J(M) \cap ASp_J(M) = ASp_J(M) \cap O(M) = O(M) \cap AH_J(M).$$

Proof. If $f \in AH_J(M) \cap ASp_J(M)$, we have $d_m f \circ J_m = J_{f(m)} \circ d_m f$

and $^{\tau}(d_m f) \circ J_{f(m)} \circ d_m f = J_m$, $m \in M$. It follows that $^{\tau}(d_m f) d_m f \circ J_m = J_m$, i.e. $^{\tau}(d_m f) d_m f = 1$, $f \in O(M)$. Analogously, we can see that

$$ASp_J(M) \cap O(M) \subset AH_J(M) \quad \text{and} \quad O(M) \cap AH_J(M) \subset ASp_J(M).$$

In view of the three inclusions obtained the proposition is proved.

Remark. In the case of almost Kähler manifold the almost symplectic transformations are symplectic and $ASp_J(M) = Sp_J(M)$. In this case Proposition 3 has been given in [9].

The group $U_J(M) := AH_J(M) \cap O(M)$ is considered a group of almost holomorphic isometries of M [8]. In the case of compact M, the group $U_J(M)$ is a Lie group of transformations. Since $U_J(M) \subset ASp_J(M)$, we get that the fundamental 2-form Ω of M is invariant under the action of $U_J(M)$ [8].

2.3. Deformations of Almost Hermitian Structures

A kind of deformations of (M,g,J) will be developed here via the notion of the common deformation of the almost complex structure J and the netric g. More precisely, we shall consider the $AH(2\mu, \mathbb{R})$-structure and the $O(2\mu)$-structure on M as well as their common deformations. In this case, the propositions of 2.1. are as follows:

Let us set

$$\Gamma_{AH(2\mu,\mathbb{R})}[t] = \Gamma_J[t], \quad \Theta_{AH(2\mu,\mathbb{R})} = \Theta_J$$

and

$$\Gamma_{O(2\mu)}[t] = \Gamma_g[t], \qquad \Theta_{O(2\mu)} = \Theta_g.$$

In fact, $\Gamma_J[t]$ is the sheaf of germs of almost holomorphic maps and $\Gamma_g[t]$ the sheaf of germs of isometries on M. Consequently, Θ_J is the sheaf of germs of infinitesimal automorphisms of J and Θ_g the sheaf of germs of the Killing vector fields. Then we have

PROPOSITION 1'. *The germs of deformations (in the considered sense) on compact almost Hermitian manifold* (M,g,J) *may be identified with the elements of the intersection* $H^1(M,\Gamma_J[t]) \cap H^1(M,\Gamma_g[t])$.

PROPOSITION 2'. *If* (M,g,J) *is a compact almost Hermitian manifold, then we have that*

$$\{rf_{ij}\} \in H^1(M,\Theta_J) \cap H^1(M,\Theta_g)$$

is an infinitesimal deformation determined by the transition functions $\{f_{ij}\}$ *of the* (y)*-coordinates of the common deformation of* J *and* g.

Let us notice that in view of the inclusion

$$H^1(M,\Gamma_{AH(2\mu,\mathbb{R})\cap O(2\mu)}[t] \subset H^1(M,\Gamma_J[t] \cap H^1(M,\Gamma_g[t]),$$

it follows that if $H^1(M,\Gamma_J[t]) \cap H^1(M,\Gamma_g[t]) = \emptyset$, then there are no almost holomorphic isometries on the compact (M,g,J). It is interesting to answer the following question: If

$$H^2(M,\Theta_J) \cap H^2(M,\Theta_g) = 0,$$

does every element of $H^1(M,\Theta_J) \cap H^1(M,\Theta_g)$ define a germ of deformation ?

2.4. The Case of Almost Kähler Structures

Let $\tilde{\omega}: M \to D$ be a deformation of the almost Hermitian manifold (M,g,J), i.e. a family (M_t,g_t,J_t) is given with $(M_o,g_o,J_o) = (M,g,J)$, $0 \in D \subset \mathbb{R}$. Then a family of fundamental 2-forms Ω_t appears. The given deformation is said to be an almost Kähler one if Ω_t is closed for each $t \in D$.

If there exists a deformation of (M,g,J) in the considered sense, then by the condition (B) (see 1.3.) we obtain

$$\Omega_j(y_j,t) = \Omega_j(y_j,0) \quad \text{on each chart } U_j \subset M.$$

In other words, each common deformation of an almost Kähler manifold is an almost Kähler deformation.

In the case of the existence of almost holomorphic isometries on a compact almost Kähler manifold it is obvious that the deformations mentioned do exist, as it follows from Proposition 1' of 2.3.

3. EXAMPLES ON THE MANIFOLD OF THURSTON-ABBENA

3.1. The Manifold of Thurston-Abbena

This manifold can be described as a factor-group of the Lie group $H \times S^1$, where H is the Heisenberg group, S^1 the circle and I the subgroup of $H \times S^1$ composed by the matrices with integer coefficients. We have

$$H \times S^1 = \left\{ \begin{bmatrix} 1 & x & z & 0 \\ 0 & 1 & y & 0 \\ 0 & 0 & 1 & 0 \\ 0 & 0 & 0 & e^{2\pi i t} \end{bmatrix} := m_{xyzt} : x,y,z,t \in \mathbb{R} \right\},$$

$$\Gamma = \left\{ \begin{bmatrix} 1 & p & r & 0 \\ 0 & 1 & q & 0 \\ 0 & 0 & 1 & 0 \\ 0 & 0 & 0 & 1 \end{bmatrix} : p,q,r \in \mathbb{Z} \right\}.$$

R e m a r k. Sometimes the matrix group $H \times S^1$ is regarded as cylindrical Heisenberg group (cf. E. ARNAUDOVA, S. DIMIEV, and K. PETROV: 'Potentials of the fundamental 2-form of the almost complex cylindrical Heisenberg group', in: *Complex analysis and applications, Proceedings, Varna 1985*, Publ. House Bulg. Acad. Sci., Sofia 1985, pp. 47-60).

On $H\times S^1$ we have the following two charts: $(U_o, x_o, \mathbb{R}^4)$ and $(U_1, x_1, \mathbb{R}^4)$, where

$$x_k(m_{xyzt}) = (x_k^1, x_k^2, x_k^3, x_k^4), \quad k = 0,1,$$

$$U_o = \{m_{xyzt} : x_o^4 \equiv t(\mathrm{mod}\ 1),\ x_o^4 \in (-\tfrac{1}{2}; \tfrac{1}{2})\},$$

$$U_i \neq \{m_{xyzt} : x_1^4 \equiv t(\mathrm{mod}\ 1),\ x_1^4 \in (0;1)\}.$$

Let us denote by 1_p the left-hand side translation on $H\times S^1$ by step p. If o is the unit of $H\times S^1$, $d_o 1_p$ is the differential of 1_p in o. Its matrix representation is denoted by $A(p)$:

$$A(p) = \begin{bmatrix} 1 & 0 & 0 & 0 \\ 0 & 1 & 0 & 0 \\ 0 & x & 1 & 0 \\ 0 & 0 & 0 & 1 \end{bmatrix}.$$

Let $e = (e_1, e_2, e_3, e_4)$ be a base of the Lie algebra g of the left-hand side invariant vector fields on $H\times S^1$. We have

$$e(p) = \frac{\partial}{\partial x_j}\Big|_p A(p), \qquad e^*(p) = d_p x_j\ A^{-1}(p),$$

where $e(p) = (e_1(p), e_2(p), e_3(p), e_4(p))$ is the base of g in the point $p\in M$.

Now, let us take the factor-group $H\times S^1/\Gamma$ and let $\pi : H\times S^1 \to H\times S^1/\Gamma$ be the canonical projection. We take the induced almost complex structure $\tilde{J}$, i.e. $\tilde{J}\tilde{X} = \tilde{J}\pi_* X = \pi_* JX$, where $\tilde{X} = \pi_* X$, $X \in g$, and the induced metric $\tilde{g}(\tilde{X},\tilde{Y}) = g(X,Y)$ (locally).

3.2. Examples of Deformations

Let us take on $H\times S^1$

$$\Lambda(0) = \begin{bmatrix} 0 & 0 & -1 & 0 \\ 0 & 0 & 0 & -1 \\ 1 & 0 & 0 & 0 \\ 0 & 1 & 0 & 0 \end{bmatrix}, \quad \gamma(0) = 1_4 \quad (1_4 \text{ is the unit } 4\times 4 \text{ matrix}),$$

and J defined by $J(e) = e\Lambda(0)$, $g({}^{\tau}e, e) = 1_4$, i.e. the matrix $g({}^{\tau}e,e)$ of the metric is the unit 1_4. First, we define a deformation on $H\times S^1$ by coordinate functions of ϕ_j:

$$\phi_j(x_j, t) := (x_j^1, (1-t)x_j^2, (1-t)x_j^3, x_j^4), \quad j = 0,1.$$

This deformation induces a trivial deformation on $H\times S^1/\Gamma$ because

$df_{ij}(y_j,t)/dt = 0$.

Next, the example gives a non-trivial almost Kähler deformation. Here $\Lambda(0)$ and $\gamma(0)$ are the same as above, but

$$\phi_j(x_j,t) := (x_j^1 + tx_j^4,\ (1-t)x_j^2,\ (1-t)x_i^3,\ x_j^4),\quad j = 0,1.$$

This deformation induces a non-trivial deformation on $H\times S^1/\Gamma$. Notice that the common deformations exist for $D_x\phi_j(x_j,t)$ does not depend on j.

3.3. Other Remarks

There are many examples of compact symplectic manifolds which admit no Kähler metric, as it has been shown by L.A. Cordero, M. Fernandez and A. Gray [2].The study of deformations can be extended on the basis of some of these examples. Of course, the problem of the stability of the considered deformations is fundamental (on Kähler manifolds the famous stability theorems hold [6]).

References

[1] ABBENA, E.: 'An example of an almost Kähler manifold which is not Kählerian', *Bolletino U.M.I.* (6), 3-A (1984), 303-393.

[2] CORDERO, L.A., FERNANDEZ, M., and GRAY, A.: 'Variétés symplectiques sans structures kählériennes', *C.R. Acad. Sci. Paris* 301 (1985), 217-218.

[3] CORDERO, L.A., FERNANDEZ, M., and M. de LEON: 'Examples of compact non-Kähler almost Kähler manifolds', *PAMS*, to appear.

[4] THURSTON, W.P.: 'Some simple examples of symplectic manifolds, *PAMS* 55 (1976), 467-468.

[5] KODAIRA, K., NIRENBERG, L., and SPENSER, D.C.: 'On the existence of deformations of complex analytic structures', *Ann. Math.* 68 (1958), 450-459.

[6] MORROW, J. and KODAIRA, K.: *Complex manifolds*, Holt, Rinchart and Winston, INS, 1971.

[7] GRIFFITS, P.A.: 'Deformations of G-structures', *Math. Ann.* 155 (1964), 292-315.

[8] DIMIEV, S.: 'Applications presque symplectiques', *C.R. Acad. Sci. Paris* 290 (1980), 161-164.

[9] DIMIEV, S. and PELOV, Ts.: 'Almost pluriharmonic functions and symplectic mappings on generalized Kähler manifolds', Seminar on Deformations Łódź-Warsaw 1982/84, *Lecture Notes in Math.* 1165 (1985), 108-121.

INDEX